The Commission of the European Communities

DECOMMISSIONING OF NUCLEAR POWER PLANTS

This conference was organized by the
Commission of the European Communities,
Directorate-General Science, Research and Development,
Nuclear Fuel Cycle Division
Brussels

PROGRAMME COMMITTEE

Members: A BERTINI
A CREGUT
W DIEFENBACHER
B HUBER
J MARSHALL
F MOTTE

Publication arrangements: D NICOLAY
Local organization : R LINSTER
Scientific secretary : K H SCHALLER

DECOMMISSIONING OF NUCLEAR POWER PLANTS

Proceedings of a European Conference
held in Luxembourg, 22–24 May 1984

Edited by
K H SCHALLER
B HUBER
Commission of the European Communities
Directorate-General Science, Research and Development, Brussels

Published by Graham & Trotman Ltd. for the Commission of the European Communities

Published in 1984 by Graham & Trotman Ltd.,
Sterling House, 66 Wilton Road, London SW1V 1DE, UK

Publication arrangements by Commission of the European Communities,
Directorate-General Information Market and Innovation, Luxembourg

EUR 9474

Softcover reprint of the hardcover 1st edition 1984

ISBN-13: 978-94-010-8983-8 e-ISBN-13: 978-94-009-5628-5
DOI: 10.1007/978-94-009-5628-5

INTRODUCTION

The European Community's programme on the decommissioning of nuclear power plants has the prime objective of developing effective techniques and procedures for ensuring the protection of man and his environment aginst the potential hazards of nuclear power plants that have been finally shut down.

The Commission of the European Communities organized this international conference to present results achieved during the past five years of the programme.

It also provided an opportunity for discussions amongst experts from Member States of the European Community and participating scientists from outside the Community, on the issues and options for future research.

The conference was limited to invited papers, presented by scientists involved in European Community contract studies. The topics covered by the conference and in the proceedings are :

- characterization of the radioactivity associated with nuclear power plants that have been finally shut down;
- implications of maintaining shut-down nuclear power plants for long periods before dismantling;
- decontamination of metal and concrete surfaces for decommissioning purposes;
- dismantling of nuclear power plant components and structures;
- conditioning and packaging of the radioactive waste arising from the dismantling;
- systems for remotely controlled decommissioning operations;
- aspects related to the classification of solid decommissioning waste, namely measurement of very low-level radioactivity and evaluation of radiological consequences;
- nuclear power plant design features facilitating decommissioning;
- experience with industrial-scale decommissioning operation.

The proceedings also report the discussions on the papers as well as the results of two technical panels and of the concluding panel.

ABBREVIATIONS

A-A Neyrtec	Alsthom-Atlantique Neyrtec, Grenoble
ACEC	Ateliers de Constructions Electriques de Charleroi
AMN	Ansaldo meccanica nucleare SpA, Genova
BMFT	Bundesministerium für Forschung und Technologie, Bonn
BSC	British Steel Corporation, London
CEA	Commissariat à l'énergie atomique, Paris
CEC	Commission of the European Communities, Brussels
CEGB	Central Electricity Generating Board, London
CEN/SCK	Centre d'étude de l'énergie nucléaire / Studiecentrum voor Kernenergie, Brussels
CRF	Centro ricerche Fiat, Orbassano (Torino)
D&W	Dyckerhoff & Widmann AG, München
EES	Electrowatt Engineering Services (UK) Ltd, London
ENEA	Comitato nazionale per la ricerca e per lo sviluppo dell'energia nucleare e delle energie alternative, Roma
Fiat TTG	Fiat termomeccanica-nucleare e turbogas, Torino
GKSS	GKSS Forschungszentrum, Geesthacht
GST	Gesellschaft für Systemtechnik mbH, Essen
Imp. College	Imperial College of Science and Technology, London
JRC	Joint Research Centre of the Commission of the European Communities
KEMA	NV tot Keuring van Elektrotechnische Materialen, Arnhem
KGB	Kernkraftwerke Gundremmingen Betriebsgesellscahft mbH
KWL	Kernkraftwerk Lingen GmbH
KWU	Kraftwerk Union AG, Erlangen
NIRA	Nucleare italiana reattori avanzati, Genova
NIS	Nuklear-Ingenieur-Service GmbH, Hanau
NNC	National Nuclear Corporation Ltd, London
Noell	Georg Noell GmbH, Würzburg
NRPB	National Radiological Protection Board, Chilton, Didcot
Nucleco	Nucleco SpA, Milano
PREAG	Preußische Elektrizitäts AG, Hannover
RWE	Rheinisch-Westfälisches Elektrizitätswerk AG
Salzgitter	Salzgitter AG, Salzgitter
Séri	Séri Renault Ingéniérie, Bois d'Arcy
Siempelkamp	Siempelkamp Gießerei, GmbH & Co., Krefeld
TH Darmstadt	Technische Hochschule Darmstadt
TN	Transnuklear GmbH, Hanau
Transnubel	Transnubel, Société belge de transports nucléaires, Brussels
TW	Taylor Woodrow Construction Ltd, Southall
U Hannover	Universität Hannover
UKAEA	United Kingdom Atomic Energy Authority, London
UK-DoE	Department of the Environment, London
U Warwick	University of Warwick, Coventry

CONTENTS

OPENING SESSION

Chairman : S. FINZI, CEC, Directorate-General Science, Research and Development

Secretary : K.H. SCHALLER, CEC, Directorate-General Science, Research and Development

Introduction by the Session Chairman

Welcoming address
P. HELMINGER, Secretary of State to the Ministry for Foreign Affairs, Grand Duchy of Luxembourg

Opening speech :
The Commission's research action programme on the development of nuclear fission energy
CEC, Directorate-General Science, Research and Development and Joint Research Centre

Decomissioning of nuclear power stations in Community countries - Carried out and projected
A. CREGUT, CEA and A.R. GREGORY, CEGB

The European Community's programme of research on the decommissioning of nuclear power plants : objectives, scope and implementation
B. HUBER, CEC, Directorate General Science, Research and Development

Factors to be considered in selecting a decommissioning strategy
A.R. GREGORY, CEGB, and A. CREGUT, CEA

Summary of discussion

INTRODUCTION BY THE SESSION CHAIRMAN

S. FINZI
CEC, Brussels, Belgium

I should like to thank the Secretary of State, on behalf of the Commission of the European Communities and also on behalf of the participants to the Conference, for his very important message because it indicates to us, technical people, the points of view and the concerns of the political authorities.

I am pleased and honoured to chair this Opening Session which presents in a way the frame of the conference, introducing a first analysis of the problems connected with the decommissioning of nuclear plants in Europe, and presenting the research programme of the Community.

The problem of decommissioning of nuclear installations has become more important in the last years, not only because a number of nuclear plants became obsolete and their operation has been stopped, but also as a result of the special attention given to environmental problems in our society in the last years.

The ways and means for the decommissioning of a nuclear plant are numerous, depending on the particular condition of the plant itself and its site. The strategy to be chosen concerning the timing and the measures of interventions, depends on different factors (technical, economical, social, political), not all quantifiable, which may render the decisions rather complex.

As far as the technical and economic aspects are concerned - if it is true that the weight of decommissioning on the cost of electricity produced is negligible -, it is nevertheless clear that the financial burden of the individual decommissioning operation is not trivial. Efforts have to be made to optimize these operations, taking in due account the very important elements of human and environment protection.

As the moment of this conference is tantamount to the end of the first pluriannual programme of the Commission of the E.C. and the beginning of a second pluriannual programme on Decommissioning of Nuclear Installations, the Deputy Director General, Dr. D.H. DAVIES, will present in his opening speech the E.C. programme on nuclear fission research which includes Decommissioning. Mr. HUBER, Head of the decommissioning programme, will then present the main achievements of the past and the plans for the future.

WELCOMING ADDRESS

P. HELMINGER
Secretary of State to the Ministry of Foreign Affairs
Grand Duchy of Luxembourg

Ladies and Gentlemen,

It gives me great pleasure, for two reasons, to be able to welcome you here in Luxembourg on behalf of the Luxembourg Government.

Firstly, as the oldest capital in Europe and the seat of several Community institutions, Luxembourg is always extremely proud to be able to provide the venue for meetings organized by the European Community. Secondly, and above all, the theme of this conference - the decommissioning of nuclear power stations - is a subject which, while it may not always make headline news, is a matter of serious concern to our Government.

The large attendance at this conference and the abilities and experience of the participants bear witness to the importance which all the Member States attach to this question.

Even though Luxembourg itself does not represent a hazard, since we have no nuclear power stations on our territory, it is very conscious of the danger which they create for Europe in general and for Luxembourg in particular. The Luxembourg Government has been made extremely wary by the proximity of nuclear power stations in neighbouring countries and has, in the interests of safety, introduced strict regulations concerning the use of nuclear power stations in general.

While strict regulations are a necessary precaution, they are not enough. We have to consider present-day technological development and, in particular, to encourage research on the development of new technologies whereby the human safety can be enhanced and our environment better protected.

Sensitive to these needs, the European Commission has become actively involved in research programmes, in particular by providing a large proportion of the funds for a research and development programme on the decommissioning of nuclear power stations.

The five-year programme, which began in 1979 and ends today, cost a total of about 10 million ECU, 4.7 million of which was financed out of the Community budget. The purpose of our meeting here in Luxembourg is to discuss the achievements resulting from five years of cooperation and to prepare the ground for a new five-year programme.

Further substantial financing is planned for the coming years, with 12.1 million ECU earmarked out of the Community budget for the period from 1984 to 1988. This gives an idea of the overall volume of the funds, which must be utilized to the full, and also of the key role which the EEC plays and must continue to play in this field, since the Member States could not undertake a programme of this scale on their own. In this period of European stagnation, this is a further example of the need to consolidate and promote Europe, while combating the disrupting forces by which it is periodically assailed.

The technological challenge which the development of nuclear power stations represents does not affect environment policy alone. The forthcoming five-year plan will be applied to nuclear installations in general and will cover not only nuclear power stations themselves but also nuclear reprocessing and enrichment plants. This is a new departure, and a very important one in view of the large amounts of low-radiation concrete and steel present in nuclear installations. It is of the utmost importance, both from the point of view of environmental protection and for economic reasons, that we should be able to re-use these enormous quantities of slightly irradiated materials. This is an opportunity which our European steel industries, which are suffering from the effects of a profound and lasting economic crisis, might well find of interest.

I feel there is no need for me to go into any details concerning the subjects to be discussed in the course of the conference, since this would be a tedious process, and in any case these are subjects which will be dealt with here in depth by specialists far more technically qualified than myself.

I trust that working conditions here in Luxembourg's municipal theatre will be fully satisfactory, as we have been at great pains to ensure the success of the conference. May I conclude, Ladies and Gentlemen, by wishing you, on behalf of the Luxembourg Government, a pleasant stay in our capital.

THE COMMISSION'S RESEARCH ACTION PROGRAMME ON THE DEVELOPMENT OF NUCLEAR FISSION ENERGY

Directorate-General for Science, Research and Development, and
the Joint Research Center
Commission of the European Communities, Brussels

Summary

For its "Framework Programme 1984-1987" the Commission has defined the major goals for a European Scientific and Technical Strategy. One of the means to reduce the energy dependance of the Community, which is an important objective, is to favour the development of nuclear fission energy. As electricity production by nuclear reactors has reached industrial maturity, the Community activities are directed mainly to safety aspects, in order to ensure the protection of workers and the general public, against hazards linked to operations in the nuclear fuel cycle.
A description of the main features of the five sub-programmes on nuclear fission energy is given below; these programmes are :
- reactor safety;
- nuclear fuels and actinides research;
- management of radioactive waste;
- safeguarding and management of fissile materials;
- decommissioning of nuclear installations.

The research and development work is carried out either by the Commission's Joint Research Center, or by organizations and companies of the Member Countries, with the Commission's financial support.

1. INTRODUCTION

The Commission has established, in a "Framework Programme 1984-1987", the major goals to which a "European Scientific and Technical Strategy" should be geared. The programme states that it is essential to improve the management of energy resources and to reduce the energy dependance of the Community. The effort is to be concentrated on the rational use of energy, the development of renewable energy sources, controlled thermonuclear fusion and the development of nuclear fission energy.

The peaceful use of nuclear fission has led to a well-developed and proven technology, and the continuation of an important nuclear energy production is an essential aspect of European energy policy. Nuclear reactors have reached industrial maturity and the entire fuel cycle, including reprocessing of spent fuel and fast-breeder reactors for a rational use of uranium resources, is well-developed.

The Community activities are directed mainly to safety aspects, in order to ensure the protection of workers and the general public against hazards linked to operations in the nuclear fuel cycle. By helping to harmonize national approaches to safety, Community action also assists in promoting industrial competitiveness and favours international cooperation.

The action programme on the development of nuclear fission energy (1) is sub-divided into five sub-programmes, which will be described in more detail below, with research on reactor safety being the most important part, as far as funding is concerned. The research and development work is carried out by the Commission's Joint Research Center, as part of its 1984-1987 research programme (2), designated as "direct action", as well as by organizations and companies of the Member Countries with the Commission's financial support ("cost-shared action"). The appropriations for the direct action research on fission (352 million ECU for 1984 to 1987) have been approved by the Council. The state of the cost-shared action programmes will be discussed under each separate chapter.

It should be underlined that sub-programmes including direct and cost-shared action research, are closely coordinated.

2. REACTOR SAFETY

Limited acceptance by the public for operation of nuclear power plants is linked with the concern on safety. Even if the operational safety record is excellent for European reactors, the TMI-2 reactor accident has nourished mistrust in designers, operators and owners of nuclear power plants. The continuing research on safety aspects is aimed at improving the prevention of accidents and the operational reliability of the plant, limiting the consequences of accidents and developing calculation tools for the analysis of events of very low probability.

2.1. JRC research

Most of the activities of the foregoing four-year programme (1980-1983), are to be pursued. Briefly, these projects were :

- Reliability and Risk Evaluation, with two main parts, the European Reliability Data System (implementation of a centralized data bank that provides the information needed for risk assessment application to Light Water Reactors), and the LWR Accident Sequence Analysis (development of methods to improve risk analysis).
- LWR Loss of Coolant Accident Studies are based on simulation in a blow-down investigation facility (LOBI) of the behaviour of the primary cooling system of a large PWR with respect to its thermohydraulic behaviour during a loss of coolant accident caused by pipe ruptures.
- Primary System Integrity with optimization of methodologies for defect detection and sizing in LWRs, and development of methods allowing a reliable prediction of the life-end of reactor components and structures, including investigation of the behaviour of stainless steel under high temperature low cycle fatigue and creep conditions (mainly for Liquid Metal Fast Breeder Reactors); an important feature of this project is the PISC-II exercise (Programme for Inspection of Steel Components), devoted to the assessment of the performance of non-destructive testing techniques used by different countries.
- LWFBR Core Accident Initiation and Transition Phase, with studies on liquid metal boiling and fuel-coolant interaction, and the development of a European Accident Code, which is a modular system of computer programmes to study the different phases of postulated hypothetical accidents.
- LMFBR Post Accident Disassembly Phase, with computer code development and assessment of transient multi-phase flow, after a core-disruptive accident, and of dynamic structure loading and response, improvement in the description of the dynamic mechanical properties of materials submitted to fast transient deformations, and studies

- on the Post Accident Heat Removal on core debris in appropriate catcher systems after a hypothetical, severe accident.

More details can be found in the annual status reports (3). It should be noted that the SUPER-SARA project, an in-pile experiment simulating the behaviour of light water reactor fuel in the event of coolant loss, had to be abandoned.

In the 1984-1987 programme, 192 million ECU have been earmarked for reactor safety. The seven projects below, with comments only where modifications in comparison with the 1980-1983 activities appear, are :

- Reliability and Risk Assessment, where modelling activities may be extended to simulation of the various phases of accidents.
- LWR integrity of components and systems, with an extension to include austenitic welds inspection techniques and possibly an exercise on irradiated materials.
- Study on abnormal behaviour in core-cooling systems in LWRs, linked to a scenario of core uncovery with subsequent fuel damage and fission release; the analysis of available in-pile data, development of models, and specific experimental out-of-pile support are planned.
- Study on severely damaged fuel of LWRs, with modelling of fuel pin and sub-assembly deformation, liquification and relocation, as well as cladding and fuel fragmentation phenomena; tests in support of modelling will be conducted in the existing installations.
- Construction models relating to accidents in fast reactors.
- Study on the properties of materials and on the behaviour of structures in fast reactors.
- Evaluation of a vibrating table, devoted to the feasibility study and possibly design of a large-capacity vibration table.

2.2. Cost-sharing programme

The cost-sharing programme on the Safety of Thermal Water Reactors (1979-1983), with a budget of 6.3 million ECU, supported research in the following fields :

- Area A : Loss of Coolant Accident and behaviour of the Emergency Core Cooling System.
- Area B : Protection of nuclear power plants from an external gas cloud explosion.
- Area C : Release and distribution of radioactive fission products in the atmosphere following a reactor accident.

The programme was complementary to the current national programmes and the JRC programme, and achieved its aim of improving collaboration between national laboratories of the Member States and encouraging exchange of information.

For the period 1984-1987, the Commission proposed a more ambitious programme, as well as an extension to fast breeder reactors (4), with a Commission contribution to research and development projects of 74.6 million ECU and a total budget of 81.3 million ECU. The LWR section included studies on human factors and man/machine interaction, protection against gas-cloud explosions, mechanical and material problems for steel components, thermohydraulics and fuel degradation during LOCA, problems arising due to a release of hydrogen during, and as a consequence of, an accident, source term caused by fission products in severe accidents, atmospheric dispersion of fission products after an accident, and methods used for probabilistic risk assessment.

The section on LMFBRs deals with instrumentation, control and protection, transient analysis, integrity of components and structures, safety aspects of sodium technology, fuel behaviour and fission product transport in severe accidents, as well as movement and interaction of molten materials.

2.3. Action on technological problems of nuclear safety

The Council Resolution of July 22, 1975, asked for a continuous action to perform comparison, analysis and synthesis in the field of coordination of research and harmonization of safety methodologies, codes and standards for LWRs and LMFBRs. Until 1980, only expenses for the Commission staff were covered. From 1980 on, about half of the annual budget, which is of 1 million ECU for 1984 is spent for contractual work, supporting analysis and synthesis studies on codes and standards for Fast Breeder Reactors.

3. NUCLEAR FUEL AND ACTINIDES RESEARCH

Research in this field is done only as direct action at the European Institute for Transuranium Elements in Karlsruhe, part of the Joint Research Center, which is specialized in studies on plutonium and higher transuranium elements. In the preceding multi-annual programme, research was aimed at fast breeder reactor fuel development and fundamental research on actinides.

The 1984-1987 programme provides 66 million ECU for this work. Within the programme, the "Operation Limits of Nuclear Fuels" are being explored by optimizing so-called advanced (i.e., carbide and/or nitride) fast breeder fuel materials and by studying thermophysical and thermodynamic properties of reactor materials at very high temperatures, as they may occur in off-normal situations. Another subject of investigation is the "Transient Behaviour of Oxide Fuels", under which heading the response of nuclear fuel to sudden power changes and the release of fission products under severe fuel damage conditions are systematically being studied. "The Safety of the Nuclear Fuel Cycle" is another field of concern, where research is directed towards the formation of actinides under neutron irradiation and the development of methods for the safe handling of plutonium and plutonium compounds. Furthermore, pilot studies deal with carbide head-end reprocessing procedures and key problems of oxide fuel reprocessing. About 30% of the Institute's research capacity are devoted to basic "Actinide Research", where highly pure samples, also single crystals, of actinide metals and compounds are prepared for measurements of physical and physico-chemical parameters to be carried out at the Institute and in other European laboratories with the aim of elucidating the electronic structure, clarify the binding properties and search for possible applications of these materials.

4. MANAGEMENT OF RADIOACTIVE WASTE

Treatment and disposal of low-level radioactive waste has been safely practised by the Community countries for several decades; for waste containing significant amounts of long-lived isotopes, only intermediate storage has been provided. As the quantities of waste are rapidly increasing, improvements in management and disposal of wastes, including alpha bearing and high-level waste from reprocessing of fuel elements, are needed. The following table gives the estimated quantities of treated and conditioned radioactive waste arising until the year 2000, in the Community countries (5) :

Type of waste	Total quantity of waste per 5-year period (m^3) 1981-1985	1986-1990	1991-1995	1996-2000	Total (m^3) 1981-2000
Low-level	268,600	299,600	335,600	359,200	1,263,000
Medium-level	43,350	57,750	86,050	108,550	295,700
Alpha-waste	15,400	16,800	22,750	23,950	78,900
High-level	190	580	1,700	2,230	4,700

The Community actions have been and are devoted to the characterization of conditioned waste with a view to disposal and the improvement of processes, the study of performance of the various geological disposal options and the corresponding confinement methods as well as the assessment of disposal safety. A twelve-year (1980-1992) Community action plan for radioactive waste ensures concertation in the Community, as well as the orientation and the continuity of the activities. The overall programme and the objectives of the Community programme, have been described extensively elsewhere (6), as well as the main lines of research in the past (7).

4.1. Cost-sharing action

The current programme 1980-1984 (8) with total commitments of 43 million ECU is already the second five-year programme. It covers :

(a) Work to solve certain technological problems involved in the processing, storage and disposal of radioactive waste.

Processing :

- immobilization of low and medium-activity waste; development of processes and operation of pilot installations;
- conditioning of high-activity waste : fuel claddings and residues from dissolvers;
- processing of medium-activity liquid waste;
- processing of waste contaminated by alpha emitters;
- examination and evaluation of high-activity solidified waste;
- immobilization and storage of gaseous waste.

Storage and disposal :

- burial of low-activity solid waste at shallow depth;
- storage and disposal in geological formations;

(b) work to define the general framework for the projects relating to the storage and disposal of radioactive waste :

- evaluation of processes, criteria for the launching of experiments and strategies for waste management;
- studies of the legal, administrative and financial aspects of waste management.

Progress of work is regularly given in the annual reports (9). The proposal for "an action programme on the development of nuclear fission energy (1984-1987)", announces that the Commission will propose a third shared-cost programme (1985-1989) on completion of the current (1980-1984) R&D shared-cost programme on radioactive waste management and storage.

The 1980-1984 programme has allowed to make important strides and to obtain important results both in the field of the waste treatment / conditioning and disposal. On the whole, the results confirm the interest of waste disposal in geological formations; they need to be confirmed, in real conditions, by means of experimental underground disposal projects.

The Commission has prepared the third programme (1985-1989) which takes into account the current JRC programme and is aimed, to a large extent, at validating the basic information acquired so far. To this end, experiments approximating real conditions and the construction of experimental underground disposal facilities, are foreseen. The projects concerning the underground facilities have been proposed by various bodies in the Member States with the support of those Member States and will be open to Community cooperation.

The first part, waste management studies and associated R&D actions would be sub-divided into six tasks :

- System studies, to characterize management schemes for each category, orientate R&D, and integrate work into coherent schemes.
- Improvement of waste treatment and conditioning technologies, aimed at minimization of volume and activity, and promotion of conditioning processes.
- Evaluation of conditioned waste and qualification of engineered barriers, with assessment of long-term behaviour of multi-barrier systems and establishment of the product quality control.
- Research in support of the development of disposal facilities : shallow burial and geological disposal studies, with assessment of long-term behaviour for predictive modelling and development of disposal facilities.
- Safety of geological disposal, covering the evaluation of isolation systems and of radiological risks.
- Joint evaluation of waste management policies, including elaboration of waste management and disposal criteria in cooperation with international organizations competent in this field, and evaluation of possible approaches to waste disposal at Community level.

A second part, construction and/or operation of underground facilities open to Community joint activities would include, for the time being, the pilot underground facility in the Asse salt mine (FRG), the pilot underground facility in the argillaceous layer under the Mol (Belgium) nuclear site, and an experimental underground facility in France, at a site to be determined.

4.2. JRC Programme

In the preceding research programme, research on radioactive waste was part of the activity in the field of Nuclear Safety and the Fuel Cycle, under the heading "Safety of Nuclear Materials". The central theme of the JRC activity was the study of the long-term safety of radioactive waste management. Research was carried out essentially in three directions :

- the evaluation of the long-term risks linked to geological disposal of radioactive waste, carried out both by theoretical assessment studies and by experimental activities aimed at providing the necessary data base (risk evaluation);
- the study of possible modifications of the fuel cycle to keep hazardous radionuclides at the lowest possible level in waste (actinide separation);
- the study of analytical control procedures for actinides in waste (actinide monitoring).

Progress is regularly described in the annual reports (10). The activities have been strictly coordinated with the cost-sharing programme on waste management, and links between these activities are strenghtened by joint coordination of important research actions, such as PAGIS (Performance Assessment of Geological Isolation Systems) and MIRAGE (Migration of Radioisotopes in the Geosphere).

The overall programme objectives were maintained for the 1984-1987 programme, which provides 49 million ECU for this activity. In particular it is intended :

- To continue, at a reinforced level of staff and budget, the activity on long-term safety assessment of geological disposal in both continental and submerged geological formations.
- To maintain the on-going activities on development of non destructive monitoring methods for actinides in waste and possibly extend them to the development of non-destructive methods for physical characterization of high activity waste.
- To continue, at a reinforced level, the activities on actinide separation from medium level waste, taking advantage of the new hot-cell installations of the ADECO complex, and including the cost-benefit evaluations of alternatives under study.

The activities mentioned above have been structured into three projects :

- Waste management and the fuel cycle deals with all aspects linked to fuel cycle installations, and aims at minimizing alpha-waste arising and improving their quality in relation to final disposal.
- Safety of waste disposal in continental geological formations includes both theoretical evaluation activities and linked experimental activities, aiming at providing a long-term safety assessment of waste disposal in continental geological formations.
- Feasibility and safety of waste disposal in deep oceanic sediments; the particular character of sub-seabed disposal made it advisable to organize the activities related to this option as a separate project, although the project goals are similar to those of the second project.

While the previous programmes were essentially executed at the Ispra Establishment, in the 1984-1987 programme the Transuranium Institute of JRC Karlsruhe will contribute to projects 1 and 2 of the programme.

5. SAFEGUARDING AND MANAGEMENT OF FISSILE MATERIALS

A reliable safeguard system for nuclear materials in circulation in Europe, is among the fundamental elements which condition the acceptance of the development of nuclear energy and is one of the means to respond to concern on proliferation of nuclear weapons. As an independent European centre, the JRC plays the role of reference laboratory for the analysis of samples taken during safeguarding inspections. In addition to analytical verification work, the JRC has given, and will give in future, technical assistance to the EURATOM Safeguard Directorate; it advises also plant operators on the implementation of their obligations in the frame of the EURATOM and Non-Proliferation Treaties, or other agreements.

The assistance to the Safeguard Directorate includes the adaptation and checking of complete systems for scientific installations.

The R&D programme 1984-1987, with appropriations of 45 million ECU as a direct action only, continues in sub-dividing activities into three projects :

- Methods and instruments for the determination of fissile materials and for containment and monitoring, has as main activities the evaluation of nuclear material assay, containment and surveillance methods and procedures, including the establishment of validation and training laboratories, and the development of methods and instruments with industrialization of instruments and devices.
- Processing, transmission and evaluation of safeguard data, with research on the nuclear material accountancy data management, and the automatic processing, transmission and evaluation of inspection data, produced by nuclear material assay and containment and surveillance instrumentation.

- Integration of safeguard activities, covering the study of safeguard practices the development of a conversational simulator for the analysis and assessment of safeguard practices, the application of EXPERT system techniques to safeguard studies, and the analysis of safeguard procedures through the fuel cycle.

6. DECOMMISSIONING OF NUCLEAR INSTALLATIONS

The decommissioning of nuclear installations, which have been definitively shut down after having reached design-life or being economically or technically obsolete, has been limited for the time being to relatively small installations. In order to prepare a management system for decommissioning, which will ensure the safety and protection of man and the environment against the potential hazards involved in these operations, a cost-sharing action has been set up by the Commission. A first programme on the "Decommissioning of nuclear power plants" produced valuable information, as documented in the annual reports (11).

A second cost-sharing programme (1984-1988), enlarging the scope from nuclear power plants to nuclear installations in general, with appropriations of 12.1 million ECU, has been approved by the Council (12). Continuing and extending the activities of the first programme, three parts are foreseen. In a first section, there are seven research and development projects :

- long-term integrity of buildings and systems;
- decontamination for decommissioning purposes;
- dismantling techniques;
- treatment of specific waste materials : steel, concrete and graphite;
- large containers for radioactive waste produced in the dismantling of nuclear installations;
- estimation of the quantities of radioactive waste arising from decommissioning of nuclear installations in the Community;
- influence of plant design features on decommissioning.

A second section is devoted to the identification of guiding principles :

- certain guiding principles in the design and operation of nuclear installations with a view to simplifying their subsequent decommissioning;
- guiding principles in the decommissioning of nuclear installations which could form the initial elements of a Community policy in this field.

The final section is mainly concerned with the testing of new techniques within the framework of large-scale decommissioning operations undertaken in Member States. Besides, the exchange of information on such operation would be promoted, for example, by delegating experts from other Member States.

REFERENCES

1. Communication from the Commission to the Council (1983). Research action programme on the development of nuclear fission energy (1984-1987). COM(83)300 final.
2. Council (1983). Council decision of December 22, 1983, adopting a research programme to be implemented by the Joint Research Centre for the European Atomic Energy Community and for the European Economic Community (1984-1987). OJ No. L 3, January 5, 1984, p. 21.
3. Commission (1983). 1982 annual status report - Reactor Safety. EUR 8519.
4. Commission (1983). Proposal for a Council decision adopting a research programme on reactor safety (1984-1987). COM(83)299 final.
5. Commission (1983). Analysis of the present situation and prospects in the field of radioactive waste management in the Community. Part of COM(83)262 final.
6. Orlowski, Cricchio, Girardi. The programme of the CEC and the objectives of the Community in the field of radioactive waste. IAEA-CN-43 /59, International Conference on Radioactive Waste Management, Seattle, WA, USA, May 16-20, 1983.
7. Orlowski, Bresesti. Research and development action of the CEC in the field of radioactive waste management. IAEA-CN-42/424, International Conference on Nuclear Power Experience, Vienna, September 13-17, 1982.
8. Council (1980). Council decision of March 18, 1980, adopting a programme on the management and storage of radioactive waste (1980-1984). OJ No. L 78, March 25, 1980, p. 22.
9. Commission (1983). Research and development on radioactive waste management and storage. Annual Progress Report 1982. EUR 8843, Harwood Academic Publishers GmbH.
10. Commission (1983). Safety of nuclear materials, 1982 annual status report. EUR 8521.
11. Commission (1983). The Community's research and development programme on decommissioning of nuclear power plants. Third annual progress report (year 1982). EUR 8962.
12. Council (1984). Council decision of January 31, 1984, adopting a research programme concerning the decommissioning of nuclear installations. OJ No. L 36 of February 8, 1984, p. 23.

DECOMMISSIONING OF NUCLEAR POWER STATIONS IN COMMUNITY COUNTRIES CARRIED OUT AND PROJECTED

A. CREGUT (CEA) - A.R. GREGORY (CEGB)

INTRODUCTION

The present stage of nuclear energy and the special nature of this industry explains the interest shown in the decommissioning of nuclear power plants.

The number of nuclear plants which have been permanently shut down or which will be shut down in the coming years in Community countries and on the fate of which a decision will have to be taken explains the need for knowledge and experience felt by those responsible for the decommissioning of these units.

To date, the many small units and few larger plants which have been definitively shut down have been handled on an ad hoc basis.

In point of fact, however, the decommissioning of large plants such as nuclear power stations merits a different approach requiring the introduction of measures and procedures allowing them to be dealt with efficiently; this efficiency would imply concern for optimum economy of operations while respecting the safety and protection rules inherent in nuclear energy.

Consequently, plant owners require :

- the tactical and policy elements to guide them in their decisions and choices;
- efficient tools, equipment and processes which meet their needs;
- information gained from experience of decommissioning already carried out which would provide them with a verified background knowledge when dealing with problems.

It should be emphasized that while the choice of the level of decommissioning and the method used to achieve this are based and depend mainly on the national safety and waste management legislation which applies in each State, knowledge of work carried out in other countries and the results achieved will nevertheless be an important element to be taken into account when taking decisions on decommissioning.

Since decommissioning experience to date has not made it possible to draw up codes and guidelines, it is important to review the work carried out by Community countries in particular on the decommissioning of nuclear power plants.

The following paper does not claim to be exhaustive or to make value judgements. Its aim is to list the nuclear power stations shut down in Community countries, to outline the decommissioning levels selected in each case and to underline some interesting aspects of the technical options.

In conclusion it will review what appeared to be the difficulties common to the various projects.

COMMENTS

- The following paper presupposes a knowledge of the definitions of the various levels of decommissioning;
- Although the subject concerns the decommissioning of nuclear power stations, the authors have included other reactors on which current or planned studies or operations include elements of significance which could apply also to power stations.
- The paper is based mainly on the background document containing the communications presented by experts from the various countries at meetings held at Windscale on 10 and 11 May 1983 and on some additional information which the experts kindly made available to the authors.

1. POWER STATIONS WHICH HAVE BEEN SHUT DOWN

1.1 Level of decommissioning selected

The enclosed table entitled 'Nuclear power stations within the EEC which have been definitively shut down' lists the various power stations where decommissioning is underway or planned, stating in each case the type (line), capacity, length of time in operation and decommissioning level selected. From this it can be said that :

- the choice of decommissioning level is independent of the reactor type;
- the level selected is generally level 2 (mothballing in a safe state for a period of 20 to 30 years) or upgraded level 1 (examples : G1, Chinon A1 in France; Gundremmingen and Lingen in the Federal Republic of Germany; Garigliano in Italy);
- clearly for those responsible the concern to put their plants in a safe state is accompanied also by concern for the economy of the project, as a result of which they try to reduce maintenance and surveillance costs and to undertake only the minimum of dismantling work;
- clearly also in many cases the lack of solution for the definitive storage of waste and the lack of effective and tried technical methods explain the desire to postpone complete dismantling of the unit until later.

On the other hand, it is clear that the main countries concerned wish to demonstrate that it is possible to dismantle a nuclear power station entirely. This is case of G2 in France, the KKN and FR2 reactor in the Federal Republic of Germany and the WAGR in the United Kingdom.

These operations are treated as demonstration-type pilot projects. Planning covers many years (between 7 and 12 years) and the planned operations are based on programmes currently underway to develop equipment and processes for cutting, decontamination, etc.

To sum up, it can be said that today the concern for sound financial management, the need for more efficient technical methods, and for confirmation of the value of solutions to be adopted, and the advantage of being able to store waste definitively without passing through intermediate storage, mean that :

- complete decommissioning is delayed,
- the necessary technical resources are developed,
- appropriate solutions are sought on the basis of the experience gained in completely dismantling several large plants.

1.2 Specific technical elements

The second concern for those responsible for a decommissioning project is to work out the tactical approach to be adopted in order to attain the objectives corresponding to the decommissioning level selected. The approach is based on a set of technical options in all the fields affecting decommissioning operations (decontamination, treatment, packaging and storage of waste, cutting tools, operation by remote control, etc.). For this reason it is important for the heads of projects to be informed of the current state of the art. This is what will be done at this conference by presenting the results of the Community five-year plan.

It is important for the heads of projects to know the choices or options adopted in the various Community countries when decommissioning plants.

Below the authors present some of the ideas or conclusions which emerge from studies on the decommissioning of reactors.

GUNDREMMINGEN

This project, which is concerned initially only with the dismantling of circuits and turbine house equipment, has led to the development of electrolytic decontamination.

The tests on 100 tonnes of pipes and turbine equipment have shown the effectiveness of this process, which makes it possible to recover 90% of the contaminated waste.

FR2

Although this is not a power reactor, the studies carried out are interesting because they allow the possibility of extrapolation to the Biblis A power station.

The important points which arise from these complete studies of the dismantling of the reactor block are :

- the choice of cutting under water and in situ for equipment with high radioactivity;
- the maximum possible utilization of mechanical cutting, in particular for substantial thicknesses, the plasma arc being reserved for small thicknesses;
- the need to set up an extensive ventilation and containment system to carry out the work;
- the choice of a central rotating mast, equipped with two separate arms, one for the manipulator and the other for the tool holder.

LINGEN

The measures taken whereby only the reactor building, the link building and the waste treatment building and their equipment were left standing tend to make for maximum ease of monitoring and surveillance. They demonstrate the need for installing a ventilation system adapted to the new situation.

KKN

Following a period of mothballing at level 2, it is now planned to dismantle this power station completely. Two points should be underlined : the choice of the type of dismantling equipment involving a separation of tool-holding machinery and handling machinery and the quantity of equipment to be created in order to carry out the operation successfully (buffer storage, handling cells, cutting and monitoring cells; transfer lock, special drainage and ventilation system). Particular note should also be taken of the cutting of the biological shield concrete using explosives.

GARIGLIANO

A decision was taken to leave this plant at level 1. Nonetheless, items of technical interest are the rotating machine for the mechanical decontamination of 24" pipes and the pipe and tube cutting machine which were used on this plant, as well as the various decontamination processes used (hydro-laser, chemical and mechanical).

CHINON A1

As regards this type of plant mention can be made of the advantage of separating the reactor and circuit section into several areas, thereby reducing, during the period at level 1, the disadvantages represented by the phenomena of natural convection and condensation at certain particular points of these spaces.

G2 MARCOULE

Like that of the KKN this project provides for the total dismantling of the plant.

Two technical difficulties are worthy of note : the problems posed by the prestressing of the reactor vessel and what to do with the graphite internals on which a policy decision has not yet been taken.

Note also the size of tool-holding and manipulating equipment, which is linked to the size of the reactor and the number of handling operations to be carried out.

WAGR

For this unit total dismantling is planned. As in the case of G2 the study shows the importance of handling and remote-controlled equipment; efforts at developing methods of thermal cutting (oxy-propane torch and plasma arc cutting).

OTTO HAHN

Although this is not a nuclear power station and the core structures could be stored temporarily en bloc; the dismantling which was carried out up to level 3 is worthy of note. In particular it shows the number and extent of the measurements which accompany dismantling and storage operations.

1.3 Aspects common to all decommissioning projects

Analysis of the various documents presented on the decommissioning of Community power stations which have been definitively shut down, gives rise to the following comments concerning aspects common to all projects :

PROCEDURE

Whatever the type of plant, the level of decommissioning proposed and the methods chosen to attain this, the procedure for obtaining the endorsement of national safety authorities was very lengthy.

It is true that the situation was a new one requiring the introduction of ad hoc rules or, where the legislation had already been adapted, adequate monitoring to get the system underway.

CHOICE

With respect to the level of decommissioning, the methods used and the storage of waste, those responsible had to carry out extensive studies and numerous tests before making their choice. Again this was due to the lack of precedents.

TECHNOLOGY

Three important points must be borne in mind :

a) In almost all cases, the equipment and fittings necessary for cutting and manipulation had to be designed especially for the project in question.
Even where existing equipment could be adapted, the changes which had to be carried out were considerable.

b) <u>Cutting tools and decontamination processes</u>
Here again it was obviously necessary to check the validity of the choice and where necessary perfect the tool and the process.

c) <u>Preliminary work before dismantling operations</u>
The importance of the preparation work before beginning dismantling operations should also be emphasized.
For example, ventilation systems have to be changed or adapted, means of decontamination planned, temporary containment installed and in certain cases also supplementary biological shielding.

RADIOLOGICAL PROTECTION

Knowledge of all aspects of the radioactivity contained in the plant (quantity, distribution, types of physical-chemical fixation etc) is an essential factor on which depend the choice of decontamination process, cutting methods and the method of packaging and storing of waste; indeed it is to some extent on this knowledge that the approach to decommissioning will depend.

It must be emphasized that this knowledge involves precise calculations and numerous monitoring measurements.

2. GENERAL CONCLUSIONS

While it is true that experience acquired to date does not permit a definitive policy and codes and guidelines to be drawn up, it nonetheless makes a significant contribution towards decommissioning options and techniques. Above all, this experience is already adequate to permit the introduction of a true research and development programme covering the whole range of the technical requirements of decommissioning.

There is an obvious need to bring the decommissioning of some large units to a successful conclusion as rapidly as possible. These demonstration decommissioning cases should make it possible to set up and establish the necessary industrial infrastructure to satisfy the increased national and international decommissioning requirement, which are likely to be very substantial from now on.

I – NUCLEAR POWER STATIONS

Name	Country	Line	Capacity MW(e)	Exploitation	Decommissioning		
					Level	Envisaged time scale	Comments
GUNDREMMINGEN KRB- A	FRG	BWR double cycle	250	1966) 1977) 10 years	II	1987	
LINGEN KWL	FRG	BWR Superheat fuel	250	1968) 1977) 9 years	II/ 30 years		
GARIGLIANO	Italy	BWR	160	1965) 1978) 13 years	I/ 30 years		
CHINON I	France	GCR	70	1963) 1973) 10 years	I/ 30 years		
G2 Marcoule	France	GCR	40	1958) 1980) 22 years	I/ to date	II 6 years III 20 years	
AGR Windscale	UK	AGR	30	1963) 1981) 18 years	III	1987-1993	
KKN Niederaichbach	FRG	GC HWR	100	1972) 1974) 2 years	III	1984-1990	
II NUCLEAR REACTORS							
			MWth				
G1 Marcoule	France	GCR	20	1956) 1968) 22 years	II/ 30 years		
EL3 Saclay	France	HWR	18	1957) 1979) 22 years	II/ 30 years		
OTTO HAHN	FRG	PWR	10	1969) 1979) 10 years	III		
FR2 Karlsruhe	FRG	HWR	44	1961) 1981) 20 years	II/ 1968		

THE EUROPEAN COMMUNITY'S PROGRAMME OF RESEARCH ON THE DECOMMISSIONING OF NUCLEAR POWER PLANTS: OBJECTIVES, SCOPE AND IMPLEMENTATION

B. HUBER
Commission of the European Communities
Brussels, Belgium

Summary

The European Community's research activities on the decommissioning of nuclear installations are aimed at developing effective techniques and procedures for ensuring the protection of man and his environment against the potential hazards from nuclear installations that have been withdrawn from service.
The first five-year (1979-1983) programme of research on the decommissioning of nuclear power plants has comprised seven R&D projects concerning the following areas:
- maintaining disused plants in safe condition;
- surface decontamination for decommissioning purposes;
- dismantling techniques;
- treatment of the main waste materials arising in decommissioning, i.e., steel, concrete and graphite;
- large containers for decommissioning waste;
- arisings and characteristics of decommissioning waste;
- plant design features facilitating decommissioning.

The research work was carried out by organizations and companies in the Member States under 51 research contracts, most of them cost-sharing.
The Commission is now launching a new five-year (1984-1988) programme of research on the decommissioning of nuclear installations.

1. INTRODUCTION

Production of electricity from nuclear energy, having started about thirty years ago, is now meeting a fifth of the total electricity requirements of the European Community, with more than hundred power reactors operating. Ten nuclear power plants, all of them small or medium sized, have already been finally shut down, but so far, none has been decommissioned to the final stage.

In 1979, the European Community started a first five-year programme of research on the decommissioning of nuclear power plants. Having completed this programme, the Commission organized this conference in order to present and discuss the results obtained. These discussions may also throw new light on some of the questions which remain to be answered under the new five-year (1984-1988) programme of research on the decommissioning of nuclear installations, which is now being started up.

The purpose of this paper is to give an overview of the objectives, the scope and the implementation of the Community programme on decommissioning, as a general introduction to the topical papers which will be presented at this conference.

2. OBJECTIVES OF THE PROGRAMME

The general objective of the programme has been set out by the Council of the European Communities, when it adopted the programme in 1979, considering: "Certain parts of nuclear power plants inevitably become radioactive during operation; it is therefore essential to find effective solutions which are capable of ensuring the safety and protection of both mankind and the environment against the potential hazards involved in the decommissioning of these plants" (1).

The radioactivity associated with a nuclear power plant, once the plant has been finally shut down, decays according to strict laws and there is no practical way to influence the amount of radioactivity subsisting at a given time. Consequently, protection against the potential hazards due to this radioactivity means to isolate the latter from man and his environment by appropriate barriers. Since strong radioactivity barriers are already existing in every nuclear power plant, in-situ storage of the radioactive components is a possible means to isolate the radioactivity till the initially dominating shorter-lived radionuclides have died away. Eventually, however, the remaining radioactive material is to be transferred to a final radioactive waste repository, because its isolation in the very long term is best achieved there, and in order to make it possible to release the plant site for other uses. This transfer, which involves the breaking-up of the existing radioactivity barriers and the creation of new ones, is a critical phase in decommissioning. Accordingly, the major part of the Community programme has been devoted to the operations required in this phase, i.e. dismantling and decontamination of plant components, and treatment and packaging of the arising waste materials.

The main effort in the programme has been aimed at providing efficient techniques and procedures for decommissioning, considering radiation protection, restriction of radioactive waste arisings, and cost effectiveness. Though a number of conventional techniques appear to be applicable in decommissioning, there is a need to improve their performances and to optimize and characterize them with a view to the special requirements due to radioactivity, e.g., control of airborne contamination and remote operation.

Another aim of the programme has been to provide basic information needed in decommissioning, such as data on the distribution, the level and the radionuclide composition of the radioactivity associated with nuclear power plants finally shut down.

A third aim has been to identify and develop plant design features facilitating decommissioning.

3. SCOPE OF THE PROGRAMME

The subjects of the programme are listed in Table I. Figure 1 illustrates how the various R&D Projects relate to successive steps of decommissioning (since this is a simplified scheme, it does not reflect all possible procedures). Concerning radioactive waste management, only aspects and procedures that are specific to the waste arising from the dismantling of nuclear power plants have been investigated under the programme on decommissioning; other procedures, in particular waste disposal methods, are being developed in the framework of the Community Programme on Radioactive Waste Management.

The scope and the particular objectives of the various projects are outlined in the following.

TABLE I. SUBJECTS OF THE PROGRAMME 1979-1983

A. Research and development projects:
- Project N° 1: Long-term integrity of buildings and systems;
- Project N° 2: Decontamination for decommissioning purposes;
- Project N° 3: Dismantling techniques;
- Project N° 4: Treatment of specific waste materials: steel, concrete and graphite;
- Project N° 5: Large transport containers for radioactive waste produced in the dismantling of nuclear power plants;
- Project N° 6: Estimation of the quantities of radioactive wastes arising from decommissioning of nuclear power plants in the Community;
- Project N° 7: Influence of nuclear power-plant design features on decommissioning.

B. Identification of guiding principles

3.1. In-situ storage of plants

In-situ storage appears to be the preferred procedure for the commercial nuclear power plants that have already been finally shut down. The implications of storage over extended time periods, ranging from several decades to about a hundred years, have been investigated under this programme.

Measures to put plants into conditions assuring safety essentially through passive means and with a minimum of continued surveyance and maintenance have been studied. The operations required are not technically demanding, but they need to be carefully designed and planned in order to minimize radiation exposures and later interventions. Plant systems which are to be maintained in operation or in operable conditions (to serve in the dismantling of the plants) were identified.

The mode and pace of degradation of various materials as they exist in nuclear power plants have been studied. This involved inspections of plants, to identify critical areas, and examinations of samples. Measures to prevent or control degradation have been proposed.

These studies showed no problem which would make the feasibility of in-situ storage of finally shut-down nuclear power plants doubtful. The radiological consequences and costs of in-situ storage have been estimated. It certainly allows to reduce the occupational radiation exposure during dismantling of the plants. To back up these studies, the experience being gained with the plants that have already been finally shut down should be continuously analysed.

3.2. Decontamination for decommissioning purposes

Surface contamination represents only a minor amount of radioactivity compared with the internal activation by neutron exposure of the components near the reactor core, but it is spread over a wider range of components and more readily releasable. Therefore, surface contamination is an important source of low-level radioactive waste and a concern for radiation protection in decommissioning operations.

Surface decontamination, i.e. removal of the radioactive surface layer from a plant component, facilitates the further handling and processing of the component. The removed radioactive substance should be concentrated into a small volume of secondary waste that is easy to condition for

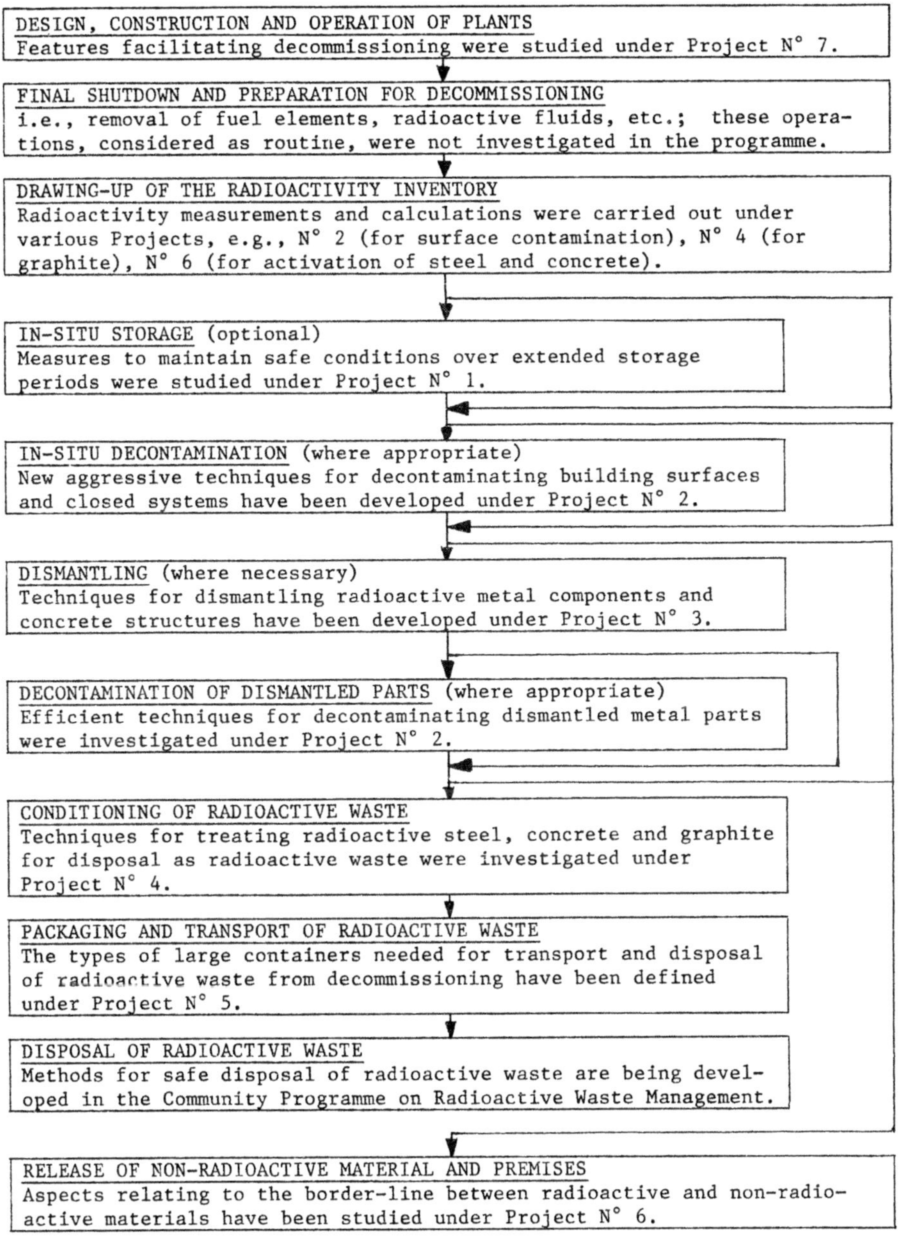

Figure 1. Relation of the various R&D Projects to successive steps of decommissioning (simplified scheme; the ramifications indicate that different plant components may follow different paths)

disposal. If the decontamination is efficient enough, it makes it possible to consider the treated component as non-radioactive material and is then a way to reduce the overall radioactive waste volume and to recover valuable raw materials.

The importance of decontamination for the purpose of unrestricted release of material may be illustrated by the following estimates, which are based on a large Pressurized Water Reactor and on present-state decontamination technology (2). Of a total mass of 11700t of radioactive components, 4500t of components, exposed to airborne contamination only, were considered decontaminable. Of the remaining mass of 7200t,

- 12% are components considered border-line cases,
- 73% are components contaminated only that were considered non-decontaminable,
- 15% are components internally activated and, therefore, non-decontaminable.

Decontamination for the purpose of unrestricted release is not yet a common procedure, and for the components considered border-line cases the available information is not sufficient to decide whether decontamination is appropriate, taking into account the attendant radiation exposure, secondary waste production and cost. A low level and penetration depth of the contamination and a low surface to volume ratio of the component are characteristics which are favourable for decontamination. The range of components for which decontamination for unrestricted release is feasible is being enlarged by the development of more efficient decontamination procedures, which may be more aggressive than those currently used in operating plants, since weakening of the components treated can be accepted.

The following techniques have been developed for the decontamination of metals:

- techniques using chemically aggressive decontaminants in liquid and in gel-like form;
- electrochemical techniques;
- hydromechanical techniques, i.e., high-pressure water lance and erosion by cavitation.

For the decontamination of concrete, flame scarifying has been developed. With this new technique, which is based on rapid heating using an oxyacethylene torch, thin surface layers (about 1 mm in each pass) can be removed from concrete structures.

Both electropolishing and flame scarifying have been successfully tested at a significant scale in a shut-down Boiling Water Reactor plant.

Also in the framework of this Project, valuable information has been obtained on the distribution, the radionuclide composition, the chemical nature and the penetration depth of surface contamination in nuclear power plants.

3.3. Dismantling techniques

In the decommissioning of nuclear power plants, large radioactive components and structures must be dismantled. Important examples are:

- the highly radiating reactor internals of light-water reactors, which are made of stainless steel with thicknesses up to a hundred millimetres;
- the reactor pressure vessels of light-water reactors, which consist of stainless steel clad carbon steel, with thicknesses up to 300 mm in the wall and up to 600 mm at the vessel flange;
- biological shielding structures, consisting of reinforced concrete of several metres thickness.

The dismantling of substantially radioactive components must be carried out remotely with adequate radiation shielding for personnel protection. Where possible, segmenting may be performed under water.

There exist various conventional dismantling techniques which may be employed in decommissioning. Generally speaking, thermal segmenting techniques tend to achieve high cutting rates but pose problems due to the production of fine particles. Mechanical techniques tend to be time-consuming and to need heavy manipulators.

Various thermal and mechanical cutting techniques have been tested on components made of stainless steel, carbon steel and concrete, using existing equipment. These tests have served to optimize process parameters, to assess and compare cutting performances and by-product characteristics, and to examine the suitability of filter systems.

Three thermal cutting techniques have been developed in particular, i.e., plasma cutting (including underwater operation), plasma-oxygen cutting and laser cutting.

The combined plasma-oxygen torch is aimed at cutting reactor pressure vessels from the inner, stainless steel clad side. With oxygen burning alone, the cut could not be started at the clad face, whereas the plasma torch alone would not achieve the necessary cutting depth. With the combined torch developed, clad plates with thicknesses up to the target value of 600 mm were successfully cut. However, the procedure produces a large amount of fine particles (<0.3 µm) and the filter combinations tested, while showing good efficiency, failed to reach sufficient service time.

Laser cutting is a new segmenting technique which is expected to benefit from the development of laser for other applications. Its performances in cutting steel have been substantially enhanced by the use of an oxygen jet to remove the molten material from the kerf.

Intergranular fissuration has been studied as another new method for segmenting metal components, which would avoid the production of fine particles.

For the dismantling of radioactive reinforced concrete structures, a diamond-tipped circular saw has been developed which achieves substantially higher cutting rates than equipment available before. In a recent test of the saw, a one metre cube was cut out from a vertical reinforced concrete wall. As an alternative to sawing, explosive techniques have been investigated with a view to employ them with a high degree of control.

Systems for remotely controlled decommissioning operations have been reviewed in order to identify the areas needing future research.

3.4. Treatment of waste materials

The waste from the dismantling of nuclear power plants consists mainly of steel, concrete and - for gas-cooled reactors - graphite. Methods for treating these materials with a view to disposal or, for steel, possibly reuse have been investigated.

One of these methods is melting of steel, which offers several advantages with regard to waste disposal, i.e. reduction of the bulk volume, reduction of the surface exposed to corrosion, immobilization of surface contamination by incorporation into the base metal. On the other hand, the possible decontamination effect of melting is important with a view to the recycling of steel.

Melt tests using low-level radioactive steel from different reactor types have been carried out in order to determine the distribution of the various radionuclides on the metal product, the slag and the off-gas. An installation for melting radioactive steel waste, which can be repeatedly mounted and operated within the controlled area of nuclear power plants finally shut down, is being designed. Experiments aimed at separating cobalt from steel, using special melting processes, failed to achieve separation factors high enough to encourage further work in this direction.

Coating techniques using epoxy or polyester resins to durably fix surface contamination on metal and concrete have been developed. The immobilization of concrete dust with silicate solutions has been investigated.

Various management modes have been assessed for graphite waste, considering incineration, sea disposal, shallow land burial and deep geological disposal. This research involved in particular determination of the radionuclide inventory of Magnox Reactor and Advanced Gas-cooled Reactor graphite, conceptual studies of the incinerator flow-sheet and of waste packages, leach tests of irradiated graphite samples, and radiological assessment of the management modes considered.

3.5. Large waste containers

Radioactive waste resulting from the dismantling of major reactor components should be transported and disposed of in larger units than those currently used for other types of radioactive waste, in order to reduce the required amount of segmenting and, consequently, the radiological impact and cost of decommissioning. As a first step, a system study has been carried out, which has led to the definition of four types of waste container for further elaboration, all as large as possible within the limits set by the railway freight gauge.

3.6. Estimation of radioactive waste arisings

The low-level radioactive waste produced in the dismantling of nuclear installations will ultimately constitute a substantial part of the overall volume of radioactive waste generated by nuclear industry. Estimates of the future arisings of such waste are therefore needed for the planning of waste disposal facilities. The radioactivity associated with the waste must be known, with particular reference to long-lived radionuclides, in order to classify the various types of waste with respect to the appropriate disposal modes. This knowledge is also needed to predict the decrease of radiation levels as a function of time, which is an important factor to be considered in the timing of dismantling. This timing will in turn determine the point in time where radioactive waste from decommissioning is to be finally disposed of. Since this Project involves the definition of reference strategies for decommissioning, it is regarded as a long-term task. The work carried out under the 1979-1983 programme has been focused on two particular aspects, i.e. long-lived activation products in concrete and steel, and basic scientific information related to the border-line between radioactive and non-radioactive material.

Radionuclide contents have been measured on activated steel and concrete samples from Boiling Water Reactor plants finally shut down. In particular, the thickness of the activated inner layer of biological shields has been determined by the examination of boring samples.

A number of relevant source elements for the activation of steel and concrete are trace elements whose concentrations are not defined in material specifications. Data on these concentrations, sometimes at the ppm level, are needed for the calculation of activation. In order to explore the ranges of the relevant trace element concentrations to be expected in the materials used in nuclear power plants, non-radioactive samples from a larger number of origins were analysed. These samples included reactor-grade stainless steel and carbon steel from various manufacturers as well as concrete from nuclear power plants in various Community countries.

A methodology for evaluating the radiological consequences of the management of very low-activity steel and concrete from the dismantling of nuclear power plants has been prepared. Such a methodology is necessary if "de minimis" radioactivity levels are to be defined. The measuring techniques required for proving that a material is non-radioactive have been reviewed.

3.7. Plant design features facilitating decommissioning

The design of nuclear power plants has continuously evolved and many improvements made with a view to operation of the plants will also facilitate decommissioning. An example are improvements of radioactivity barriers, such as fuel claddings and steam generators, which reduce the contamination of the plants. On the other side, increased safety requirements tended to increase the volume of the components and structures to be dealt with in decommissioning.

The objective of this Project has been to identify and develop reasonable improvements in the design of nuclear power plants with a view to decommissioning. Though requirements related to the safety and reliability of reactor operation, as well as cost-effectiveness aspects restrict the latitude for modifying plant designs, the following specific features appeared to merit closer investigation:

- control of the cobalt and niobium content of reactor steels;
- cobalt-free materials to substitute cobalt alloys used, e.g., in valve seatings;
- coatings, in particular removable ones, to protect concrete against contamination;
- reactor shielding design features facilitating dismantling;
- documentation system for deferred decommissioning.

3.8. Identification of guiding principles

The identification of guiding principles in the field of decommissioning has been considered a long-term task, since it must be based on an appropriate body of technical information. The 1979-1983 programme comprised a number of relevant studies in this regard.

Moreover, available material in the Member States that could serve as a basis for guiding principles in the field of decommissioning has been assembled. It appeared that the existing systems of authorisation and control, together with the radiological protection standards in force, already make it possible to decommission nuclear power plants on a case-by-case basis. However, no specific technical regulations on decommissioning exist. A basic uncertainty as regards final decommissioning, i.e., release of sites for re-use, lies in the absence of specified criteria for distinguishing non-radioactive material from radioactive material.

4. IMPLEMENTATION OF THE PROGRAMME

The research work has been carried out by organizations and companies in the Member States under 51 research contracts, most of them cost-sharing, which are listed in Table II. The Commission had a budget of 4.7 million ECU for the five-year programme. In the implementation of the programme, the Commission has been assisted by an Advisory Committee on Programme Management, which is composed of experts appointed by the Member States governments and of Commission officials.

Due to various reasons, the contractual research activities could not start before 1980. On the other side, a number of contracts have been prolonged on request of the contractors and are now running till mid 1984.

TABLE II. LIST OF RESEARCH CONTRACTS

PROJECT	CONTRACT N°	CONTRACTOR - SUBJECT OF RESEARCH
N° 1 LONG-TERM INTEGRITY	DE-A-001-UK	CEGB - Degradation of building plant and materials
	DE-A-002-F	CEA, IPSN - Long-term integrity of buildings and systems
N° 2 DECONTAMINATION FOR DECOMMISSIONING PURPOSES	DE-B-001-I	CNEN - Delegation of an expert to the USNRC in relation to the clean-up of the TMI-2 plant
	DE-B-002-D	Salzgitter AG - Decontamination of concrete surfaces by flame scarifying
	DE-B-003-F	Alsthom-Atlantique Neyrtec - Erosion of metal surfaces by cavitation at very high velocity
	DE-B-004-D	KW Lingen GmbH - Composition of contamination layers and efficiency of decontamination
	DE-B-005-I	ENEL - Vigorous decontamination tests of steel samples in a special test loop
	DE-B-006-D	KRB Gundremmingen GmbH - Development of economical decontamination procedures
	DE-B-007-F	CEA, CEN Saclay - Research and development on decontamination by gel-based decontaminants
	DE-B-008-F	CEA, CEN Cadarache - Metal decontamin. by chemical and electrochemical methods and by water lance
	DE-B-009-D	NIS GmbH - Economic assessment of decontamination for unrestricted release
	DE-B-010-F	Alsthom-Atlantique Neyrtec - Follow-up of contract N° DE-B-003-F: Testing of new cavitating elements
	DE-B-011-B	CEN/SCK Mol - Delegation of an expert to the USNRC in relation to the clean-up of the TMI-2 plant
N° 3 DISMANTLING TECHNIQUES	DE-C-001-D	Transnuklear GmbH - Comparative assessment of various thermal and mechanical dismantling techniques
	DE-C-002-I	AMN SpA - Plasma techniques for cutting mineral and metal materials
	DE-C-003-UK	CEGB - Large diamond-tipped circular saws for cutting reinforced concrete structures
	DE-C-004-D	Salzgitter AG - Combined plasma-oxygen torch for cutting steel pressure vessels
	DE-C-005-I	FIAT TTG - Dismantling of metal components and concrete structures using laser beam
	DE-C-006-UK	Taylor Woodrow Ltd - Explosive demolition techniques for concrete structures
	DE-C-007-F	CEA, CEN Grenoble - Segmenting of steel components using intergranular fissuration
	DE-C-008-D DE-C-009-B	GST mbH -) Review of systems for remotely con- ACEC -) trolled decommissioning operations
N° 4 TREATMENT OF SPECIFIC WASTE MATERIALS	DE-D-001-UK	NRPB - Assessment of various management modes for radioactive graphite waste
	DE-D-002-UK	British Steel Corporation - Treatment of contaminated steel waste by melting
	DE-D-003-F	Ecopol - Immobilization of contamination on metals by coating with thermosetting resins
	DE-D-004-F	Séri Renault - Removal of cobalt from steel using special melting processes

TABLE II. LIST OF RESEARCH CONTRACTS (continued)

PROJECT	CONTRACT N°	CONTRACTOR - SUBJECT OF RESEARCH
N° 4 TREATMENT OF SPECIFIC WASTE MATERIALS	DE-D-005-UK	Taylor Woodrow Ltd - Treatment of concrete with silicate solutions to prevent dusting
	DE-D-006-I	Nucleco SpA - Coating of materials to prevent corrosion, fix contamination, avoid powder formation
	DE-D-007-F	CEA, CEN Grenoble - Follow-up of contract N° DE-D-003-F
	DE-D-008-D	Siempelkamp Giesserei - Design of an installation for melting radioactive steel waste
N° 5 LARGE TRANSPORT CONTAINERS	DE-E-001-UK	UKAEA) System of large transport containers) for waste from dismantling light-
	DE-E-002-B	Transnubel) water reactors and gas-cooled) reactors
N° 6 ESTIMATION OF RADIO-ACTIVE WASTE QUANTITIES	DE-F-001-D	KWL Lingen GmbH - Activation products in the biological shield of the Lingen reactor
	DE-F-002-D	KRB Gundremmingen GmbH - Activation products in the biological shield of the KRB-A reactor
	DE-F-003-I	Nucleco SpA - Activation and radiation of the Garigliano reactor pressure vessel
	DE-F-004-UK	Imperial College - Trace element content and activation of low-alloy and stainless reactor steels
	DE-F-005-F	CEA, CEN Saclay - Determination of trace elements on concrete samples from various nuclear reactors
	DE-F-006-UK	NRPB) Methodology for evaluating radiologi-) cal consequences of the management of
	DE-F-007-F	CEA, IPSN) low-level radioactive waste from the) dismantling of nuclear power plants
	DE-F-008-D	GKSS) Review of techniques for) measuring very low-level
	DE-F-009-F	CEA, IPSN) radioactivity in relation to) decommissioning
N° 7 INFLUENCE OF DESIGN FEATURES ON DECOM-MISSIONING	DE-G-001-UK	National Nuclear Corporation - Catalogue of design features facilitating the decommissioning of AGRs
	DE-G-002-UK	Taylor Woodrow Ltd - Design features of civil works facilitating refurbishing or dismantling
	DE-G-003-D	Kraftwerk Union AG - Erosion-corrosion testing of cobalt-free materials to substitute cobalt alloys
	DE-G-004-UK	UKAEA, AERE Harwell - Control of the cobalt content of reactor-grade steels
	DE-G-005-I	Nucleco SpA - Removable coatings for the protection of concrete against contamination
	DE-G-006-F	CEA, CEN Saclay - Characteriz. and improvement of coatings protecting concrete against contamination
	DE-G-007-I	NIRA and ENEA - Evaluation of the design features facilitating decommissioning of PWRs
	DE-G-008-UK	Electrowatt Engg. Serv. - Concepts minimizing the activation of the biological shield
	DE-G-009-D	Dyckerhoff & Widmann - Biological shield design with dose-reducing effect for decommissioning
	DE-G-010-D	Technische Hochschule Darmstadt - Documentation system for decommissioning of nuclear power plants

Emphasis has been laid on the exchange and dissemination of the results obtained. Beside this conference, this has been done through regular meetings of people working in the same area and through publication of annual progress reports of the programme (3)(4)(5) and of final reports of individual research contracts.

5. OUTLOOK

A new five-year (1984-1988) programme of research on decommissioning (6) is now being started up. It has a more general scope than the first programme, i.e., decommissioning of nuclear installations.

In the new programme, the particular techniques being developed should be increasingly integrated into overall decommissioning schemes extending from the start of the decommissioning operations to the ultimate disposal of the radioactive waste, account being taken of the waste disposal methods that already exist or are being developed in various member countries.

In the course of the progressive development of new techniques, ever greater importance will attach to the testing of these techniques under representative conditions, in particular the presence of radioactivity, and at large scale. Because of the importance of this subject, it has been added to the new programme as a separate section.

The Community research activities on decommissioning represent a long-term effort but they should not become open-ended. Their logical copestone would be the decommissioning of a few typical large nuclear installations as pilot operations.

REFERENCES

1. Council (1979). Council Decision of 27 March 1979 adopting a research programme concerning the decommissioning of nuclear power plants. OJ N° L 83, 3.4.1979, p. 19.
2. LORCHER, G. and PIEL, W. (1983). Dekontamination von Komponenten stillgelegter Kernkraftwerke für die freie Beseitigung. EUR 8704.
3. Commission (1981). The Community's research and development programme on decommissioning of nuclear power plants. First annual progress report (year 1980). EUR 7440.
4. Commission (1982). The Community's research and development programme on decommissioning of nuclear power plants. Second annual progress report (year 1981). EUR 8343.
5. Commission (1984). The Community's research and development programme on decommissioning of nuclear power plants. Third annual progress report (year 1982). EUR 8962.
6. Council (1984). Council Decision of 31 January 1984 adopting a research programme concerning the decommissioning of nuclear installations. OJ N° L 36, 8.2.1984, p. 23.

FACTORS TO BE CONSIDERED IN SELECTING A DECOMMISSIONING STRATEGY

A.R. GREGORY - Central Electricity Generating Board, UNITED KINGDOM

A. CREGUT - Commisariat a l'Energie Atomique, FRANCE

SUMMARY

The introduction outlines the scope of the work that will be carried out during the three stages of decommissioning a nuclear power station nothing that the content of each stage may vary somewhat for reactor system and country in particular whether defuelling forms an integral part of Stage 1. Some important factors to be taken into account are the natures and quantities of contaminated and neutron activated materials and the techniques required to dismantle the plant and manage the waste.

Timescales for carrying out each stage are discussed in relation to water and gas cooled reactor systems noting the benefits of delay in terms of radiological protection during decommissioning and limiting effects of radiological protection during decommissioning and limiting effects of the long lived radioactive isotopes.

It is suggested that three courses of action may be considered.

OPTION 1 - Complete Stages 1, 2 and 3 as an ongoing project within 10 to 15 years from shutdown.

OPTION 2 - Complete Stages 1 and 2 within 7 to 10 years from shutdown but defer Stage 3 up to 50 years or longer

OPTION 3 - Complete Stage 1 and defer Stages 2 and 3 for up to 50 years or longer.

In considering these options attention must be paid to the long term integrity of buildings and systems which was the subject of studies carried out by CEGB and CEA as part of the European Community Programme.

It is concluded that following final or definitive shutdown the reactors will be defuelled as soon as reasonably practicable. A decision to proceed to Stages 2 or 3 depends on safety, technical, economic, national policy and public opinion.

INTRODUCTION

1. There is general agreement that three stages of decommissioning may be identified, although the exact definition may vary from country to country to accommodate different reactor systems and policy.

2. Broadly Stage 1 decommissioning relates to the period immediately following final shutdown of the nuclear power plant. It is usually assumed that this will be a planned operation rather than precipitated by a breakdown or accident which will not be considered here, so that some small benefit may be realised by optimising the management of the last fuel charge.

3. For light water reactors defuelling is considered to be the responsibility of the operator and part of the process of shutdown. This could be completed within two or three months when, in France for example, responsibility would pass to the decommissioning team and Stage 1 would start. This stage covers the process of making the plant safe, removing operational solid wastes, sludges and resins, including those arising from the aggressive decontamination of the primary circuit, steam generators, other auxiliary and plant buildings no longer required.

4. Gas cooled reactor defuelling is, however, a more major task and will probably extend over several years so that in United Kingdom it is included in Stage 1 and the other tasks will be carried out in parallel with defuelling so that responsibility rests with the operator.

5. Defuelling and all other work associated with Stage 1 will be essentially be an extension of the operating mode and may be carried out using existing plant and facilities by either operating or decommissioning staff as appropriate to the LWR or GCR.

6. Stage 2 decommissioning has the objective of dismantling all radioactive and non-radioactive plant and buildings external to the biological shield. The radioactive, mostly contaminated, waste material is removed off site for disposal or storage in accordance with the appropriate regulations and the residual reactor structure is sealed off until Stage 3 commences.

7. The principal primary coolant circuit contaminated plant includes for PWR the steam generators, coolant pumps, primary circuit pipework and pressuriser. For BWR it includes, main pipework, turbine, condenser feedheating train. Both systems have extensive auxiliary, radwaste and fuelling plant and buildings that will also be contaminated to some extent.

Inactive plant removed at Stage 2 includes generators, main and auxiliary electrical equipment and other plant and buildings no longer needed with the exception of that for essential services to complete Stage 2 and maintain the required safety standards.

8. Although there are obvious differences between light water and gas cooled reactor systems, Stage 2 is characterised firstly by the ability to dismantle the plant using built-in facilities in conjunction with brought-in engineering plant and equipment. This is readily available and well proven needing only little modification to match the specialised tasks. Secondly, the active waste arising, though significant in terms of mass will not generally require elaborate or remote handling techniques. It should, however, be noted that GCR boilers and auxiliary circuits tend to be relatively less contaminated than LWR and in consequence Stage 2 does not represent a major part of the total GCR decommissioning costs.

9. Stage 3 is the removal of the reactor itself with its biological shield, or pre-stressed concrete pressure vessel and the final clearance of the site rendering it safe for future use. This is the most difficult stage because the activated fixed reactor structure demands the use of specialised remote operated systems and, although suitable techniques are being developed, the detailed design, planning and proving of the complete operation will be very demanding. It must be noted also that over 90% of the total radioactive inventory both at 10 years and 100 years from final reactor shutdown is associated with these structures. It is, therefore, the most expensive part of the overall task.

TIMESCALES

10. The assessed timescales for each stage of decommissioning vary to some extent for each reactor system and individual site. However, taking account of the time required to discharge fuel from reactors and allow it to cool in the ponds, the availability of transport flasks and reprocessing or storage capability, Stage 1 might take from 2 to 5 years from shutdown to complete. Some Stage 2 work could be undertaken at this stage, such as delagging and dismantlement of some conventional plant items provided this work did not diminish the high safety standards required for nuclear sites. It would also be necessary to ensure essential services and facilities to carry out Stage 2 and 3 are retained.

11. Stage 2 would be completed immediately after Stage 1 in 3 to 5 years to be followed by Stage 3 taking a little longer. Thus total clearance could be achieved in a range of about 8 to 15 years from shutdown without undue pressure. It might, however, be cheaper to take advantage of natural decay of radioisotopes which would lead to some benefit in terms of radiological protection of decommisioning staff, packaging and transport of waste and/or its storage or final disposal.

12. Stage 2 timescale will be set by the tasks of decontamination especially of the PWR primary and auxiliary circuits, removal of thermal insulation from boiler and steam generator circuit and subsequent dismantlement of these plant items.

Overall the total activity represents less than 1% of the total inventory and is associated with some 10,000 t of material mostly boiler or steam generators and concrete surfaces. Dismantling can be facilitated by ensuring surfaces, especially concrete, which can lead to a significant quantity of active waste are suitably treated during construction and by ensuring that access to the primary circuit is provided to allow aggressive decontamination following final shutdown.

13. Turning to Stage 3 at about 10 years from shutdown PWR total activity will be dominated by the gamma emitting isotope Cobalt 60 associated with the internal core structure. There will be also a substantial contribution from the Xray emitter Iron 55 and Beta emitter Nickel 63. For gas cooled reactors the total activity will be dominated by Iron 55 but Cobalt 60 will also be very important. Generally total activity is important in relation to the limits set by waste disposal or storage. However, with respect to dismantling operations Cobalt 60 which has half life of about 6 years, must be considered. Generally dose rates will be far too high to permit dismantling operation to be carried out on the reactor structures using direct labour and remote methods will be essential up to about 80 years from shutdown. Thereafter there is some prospect of limited manual contribution, especially for to gas cooled reactors which could reduce the use of expensive, remote handling methods. It may also be reasonable to assume advances in technology and experience of decommissioning work elsewhere could facilitate and reduce costs.

14. Beyond about 100 years from shutdown PWR total activity will be dominated by Nickel 63 with a half life of 100 years and will thus be important in the context of long term low level waste disposal but access will depend on the gamma activity from Niobium 94 with a half life of 20,000 years and to a lesser extent Silver 108m, half life of 97 years. The most important point to note however is that there will be little advantage in radiological terms to be gained by delaying Stage 3 any longer.

OPTION 1 : Complete Stages 1, 2 and 3 as an ongoing project within 10 to 15 years from shutdown.

15. This would probably be the most difficult and expensive because the high radiation levels would necessitate the extensive use of remote methods and shielding for dismantling the reactor and packaging waste. There would be greater limitation on the transport of radioactive waste resulting in less efficient pay loads and more packages and on its storage or ultimate disposal. The advantages in taking this option would lie in the early release of the whole site for re-development as a nuclear power station, or other use, and in the ability to use facilities and services already on site. Also the need to maintain buildings and services for an indefinite period would be avoided.

OPTION 2 : Complete Stages 1 and 2 but defer Stage 3 for a further 10 years to 100 years.

16. The advantages of this option would be the ability to remove most of the inactive plant and buildings and the relatively low active contaminated plant using generally available methods and equipment leaving the greater part of the site clear. The bulk of the radioactivity would remain secure within the biological shield and pressure vessel to decay over a long period up to 100 years. This would permit re-development of the site for a further nuclear power station or other purpose. However, the real advantage as far as gas cooled reactors were concerned would lie in the good prospect of being able to dismantle the

greater part of the reactor structure and biological shield or pressure vessel using manual support. This would give a substantial cost reduction for Stage 3 but would be offset by the accumulated cost of surveillance and maintenance over a long period.

17. The disadvantages of this option would be the need to monitor and protect the residual structure for this long period especially with respect to corrosion and the accumulated cost could be significant, especially if there were no other power stations nearby to provide a base for security and monitoring staff. Also when final clearance was eventually undertaken the possible advantages to be gained from improved technology could be offset to an extent by more stringent regulations and the costs of re-establishing site services.

OPTION 3 : Complete Stage 1 and defer Stages 2 and 3 for 50 years or longer.

18. This is apparently favoured in some countries on the grounds that the whole task can be tackled at the one time and there is no problem or cost involved in re-establishing the site services and accommodation. Most of all advantage gained from radioactivity decay would be realised and some improvement of technology would be available. However, to benefit from this all buildings and plant required to contribute to Stage 3 decommissioning would have to be maintained in good working order for the period of deferment and this could be very expensive, especially in relation to most forms of lifting appliances which could in the event be found to be no longer safe to use. Also for water cooled reactors decontamination would become increasingly more difficult.

OTHER FACTORS

Public Opinion

19. Account also would have to be taken of public interest in the visual amenity which could result from removal of the structure at an early date and enjoyment of the use of the site, especially if it were situated in an area of natural beauty, such as a coastline or river. There could be pressure also to remove the residual building because of its actual or apparent threat in terms of radiation hazard to the public. It might be argued that such structures were extremely robust and could neither deteriorate significantly by chemical process or erosion, certainly not to the extent of collapsing nor could they catch fire. However, the costs ensuring a possibly extreme standard of security could lead to a decision to dismantle on a relatively short timescale.

Costs

20. In choosing which option should be taken attention has to be given to the overall costs and incidence of expenditure. Without applying discounted cash flow, DCF, there might not be significant difference between the three main options described above. This is because the cost reductions resulting from delaying Stages 2 and 3 by way of radio-active decay and technological improvement could be offset by the cost of maintaining services and site security. Alternatively, if DCF is used without allowance for inflation, OPTION 3, appears to be very attractive, but thus is less clear if the possible effects of inflation over 100

years are taken into account. Much will depend on how utilities build-up reserve funds to meet the decommissioning costs but in basic terms whenever the task is undertaken the money will have to be found.

LONG TERM INTEGRITY OF BUILDINGS AND SYSTEMS

21. A decision to delay decommissioning depends on the confidence which can be placed in the ability of plant buildings and services to conform with the regulations in order to protect people and the environment from the radioactive material in the long term and to provide support to the dismantling operation when the selected time comes. Project 1 of the Community Research and Development programme has been directed to a study of what might be necessary to achieve this and two contracts were placed, one with CEGB and the other with CEA.

CEGB - Degradation of building and plant materials over a life span of 30 - 100 years

22. Work carried out under this contract comprised a series of inspections of Magnox nuclear power stations which were associated with existing surveys of buildings at six coal fired power stations. Using this data a programme was formulated for taking samples of building and plant materials from four Magnox power stations for laboratory analysis. The age of those stations lay between 15 and 28 years. The results of these analyses coupled with in-situ non-destructive tests have led to the general conclusion that the nuclear stations, three of which are coastal sited and so subject to salt spray, are found to be in good condition and with reasonable maintenance and attention to some minor deficiencies should emain so for 50 years. Furthermore, nothing was identified which would preclude life spans in excess of 50 years.

Two materials were identified as generally causing problems. These were bitumous roofing felt and steel wall cladding whether galvanised or coated with materials. While the effective life of the felt was expected to be limited, but now essentially difficult to repair or replace, the early corrosion of the steel cladding sheets was not anticipated. Other materials and systems reported in these surveys were giving good service and such problems as there were resulted from lack of adequate maintenance or design shortcomings rather than premature material failure.

24. Work at CEGB laboratories, which is not part of the Community programme, on the long term corrosion of pressure vessel steel samples have not yet yielded any signs that cold, damp conditions could lead to doubts about vessel and core structure integrity. These tests will continue. Similarly, the long term integrity of reinforced concrete is being studied but no results which could affect a policy to defer Stages 2 and 3 have come to light.

CEA-IPSN Long term integrity of buildings and systems

26. This contract related to a study of the behaviour of buildings and systems during long term periods of shutdown. It estimated requirements for maintenance, inspection and needs for technical control to keep the

facility in a safe condition but ready to be dismantled sometime in the future. The related costs were also assessed.

The study report also identifies the information which will be needed to allow the correct choice of option to be selected. The following contribution is not intended as a summary of the study but rather to highlight some of the more significant findings.

27. Vessel, Primary and Auxiliary Circuits

For LWR's there apparently will not be any inherent difficulties in leaving the main circuits drained and filled with dry air for at least a few decades and to ensure the tightness of gaskets and stuffing boxes particularly in relation to BWR. Also it appears interesting to dismantle the ancillary circuits as soon as possible after controls have been lifted and to remove thermal insulation to facilitate surveillance.

Regarding GCR the CEA report suggests that it could be desirable to divide the very large volume of the reactor pressure vessels and main gas circuits into smaller sections so as to limit natural convection and reduce the spread of contamination.

28. Reactor Containment Buildings

These structures are not expected to pose any specific ageing problem. It will nevertheless be necessary to adapt the methods of technical surveillance and maintenance to the new situation. Also where there is prestressed concrete it will be most important to continue surveying the differential settlement and the development of stress.

29. Auxiliary Plant

This part of the report concerns the following plant items:-

- electrical circuits
- ventilation
- lifting devices
- control and instrumentation
- steam, compressed air and water circuits
- health physics equipment
- fire fighting equipment

It is important to emphasise that the major part of these plant items have a life span in the range of 25 to 35 years thus those required for Stage 1 such as ventilation, fire fighting equipment will have to be renewed after 30 years. Furthermore, to maintain those items which it is expected will be useful later on for Stages 2 or 3 in good condition raises the problem of spare parts and costly maintenance once the first renewal may occur just after final shutdown.

CONCLUSIONS

30. It appears to be agreed generally that following final or definitive, shutdown the reactors will be defuelled as soon as reasonably practicable and Stage 1 will be completed.

A decision to proceed to Stages 2 or 3 will depend on many factors including safety, technical, economic, national policy and public opinion.

Each site has its own specific characteristics that will have to be taken into account in particular whether it is suitable and needed for re-development as a nuclear power station or put to some use.

It will therefore be necessary to decide in good time, perhaps before final shutdown, on a clear strategy noting that to change course could be very expensive.

OPENING SESSION : SUMMARY OF DISCUSSION

Paper : "The decommissioning of nuclear power plants in the countries of the European Community : achievements and projects"

The Chairman, after having expressed his gratitude for the author's complete description of the decommissioning work in the EC countries, would like to know what the situation on the other side of the Atlantic is. Mr. Crégut, having underlined that some of the participants of the conference are certainly in a better position to answer this questions, points out that the preoccupations on the other side of the Atlantic are similar to the European ones; two large projects exist : one nuclear power plant in the United States is planned to be decommissioned by complete dismantling, and another in Canada will be refurbished for nuclear or non-nuclear use. In all these countries, research is pushed forward in a large number of fields; the development of cutting by the arc-saw can be taken as an example. There is a common preoccupation of all plant operators, and here he includes not only the Americans but also the Japanese, that is to prepare decommissioning carefully, without forgetting to take the costs into consideration. He points out that the main objective in building a plant is production, be it production of electricity, plutonium or basic research results; in decommissioning a plant, the only objective is security and safety. It is difficult to make people understand that money is only coming after this priority; for this reason, the economic point of view must never be forgotten and therefore there must be investment in research and development and in demonstration of methods before, and this is his strictly personal opinion, the authorities draw up rules and the decision can be taken as to what the best solution is.

Paper : "The Community programme on the decommissioning of nuclear power plants : objectives, scope and implementation"

A participant representing an international organization, recalls that measurement techniques to prove that material is not radioactive have been reviewed. He asks how successful these procedures will be in doing these measurements on a production base for large quantities of materials arising from decommissioning. He would also like to know if these measurements will be based on gamma activity values only, or if alpha activity measurements would be included. Mr. Huber indicates that this subject will be treated in more detail later during the conference. He believes, nevertheless, that such measurements can be done on a large scale; the main problem will be costs. He had formerly thought that these costs would be prohibitive, but for a recent large scale decontamination operation to unrestricted release in a German Boiling Water Reactor, which will be reported later, the costs for measurements have been in the order of 10% of total costs, which is a reasonable figure; the activity present was mainly beta-gamma activity. Melting of steel could be a way of facilitating measurements; problems arise when the geometry of pieces is complicated and distribution of contamination is not homogenous. After melting, samples could be taken from the melt.

A Swiss participant asks if the programme takes into account developments and results obtained at the TMI-2 plant. Mr. Huber explains that experts had been delegated, in the frame of the research programme, to the TMI-2 plant, in order to obtain first-hand information on work going on at the plant; the information has been made available to the experts of the member countries. The case of the TMI-2 plant is a special one. The Community programme included provision for research on decontamination in plants having experienced a major incident; in executing the programme, this part has been limited to collecting the TMI-2 experience. In fact, consequences of hypothetical accidents can be so different from case to case, that an in-depth study of procedures and measures to be taken for decommissioning is not possible on a generic base.

Mr. Crégut adds that exchange of information has to be based on reciprocity. The main difference between a normal decommissioning operation and the TMI-2 situation is that there is typically a lack of knowledge in the latter case : firstly, it has not been possible to set up a convenient and complete map of the activity inventory; a good decommissioning management is possible only knowing perfectly composition and quantity of the activity. Secondly, it is necessary to know in what form the activity is present. The lessons learnt for decommissioning engineers are that measurement devices to cope with these specific situations have to be made available, allowing the identification, location by location, of the local activity content and distribution. Another observation is that capacities for conditioning and storage of unexpectedly high quantities of radioactive waste of a very special kind have not been foreseen. But there are also positive aspects : there has been considerable progress in the application of processes and treatments of waste. His organization gave some assistance to TMI-operations, for example by providing intervention devices, small carrier units, etc. In conclusion, TMI really is an unforeseen operation and fruitful cooperation is going on. Everybody's wishes are that these efforts will allow domination of the situation as soon as possible.

The Chairman announces that a paper in Session II will be presented on the TMI-experience by the experts having been delegated to TMI.

Paper : "Factors to be considered in selecting a decommissioning strategy"

A Belgian participant asks if the arrangement of the last fuel operation cycle could already be considered as preparation of Stage I decommissioning. He mentions a US paper which considers the economic loss due to incomplete burn-up of the fuel in the last cycle, as part of the decommissioning costs. The question is if this cost has to be supported by the budget set aside for decommissioning or by the normal operation budget. Mr. Gregory states that his organization has never attempted to charge residual value or lost value of the final fuel charge to decommissioning and does not know anybody who has done it. He is nevertheless interested in such a possible US-practice, but proposes to let the decision be taken by the plant operators.

A U.K. participant recalls that the paper is supposed to deal with reactors that have not been shut-down due to break-downs or accidents. She suspects that there will be a pressure to get longer useful life out of existing reactors and that there will finally be a number of minor break-downs leading to the decision to close the plant. Without suggesting any major break-down, such smaller incidents, which do not result in a lower safety level, may present to decommissioning more unexpected problems. Mr. Gregory thinks that the most likely for ultimate shut-down is the presence of a number of small items which tend to give difficulty and which are more and more difficult to maintain. Safety has to be maintained, of course, at all times. If there is no large event, the items tending to give trouble are conventional items and, consequently, there would be no added difficulty to the decommissioning process.

Mr. Crégut adds that ageing components need more and more maintenance to a point where they become technically obsolete. The cost of maintaining them in working order becomes so high that closing is preferable. Fabrication aims at achieving a life time as high as possible. He underlines that, in any case, operation of a nuclear power plant is subject to control by authorities and licensing bodies and that there is no allowance for operation with items suspected to give even a small risk.

A Swedish participant recalls that it is understood in the paper that in Stage II, activity is concentrated inside the biological shield of the reactor. He asks if the Stage II could not also be reached by enclosing activity in the concrete containment, for example the steam generators of a BWR, since removal of these items of the primary loop could lead to considerable exposure. Mr. Gregory confirms that the UK policy would be to treat the steam generators as being part of Stage II activity, so that they would be removed during Stage II, if possible. The structure could thus be reduced, if that were suitable or necessary, leaving over 90% of the activity neatly boxed up inside the reactor.

SESSION I

RADIOACTIVITY INVENTORIES AND WASTE ARISING

Chairman : K. HÜBENTHAL, BMFT

Secretary : K.H. SCHALLER, CEC, Directorate-General Science, Research and Development

Introduction by the Session Chairman

Activation of biological shields

Activation of steel components

Estimation of radioactive waste quantities arising during decommissioning

Methodology for evaluating radiological consequences of the management of very low-level solid waste

Some examples of release of materials and sites, including measurement techniques

Summary of discussion

INTRODUCTION BY THE SESSION CHAIRMAN

K. HUBENTHAL
BMFT, Bonn, F.R.Germany

The first working session of the conference is devoted to the radioactive inventory and the quantities of waste arising in the decommissioning of nuclear power plants.

A good knowledge of the radioactive inventory is not only of importance for the estimation of the arisings in waste, but is also of decisive influence on planning and execution of the effective decommissioning activities.

The estimation of the activity inventory will be treated separately for concrete and steel in two presentations. The computation of the local neutron fluxes in the different zones certainly raises some difficulties; nevertheless, the larger uncertainties are due to the lack of knowledge on steels and concretes employed, as far as the impurities containing isotopes with high capture cross sections and leading to long-lived activation products, is concerned.

Despite these difficulties, the concordance of calculated and measured values has been found to be rather satisfactory in cases where such a comparison was feasible.

One presentation produces an estimation of the arisings of waste quantities in the Community countries, based on a number of assumptions on decontamination, limits for unrestricted release and waste conditioning; the estimation takes into account a number of reactor types and decay times. These quantities of waste are certainly important but are in the same order of magnitude as the waste quantities arising already now from normal reactor operation.

A further presentation covers a methodology for the estimation of the radiological consequences for waste disposal, which is not to be eliminated by geological disposal. It appears that the radiological consequences can be tolerated for disposal modes like shallow land disposal, disposal at municipal refuse sites and sea-dumping.

The last presentation describes some major decommissioning activities performed in France and the Federal Republic of Germany. Even if it is true that no large power plant of today's usual size has been decommissioned, it has been shown that dismantling of shut-down nuclear installations can be performed under conditions which are acceptable to the environment and the decommissioning workers.

ACTIVATION OF BIOLOGICAL SHIELDS

S. MAY, D. PICCOT
Commissariat à l'Energie Atomique, France
L. BERGEMANN
Kernkraftwerk RWE - Bayernwerk GmbH, Gundremmingen
W. HARBECKE
Kernkraftwerk Lingen GmbH, Lingen (Ems)
G. LOERCHER
NIS Nuklear-Ingenieur-Service GmbH, Frankfurt/Main

Abstract

Knowledge of the quantities of radioactive waste resulting from the dismantling of the concrete biological shields of nuclear reactors is very approximate because there is little accurate data available on the composition of the concrete. It is possible to calculate the residual radioactivity in the concrete from nuclear reactors if the amounts of elements which give rise to long-lived radionuclides, and in particular chlorine, calcium, nickel, cobalt, niobium, europium and samarium are known accurately. The highly effective techniques of neutron activation analysis were used to determine the quantities of 29 elements in 21 concrete samples from reactors in the European Community. Research was conducted on the Lingen (KWL) and Gundremmingen (KRB-A) reactors. The method employed consisted of measurements conducted on concrete samples and calculations based on the composition of the concrete determined by activation analysis. The measurements made and the results of the calculations are in reasonable agreement.

INTRODUCTION

For the purposes of making forecasts on the complete dismantling of nuclear power reactors it is necessary to know the quantities of concrete from the biological shield which have to be stored as nuclear waste. Calculation of the radioactivity of the biological shield is therefore very important. Theoretical findings must be proved by measurements. Measurements of this type are difficult and expensive since the samples must be taken by drilling from the external wall of the biological shield as far as the highly radioactive containment barriers.

In order to avoid uncertainty in the calculations and high expenditure in future projects we made both experimental measurements and theoretical forecasts. The work was carried out on the German nuclear reactors at Gundremmingen (KRB-A) and Lingen (KWL), both of the BWR type, which ceased operation in January 1977. The power of reactor KRB-A is 801 (thermal) megawatts and that of the KWL reactor 520 megawatts. The theoretical duration of operation at full power (the total power generated during the reactor's life divided by the annual power rating) is 7.34 years for KRB-A and 5.01 years for KWL.

For calculation purposes it is necessary to know the amounts of various elements in the concrete. Unfortunately, most of the elements which are most important for activation are present in very low concentrations (< 1 ppm). We therefore decided to develop and apply an analysis methodology making it possible to quantify the elements in the concrete used in the KWL and KRB-A nuclear reactors and in other reactors in the European Community. Knowledge of the concentrations of trace elements, in particular chlorine, nickel, cobalt, niobium, samarium and europium plus a major element, calcium, makes it possible to calculate the activities of radioactive elements with long half-lives generated by radio-activation in nuclear reactors.

Neutron activation analysis was the main method used to analyse the concrete. It was chosen for the analysis of elements giving rise to long-lived radionuclides because of its effectiveness in the analysis of trace quantities. Other analysis techniques were not entirely satisfactory. X-ray fluorescence, however, provided interesting results for some samples.

1. NEUTRON ACTIVATION ANALYSIS

The elements chlorine, nickel, cobalt, calcium, europium, samarium and niobium are the most important ones because they generate radionuclides with long half-lives when subjected to neutron irradiation :

$T_{\frac{1}{2}}$ ^{36}Cl	:	3×10^6 years	$T_{\frac{1}{2}}$ ^{151}Sm	:	87 years
$T_{\frac{1}{2}}$ ^{41}Ca	:	8×10^4 years	$T_{\frac{1}{2}}$ ^{152}Eu	:	12.4 years
$T_{\frac{1}{2}}$ ^{59}Ni	:	7.5×10^4 years	$T_{\frac{1}{2}}$ ^{154}Eu	:	16 years
$T_{\frac{1}{2}}$ ^{63}Ni	:	92 years	$T_{\frac{1}{2}}$ ^{94}Nb	:	2×10^4 years
$T_{\frac{1}{2}}$ ^{60}Co	:	5.2 years			

In most cases the isotopic composition of the elements studied makes it possible to choose for the study of the n, γ reaction an isotope which yields a radionuclide with a much shorter half-life than those mentioned above.

Development of the methodology and activation analysis of concrete samples were carried out at the Laboratoire d'Analyse par Activation Pierre Süe on the site of the Centre d'Etudes Nucléaires at Saclay.

We were provided with 21 concrete samples from various reactors in the European Community of masses ranging from a few grammes to a few kilogrammes.

Regardless of the size of the sample, the whole quantity was crushed for homogenization purposes by the Service des Techniques Avancées, Laboratoire des Bétons, CEN Saclay.

The amount actually analysed was about a gramme or less; hence a very homogeneous sample was needed to ensure that it was representative.

1.1 Analysis of the calcium, cobalt, europium and samarium contents

The first tests showed that these elements may be identified, after neutron irradiation, by direct gamma spectrometry, without chemical separation. The radiosisotopes measured were calcium-47, cobalt-60, samarium-153 and europium-152 (for nuclear characteristics see Table 1).

1.2. Specific analysis of chlorine, by the reaction $^{37}Cl(n,\gamma)^{38}Cl.(T\frac{1}{2}\ 37.3mn)$

It is impossible to determine chlorine directly by gamma spectrometry after neutron irradiation because of the radioactivity of the major constituents of the concrete, which masks the spectral gamma lines characteristic of chlorine-38.

An analysis method based on the radiochemical separation of chlorine-38 after irradiation was therefore needed.

1.2.1. Radiochemical separation of chlorine-38.

About 500 mg of crushed concrete was irradiated for 2 minutes in the Osiris reactor together with a standard chlorine sample in a neutron flux density of 3×10^{12} n. $cm^{-2}.s^{-1}$.

A pause of 20 minutes after irradiation was followed by alkaline fusion in a nickel crucible in the presence of sodium peroxide. After cooling, the mixture was set in water and the insoluble residues were dissolved by adding concentrated nitric acid. The carrier (hydrochloric acid) was added plus a solution of silver nitrate. The silver chloride (containing the chlorine-38) was precipitated. The precipitate was filtered, washed in water, dried and counted.

The intensity of the gamma lines from chlorine-38 (1642 and 2166.8 keV) in the sample and the standard were compared, which made it possible to perform quantitative analysis. The chemical yield of chlorine separation was measured by tracer studies with labelled calcium chloride. It amounted to 100 %; this demonstrates the validity of the method used which is also perfectly reproducible.

1.3. Specific analysis of nickel by the reaction $^{58}Ni(n,p)^{58}Co.(T\frac{1}{2}\ 71.3\ days)$

Like chlorine, cobalt-58 cannot be measured directly because of the high radioactivity of the major constituents of the concrete.

1.3.1. Radiochemical separation of cobalt-58

100 mg of pulverized concrete was irradiated for 16 hours (Osiris reactor) in a neutron flux density of 10^{14}n. $cm^{-2}.s^{-1}$.

Irradiation took place in a cadmium container which stops the passage of thermal neutrons. The standard irradiated at the same time was a small piece of pure nickel.

Chemical separation of cobalt-58 was carried out 48 hours after removal from the reactor. The irradiated concrete sample then underwent alkaline fusion with sodium peroxide in a nickel crucible. After cooling, it was recovered using water and then concentrated hydrochloric acid, to dissolve the insoluble portion. Then the cobalt carrier was added. The cobalt and β - nitroso α - naphthol complex was extracted by carbon tetrachloride. This organic phase was analysed by gamma spectrometry for cobalt-58 using the 810.3 keV gamma ray emission line.

The chemical yield of the separation was measured by counting cobalt-60 (present in the irradiated concrete) before and after chemical separation.

1.4. Specific analysis of niobium

The classic nuclear reaction ^{93}Nb(n, γ) ^{94m}Nb forms a radioisotope with a very short half-life: 6.3 minutes. Direct measurement of niobium 94m is not possible because of the high radioactivity of the other constituents of the concrete.

Another activation route is the reaction ^{93}Nb (n,2n) ^{92m}Nb. The ^{92m}Nb has a half-life of 10.2 days. But there is interference from molybdenum due to the reaction ^{92}Mo (n,p) ^{92m}Nb.

For the purposes of niobium analysis by irradiation with epithermal and fast neutrons it is therefore also necessary to measure the molybdenum content in order to allow for this interference.

1.4.1. Radiochemical separation of ^{92m}Nb.

One gramme of concrete was irradiated for 65 hours in a neutron flux density of 2×10^{14} $n.cm^{-2}.s^{-1}$ in the presence of a niobium standard. In this irradiation position the fast neutron flux is about 3×10^{13} $n.cm^{-2}.s^{-1}$. Samples and standard were placed in a special cadmium container in order to stop the thermal neutron flux. 10 days were allowed for deactivation before starting radiochemical separation.

The irradiated concrete sample underwent alkaline fusion with sodium peroxide, then the fusion residue was recovered with water and nitric acid. After addition of the carrier iron, the hydroxides (containing the niobium) were precipitated in an ammonia medium. The precipitate isolated was treated with hydrofluoric acid to eliminate the remaining silica.

The hydroxide precipitate was recovered with hydrochloric acid and the niobium was separated by chromatography on Dowex 1 x 8 resin (100 - 200 mesh) by means of elution with a fluorohydrochloric mix.

This solution was dried and recovered with a mixture of sulphuric and hydrofluoric acids and the niobium extracted by hexone. The organic fraction was then treated to coprecipitate the niobium with an iron hydroxide precipitate. For this purpose the hexone was separated by distillation and the residue mineralized with a nitrosulphuric mixture. After mineralization and recovery with water, the iron hydroxide was precipitated by means of ammonia in the presence of the carrier iron. The final precipitate was filtered, dried and analysed by gamma spectrometry using the characteristic 934 keV gamma ray emission line.

1.5. Specific analysis of molybdenum y nuclear reaction
$^{98}Mo(n, \gamma)^{99}Mo \xrightarrow{\beta^-} {}^{99m}Tc$.(Half-lives: ^{99}Mo : 66.7 hours - ^{99m}Tc : 6 hours)

After one hour's irradiation in a $10^{14} n.cm^{-2}.s^{-1}$ flux followed by two days deactivation, the concrete sample (500 mg) was subjected to radiochemical separation treatment. The concrete was dissolved by alkaline fusion, recovered with water in the presence of the molybdenum carrier then with hydrochloric acid.

The molybdenum was extracted from the solution by fixing it on Dowex 1 x 8 resin. It was then re-extracted by hydrochloric acid and precipitated by forming a complex with 1' α-benzoin oxime. This complex was filtered, washed and dried and finally analysed by gamma spectrometry for molybdenum-99 using the 740 keV gamma ray emission line.

We have determined the equivalence between the molybdenum content and the apparent niobium content from the reaction $^{92}Mo\ (n,p)\ ^{92m}Nb$. It seems that : 1 μg molybdenum results in an apparent mass of 2.16 μg niobium.

1.6. Other elements determined.

All the appropriate elements, i.e. those which could lead to the production of long-lived radionuclides, were measured. It emerged that many other elements could be analysed by neutron activation on a purely instrumental basis. The radioisotopes of these elements appear in the gamma spectra obtained by direct analysis of calcium, cobalt, europium and samarium. Their nuclear characteristics are mentioned in Table I.

We therefore thought it useful to include them in the quantitative analysis in order to obtain a more complete picture of the trace element composition of the concrete.

The following elements were determined : As, Au, Ba, Ce, Cr, Cs, Fe, Hf, Ir, La, Lu, Nd, Rb, Sb, Sc, Ta, Tb, Th, U, Yb, and Zr.

Twenty-nine elements were therefore determined by quantitative analysis in most of the 21 concrete samples from reactors in the European Community.

1.7. Analysis results

The results of analysis of the 21 concrete samples are shown in Tables II and III. Some comments must be made about the variations found in the concentrations of elements with long-lived radionuclides.

- Calcium : the calcium content varies from 3 to 25 % with an average of 13 %
- Chlorine : apart from the major elements, iron and calcium, this is the element with the highest concentrations (~ 500 ppm). The average is 109 ppm.
- Cobalt : concentration levels are relatively stable, generally below 10 ppm, with an average of 5 ppm.
- Europium : it is the rarest of the most important elements. The average value is 0.6 ppm.

- Nickel : with an average of 10.9 ppm, the distribution shows no great variation.
- Samarium : all the values are < 10 ppm, with an average of 3.2 ppm.
- Niobium : the contribution of molybdenum, which gives rise to the same radio-isotope as niobium, explains the important variations observed. The highest niobium concentration is 47 ± 11ppm. The average value for actual measurements is 16.5 ppm.

The average values of the concenrations observed lead to the following classification of the elements : Ca >>> Cl >> Nb > Co > Sm > Eu.

The other elements identified, of secondary importance, may be classified in accordance with the average concentrations observed, in the following order : Fe >>> Ba > Zr > Cr > Rb > Ce > La > Nd > As > Mo Th > Sc > U > Cs > Hf > Yb > Sb > Tb > Ta > Lu.

2. COMPARISON OF THE RESULTS OF NEUTRON ACTIVATION ANALYSIS (NAA) WITH OTHER ANALYSIS TECHNIQUES

Some concrete samples from reactors in the Federal Republic of Germany were analysed by X-ray fluorescence by Dyckerhoff and Widmann. Comparative results for some elements are given in Table IV. These comparative results call for some comments.

- Arsenic : the results of the two methods do agree to some extent, but there are marked differences (8.7 ppm → 5 ppm in the KKG sample).
- Barium : agreement is excellent.
- Calcium : the results agree except for the KWG sample.
- Cerium : same order of magnitude, but significant differences.
- Cobalt : X-ray fluorescence gives results 3 to 5 times higher than NAA.
- Chromium : very marked systematic differences : NAA gives much higher values than X-ray fluorescence.
- Iron : same comments as for chromium except for the KKG sample where agreement is excellent.
- Lanthanum : very good agreement for the KWG sample (11.8 - 12 ppm). For the other samples there are substantial differences.
- Nickel : same order of magnitude but substantial differences.
- Rubidium : excellent agreement.
- Scandium : differences of 50 % sometimes observed.
- Tantalum : for all samples X-ray fluorescence indicates a level of 5ppm while NAA provides results from 0.18 to 0.54 ppm. Very substantial divergencies between the two methods therefore.
- Thorium : same order of magnitude but discrepancies of more than 50 %.
- Uranium : same comments as for thorium.
- Zirconium : same order of magnitude but appreciable differences.

3. SAMPLING AND MEASUREMENTS IN THE KWL AND KRB-A REACTORS

In order to obtain significant results, three samples of concrete situated opposite the reactor core were taken from the biological shield by drilling from the exterior to the centre.

One sample came from the lowest level of the reactor core, the second from mid-level and the third from the upper level. Samples were taken from the material obtained at intervals based on a logarithmic scale derived from earlier calculations. The samples of concrete taken were broken into pieces, then pulverized. The pulverized samples were weighed in order to measure their specific activities and compare them with the values calculated.

In the first stage we found that the most important radionuclides were cobalt-60, cesium-134 and europium-152 which emit gamma lines that can be measured precisely. We collected the best possible data for the calculations in order to determine the masses accurately, in particular for tne elements giving rise to the radionuclides determined by activation analysis.

It is also necessary to know the quantities of other elements in order obtain good calculation results, although the comparison of theoretical calculations and measurement results is based purely on the radionuclides precipitated.

4. CALCULATION OF THE ACTIVATION OF THE CONCRETE

In order to obtain reliable information on the activities of the concrete and reinforcing steel, the calculations must be based on reliable input data. In addition to the weight of the element, the neutron flux and activation data such as absorption, cross-section and half-life must be known.

We therefore used a tired and tested NIS (AKAT-2) computer program which can calculate 70 activation reactions producing 50 radionuclides with half-lives longer than 30 days.

This program also makes it possible to consider realistic operational periods. We have good information on operational data, i.e. monthly power levels, and it is therefore possible to obtain input data on the relationship between real power and average power. This relationship was used to calculate the correct neutron flux for each period of operation. We considered 21 periods for KRB-A and 17 for KWL. In the case of KRB-A we concentrated on the calculation of neutron flux in the concrete.

Finally, we obtained a wide range of radionuclides and their activities. They are calculated for decay times of 1, 5, 10, 30, and 50 years which makes it possible to calculate total activity as a function of time and thus the advisability of dismantling the reactor after a given time.

5. COMPARISON OF ACTIVATION CALCULATIONS AND MEASUREMENTS

As we mentioned earlier, the comparison concentrated on Co-60, Eu-152 and Cs-134. There is good agreement for the two reactors in spite of a few design differences.

The comparison is shown on the curves in Figure 1.

Because of the difficulties of drilling into KRB-A, the sample was broken and we do not have full information on the exact position of each piece of concrete. This is doubtless the reason for some of the discrepancies observed.

In one case the difference was very great but it was possible to correct it by making new measurements on the sample. The difference was clearly identified as an error during measurement. This shows that serious discrepancies are always identifiable and can then be corrected.

It the case of KWL the differences observed between the values calculated and the measurements may be explained by the calculation of the neutron flux, which may vary as a function of thermalization of the neutrons in the concrete.

In the two reactors the radionuclides occur in the following order : Eu-152 > Co-60 > Cs-134. This classification is demonstrated by both calculations and measurements.

It is therefore true to say that the calculations made with reliable data (especially as regards the amounts of the elements present) are enough to provide precise information.

6. RESULTS OF EXPERIMENTS ON KWL AND KRB-A

The results obtained for the two reactors KWL and KRB-A may be useful for calculations on other reactors and the values obtained give a good idea of the phenomena occurring in the concrete of the biological shield.

The very long service lives of these reactors gave rise to little or no increase in activity. This was proved by the calculations.

An example of variation in total specific activity (by calculations) is given in Figure 2 as a function of the depth of the sample and the decay time after final reactor shut-down. The curves dip slightly less steeply for the external part of the concrete. The dotted lines give an idea of what the limit values (for disposal without restriction) might be in future.

It should therefore be noted that, after five years, only the outermost part of the concrete is below the most restrictive value (0.37 $Bq.g^{-1}$). This is not the case for the reinforcing steel (Figure 2). It would, therefore, be interesting to separate the reinforcing steel and the concrete proper.

7. CONCLUSION

Because of the effectiveness of the nuclear analysis methods it was possible to chart the amounts of elements leading to long-lived radioisotopes in the concrete of the biological shield of power reactors. Many other elements of secondary importance, however, were also determined quantitatively.

The analysis results were used as a basis for calculation of the activation of concrete in the biological shields of the KWL and KRB-A reactors. The experimental measurements carried out on the samples of the concrete shield of these reactors are in good agreement with the results of theoretical calculations based on reliable data.

It would appear, therefore, that in the future information on the concentrations of major, minor and trace elements will make it possible to calculate with adequate accuracy residual activity in the concrete of a reactor due for dismantling.

Table I : Nuclear characteristics of the elements analysed

Element	Z	A	Isotopic abundance	σ in barns	Nuclear reaction	Half-life	Gamma line (keV)
As	33	75	1	5,4	$^{75}As\ (n,\gamma)\ ^{76}As$	26,7 hours	559,3
Ba	56	130	0,001	10	$^{130}Ba\ (n,\gamma)\ ^{131}Ba$	12 days	496,3
Ca	20	46	0,0033	0,25	$^{46}Ca\ (n,\gamma)\ ^{47}Ca$	4,7 days	1296,9
Ce	58	140	0,884	0,31	$^{140}Ce\ (n,\gamma)\ ^{141}Ce$	32 days	145,4
Co	27	59	1	20	$^{59}Co\ (n,\gamma)\ ^{60}Co$	5,2years	1173,1
Cr	24	50	0,04	13,5	$^{50}Cr\ (n,\gamma)\ ^{51}Cr$	27,8 days	320
Cs	55	133	1	30	$^{133}Cs\ (n,\gamma)\ ^{134}Cs$	2 years	795,8
Eu	63	151	0,477	2500	$^{151}Eu\ (n,\gamma)\ ^{152}Eu$	12,5 years	1407,5
Fe	26	58	0,003	0,98	$^{58}Fe\ (n,\gamma)\ ^{59}Fe$	45 days	1098,6
Hf	72	180	0,364	10	$^{180}Hf\ (n,\gamma)\ ^{181}Hf$	45 days	482,2
La	57	139	0,999	8,2	$^{139}La\ (n,\gamma)\ ^{140}La$	40,2 hours	1595,4
Lu	71	176	0,026	4000	$^{176}Lu\ (n,\gamma)\ ^{177}Lu$	6,8 days	208,4
Nd	60	146	0,171	1,8	$^{146}Nd\ (n,\gamma)\ ^{147}Nd$	11,1 days	531
Rb	37	85	0,721	0,45	$^{85}Rb\ (n,\gamma)\ ^{86}Rb$	18,7 days	1076,6
Sb	51	123	0,427	2,5	$^{123}Sb\ (n,\gamma)\ ^{124}Sb$	60 days	1690,9
Sc	21	45	1	12	$^{45}Sc\ (n,\gamma)\ ^{46}Sc$	84 days	889,4
Sm	62	152	0,265	140	$^{152}Sm\ (n,\gamma)\ ^{153}Sm$	47 hours	103,2
Ta	73	181	1	19	$^{181}Ta\ (n,\gamma)\ ^{182}Ta$	115 days	1221,6
Tb	65	159	1	22	$^{159}Tb\ (n,\gamma)\ ^{160}Tb$	72 days	879,3
Th	90	232	1	7,4	$^{232}Th\ (n,\gamma)\ ^{233}Th \xrightarrow{\beta^-} {}^{233}Pa$	27,4 days	311,8
U	92	238	0,992	2,73	$^{238}U\ (n,\gamma)\ ^{239}U \xrightarrow{\beta^-} {}^{239}Np$	2,35 days	277,5
Yb	70	174	0,318	60	$^{174}Yb\ (n,\gamma)\ ^{175}Yb$	4,2 days	282,6
Zr	40	94	0,174	0,09	$^{94}Zr\ (n,\gamma)\ ^{95}Zr$	65 days	756,6

TABLE II : Results in ppm (10^{-6}) except where specified. Differences $\pm$ 10 %

Reactor	As	Ba	Ca%	Ce	Cl	Co	Cr	Cs	Eu	Fe%	Hf	Ir ppb	La	Lu
KKP-2-RFA	22,6	245	8,8	17,9	57	5,1	73	3,9	0,40	1,2	1,4	< 11	9,4	0,14
KKG (H) RFA	9,8	108	3,4	-	40	2,3	⩽ 69	0,40	0,35	61,2	0,82	< 20	6,4	0,27
KKI-2-RFA	7,2	104	11,3	10,5	87	3,8	81	1,1	0,31	0,66	0,61	< 7	7,0	0,09
KWG - RFA	4,4	235	9,9	19,1	117	4,5	86,1	2,1	0,49	0,96	1,8	< 11	11,8	0,16
KBR - RFA	1,3	420	8,1	29,5	75	3,2	122,9	1,0	0,82	1,1	3,4	< 9	18,5	0,42
KWS (GB) RFA	55,9	266	18,1	56,5	55	20	98,5	4,3	1,18	5,4	3,1	< 18	28,7	0,35
KKG (NB) RFA	8,7	343	9,4	23,7	97	6,4	68,3	1,6	0,53	2,0	2,9	< 11	12,7	0,21
NPU - RFA	4,9	271	5,8	21,6	74	3,7	81,2	1,6	0,49	1,1	3,2	< 10	12,4	0,18
KBR (H) RFA	13,1	110	4,6	-	42	4,7	<112	0,26	0,40	54,7	0,49	< 33	8,5	-
GARIGLIANO 1(It.)	5,8	380	22,2	54,7	317	7,1	49,8	5,0	1,07	1,5	1,7	< 15	31,6	0,30
GARIGLIANO 2(It.)	5,1	374	22,3	54,8	326	7,8	44,8	3,8	1,0	1,5	2,0	< 14	30,2	0,30
GARIGLIANO 3(It.)	5,1	371	21,8	55,6	493	5,4	41,8	3,1	1,03	1,3	1,9	< 11	30,6	0,35
EL2 - FRANCE	5,6	113	24,8	15,6	27	3,2	49,5	1,5	0,41	0,80	0,81	< 8	8,5	0,14
EL3 - FRANCE	6,0	150	23	17	22	3,2	39,6	1,3	0,49	0,86	1,0	< 8	10,3	0,17
TRITON - FRANCE	4,8	172	16,9	13,9	16	3,5	58,6	1,9	0,33	0,80	0,93	< 8	9,6	0,09
G1 - FRANCE	11,8	650	12,4	11	77	2,9	99,5	2,3	0,35	0,82	1,3	< 9	7,3	0,09
OLDBURY - U.K	66,6	191	23,2	19,7	208	4,9	18,0	0,66	0,50	2,1	1,3	< 12	11,5	0,16
KWL-1-RFA	-	222	4,6	46,7	52	2,7	91	0,9	0,75	0,6	1,8	< · 3	26,7	0,19
KWL-2-RFA	-	247	4,8	59,9	40	3,0	100	0,9	0,99	0,8	3,6	< 3	30,5	0,24
KWL 3-RFA	-	242	5,7	67,8	36	2,9	105,4	0,9	1,04	0,6	2,0	< 3	39,4	0,26
KRB-A-RFA	-	80	14,8	31,6	32	3,8	99,1	1,0	0,35	1,25	1,8	< 3	19,5	0,11

TABLE III : Results in ppm (10^{-6}). Differences : ± 10 % except where specified

Reactor	Mo	Nb	Nd	Ni	Rb	Sb	Sc	Sm	Ta	Tb	Th	U	Yb	Zr
KKP-2-RFA	7,1	8,8 ± 5,4	11,2	14,9	63,9	2,4	3,3	2,1	0,54	0,22	4,3	2,4	1,1	71
KKG (H)-RFA	6,2	-	-	7,2	-	2,0	3,5	1,3	-	0,29	1,3	2,3	0,77	-
KKI-2-RFA	4,8	9,0 ± 1,2	8,7	8,5	19,7	0,44	1,5	1,3	0,18	0,16	1,8	1,5	0,63	-
KWG - RFA	5,4	≤ 6	11,6	11,6	51,7	0,65	3,4	2,5	0,46	0,32	4,6	1,8	1,35	122
KBR - RFA	6,2	6,0 ± 4,4	18	6,1	62,6	0,17	4,7	3,4	0,48	0,45	5,1	2,5	1,2	189
KWS (GB) RFA	10,8	< 5,8	28,1	26	63,1	2,5	10,9	5,6	0,85	0,61	8,1	-	2,3	109
KKG (NB) RFA	5,1	4,9 ± 3,6	12,6	17,8	40,1	1,3	3,5	2,6	0,49	0,27	3,4	2,2	1,7	149
NPU - RFA	3,7	< 5,1	11,5	10,0	55,8	0,55	2,6	2,3	0,47	0,33	4,7	1,8	1,2	126
KBR (H) RFA	5,3	-	11,1	11,4	1,1	-	3,5	1,4	0,13	0,24	1,3	1,9	5,8	-
GARIGLIANO 1 (It.)	11,1	24,9 ± 10,5	32	10,0	75,1	0,52	8,2	5,8	0,41	0,57	9,4	4,8	1,2	122
GARIGLIANO 2 (It.)	11,1	17,3 ± 9,1	29,9	10,2	65,9	0,45	7,9	5,4	0,43	0,59	8,5	4,7	2,5	89
GARIGLIANO 3 (It.)	10,8	22,7 ± 9,9	28,9	8,2	53,4	0,37	7,8	5,7	0,38	0,64	9,6	5	2,7	94
EL2 - FRANCE	4,1	< 3,2	8,0	6,1	32,6	0,44	2,8	1,9	0,21	0,28	3,0	1,8	0,92	53
EL3 - FRANCE	3,9	47 ± 11	12,0	6,6	29,1	0,37	3,4	2,3	0,25	0,27	3,6	1,7	1,6	54
TRITON - FRANCE	4,5	≤ 7	11,7	8,9	43,3	0,63	2,5	1,8	0,25	0,18	2,7	1,5	0,67	83
G1 - FRANCE	5,3	≤ 3,5	7,2	13,3	49,5	0,87	2,1	1,6	0,36	0,18	2,4	2,6	0,67	66
OLDBURY - U.K	5,5	12,8 ± 5,3	12,7	11,1	17,0	7,7	3,0	2,3	0,18	0,39	2,1	2,3	1,2	77
KWL-1-RFA	3,1	10,7 ± 3,5	-	10,5	20,5	0,39	4,1	4,6	0,23	0,48	5,4	1,7	1,33	-
KWL-2-RFA	3,9	9,9 ± 3,8	-	8,9	23,9	0,57	5,0	5,7	0,34	0,60	6,3	1,8	1,87	-
KWL-3-RFA	3,8	17,3 ± 5,2	-	10,5	22,2	0,31	5,4	6,4	0,29	0,65	6,9	1,8	2,17	-
KRB-A-RFA	2,5	12,2 ± 3,5	-	11,4	26,1	0,45	3,2	2,4	0,36	0,22	4,9	1,2	0,73	-

(GB) : faced concrete (H) : Haematite (NB) : ordinary concrete

TABLE IV

Comparison of the results obtained by neutron activation (1st valve in each column) and by X-ray fluorescence (2nd value in each column ; results from Dyckerhoff and Widmann)

Results in ppm (10^{-6}) except where

specified Element	Reactor KKG-FRG	Reactor KKG-FRG	Reactor KKI-2 FRG	Reactor KKI-2 FRG
As	8,7 - 5	4,4 - 5	7,2 - 5	-
Ba	343 - 323	235 - 260	104 - 100	245 - 264
Ca %	9,4 - 8,64	9,9 - 14,0	11,3 - 12,5	8,8 - 9,3
Ce	23,7 - 32	19,1 - 16	10,5 - 10	17,9 - 10
Co	6,4 - 22	4,5 - 13	3,8 - 20	5,1 - 22
Cr	68,3 - 22	86,1 - 13	81 - 12	73 - 25
Fe %	2 - 1,79	0,96 - 0,44	0,66 - 0,28	1,2 - 0,49
La	12,7 - 28	11,8 - 12	7 - 10	9,4 - 27
Ni	17,8 - 27	11,6 - 16	8,5 - 10	14,9 - 23
Rb	40,1 - 40	51,7 - 48	19,7 - 18	63,9 - 61
Sc	3,5 - 5	3,4 - 3	1,5 - 3	3,3 - 5
Ta	0,49 - 5	0,46 - 5	0,18 - 5	0,54 - 5
Th	3,4 - 6	4,6 - 5	1,8 - 5	4,3 - 5
U	2,2 - 5	1,8 - 3	1,5 - 3	2,4 - 4
Zr	149 - 107	122 - 74	-	71 - 66

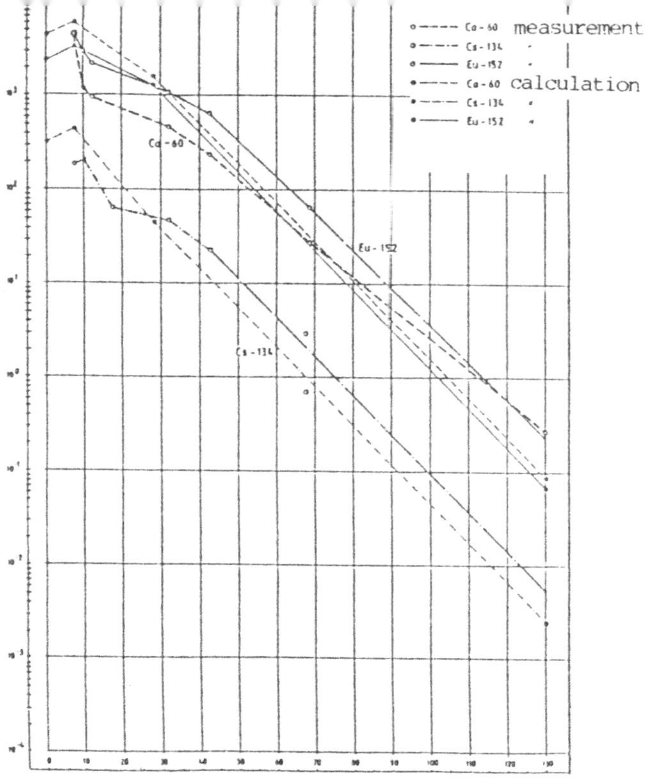

Thickness of the biological shield

Radioactivity as a function of the thickness of the concrete (measurement and calculation) KRB-A

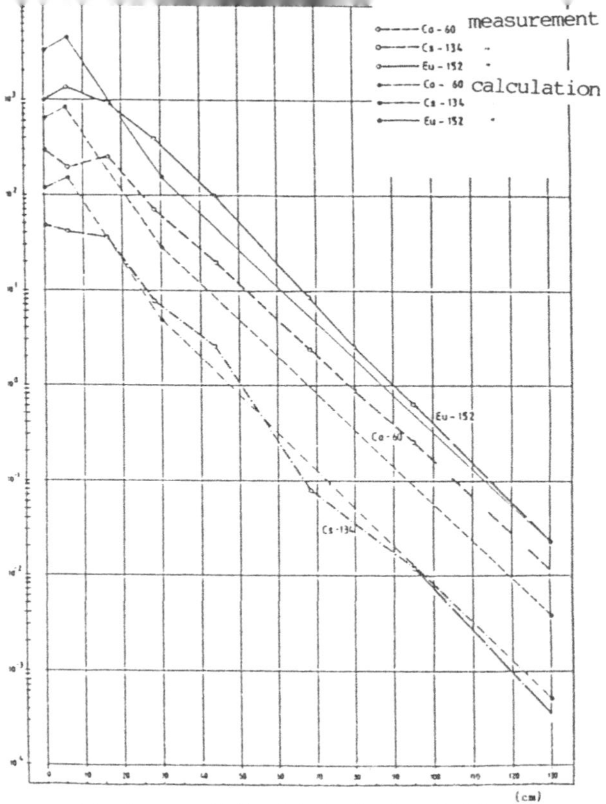

Thickness of the biological shield

Radioactivity as a function of the thickness of the concrete (measurement and calculation) KWL

Figure : 2

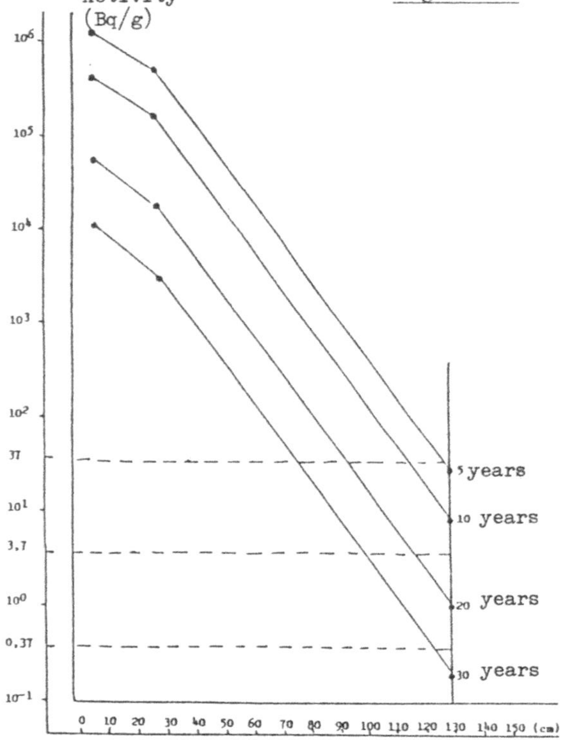

Thickness of the biological shield

Radioactivity of reinforcing steel
as a function of the thickness
of the concrete and decay time

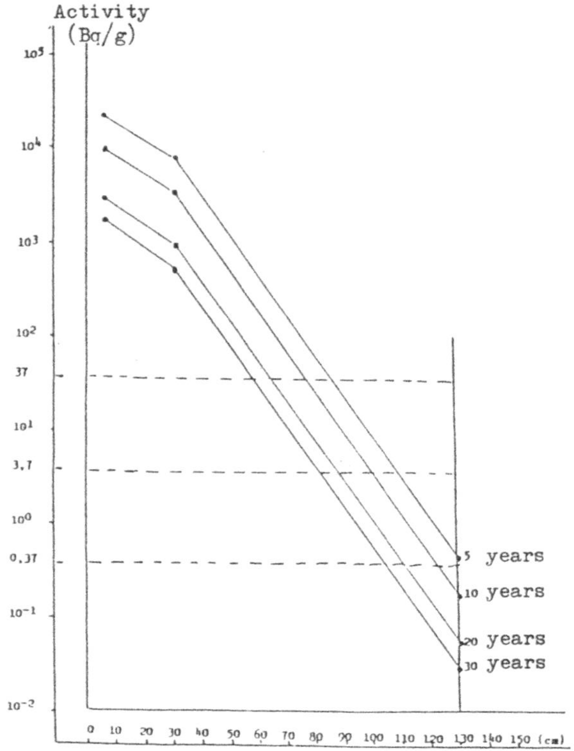

Thickness of the biological shield

Radioactivity of reinforcing steel
as a function of the thickness
of the concrete and decay time

ACTIVATION OF STEEL COMPONENTS

A.J.H. Goddard and T.D. MacMahon
Imperial College of Science and Technology,
London SW7 2BX, United Kingdom.

and

M. Peselli
Nucleco, S.p.A.
Milan, Italy.

Summary

Trace elements leading to long-lived induced gamma activity have been assessed in selected reactor steels. Concentrations of Mo, Ni, Nb, Co, Ag, Eu, Sm and Ho have been measured. Particular attention has been paid to the detection limits of various techniques. Three techniques have been applied to Nb determination and concentrations close to the detection limit of 5 ppm measured in a number of steels. Concentrations of Ag in the vicinity of 1 ppm were observed in all steels analysed. The second aspect of the work described concerns evaluation of gamma emitting nuclides in structural materials of the Garigliano BWR power plant. The trace element composition of samples of vessel steel have been measured. A series of detailed two-dimensional neutron transport calculations have been carried out and, with the aid of the operating history, the time dependent flux leading to activation of vessel, liner and thermal shield estimated. The dose rate at the centre of the reactor vessel has been calculated as a function of time following shutdown.The form of the contribution to dose rate from the vessel differs from that from the liner and thermal shield, due mainly to the need to make pessimistic assumptions of steel composition in the absence of trace element analysis for these latter two components.

1) INTRODUCTION

The timing of dismantling of reactors after final shutdown has to take into account the radiation exposure to be expected during dismantling operations. This paper is concerned with the likely external gamma exposure from activated steel components with particular reference to light water reactors, PWR and BWR. Two aspects are covered; firstly, assessment of the trace elements Co, Nb, Ag, Eu, Sm and Ho, together with Mo and Ni, for a range of reactor steels, with particular reference to potential gamma sources at long cooling times and, secondly, the estimation of the gamma activity induced in the vessel, liner and thermal shield of the Garigliano BWR power plant.

The first systematic study of the implications of trace elements in reactor steels was that of Stephens and Pohl (1). They drew attention to the long-lived activities ^{94}Nb ($T^1/_2$ = 20,300 y) and ^{59}Ni ($T^1/_2$ = 75,000 y) and showed that these activities were likely to dominate gamma dose rates external to PWR core components after about 100 years decay of the ^{60}Co activity. In a subsequent detailed study, Woollam (2) reviewed in detail the external dose rates likely to arise due to ^{59}Ni. Woollam has also shown (3) that ^{108m}Ag ($T^1/_2$ = 127 y) can also be a significant gamma source after about 100 years. Similar considerations lead to the need to set limits on the concentrations of Eu and Sm.

Various comprehensive studies of the decommissioning of representative large PWR and BWR power stations have been published (see for example references (4), (5),and (6)). These studies have included consideration of external gamma doses. Woollam (7) has published a comprehensive study of the radioactive inventory of a decommissioned MAGNOX reactor while Bibby and Goddard (8) have made a detailed study of induced gamma activities in a gas cooled reactor structure.

2. TRACE ELEMENT ASSESSMENT

Neutron activation analysis was the principal technique employed in trace element assessment and was carried out principally at the Imperial College Reactor Centre. A variety of irradiation facilities are available within and around the core ranging from in-core irradiation systems with pneumatic transfer of samples to a cyclic activation system which permits enhanced counting statistics for short-lived activities. Cadmium covered irradiation tubes are also available along the sides of the reactor core. For counting activated samples intrinsic Ge detectors and Ge(Li) detectors are linked to an automatic analysis system, and a Compton suppression spectrometer is also available. For certain elements activation analysis was not found to be the optimum technique, and inductively coupled plasma (ICP) spectrometry and flameless atomic absorption were also employed where appropriate. Secondary ion microprobe spectrometry (SIMS) was also used in performing analyses for niobium. In the case of the rare earths elements, Eu, Sm and Ho, the high activation cross-sections, particularly of Eu and Sm, give them the potential to yield significant gamma activities even with concentrations in the sub-ppm range. In the present work, in order to measure such levels of concentration, an ion exchange procedure was developed in order to concentrate the rare earths prior to activation analysis.

2.1 Samples.

The samples which have been the subject of the analyses carried out are listed in Table 1: samples 1A-D are various CrNiNb18/9 steels; sample 2 is vessel basis material; samples 3A and 3B correspond to parts of the Garigliano vessel as discussed in section 3.2; samples 4A and 4B and 5 are various AGR reactor steel types; sample 6 is a reinforcement steel; samples 7A and 7B are high niobium steels according to AISI 347; sample 8 is stock 18/8 stainless steel. The presence of significant niobium levels in samples 1A-D and samples 7A and B and 8 provide a valuable test of the methods for niobium. Selected results of the analysis are shown in Table 2.

No.	Steel Sample	Composition (Weight %)					
		Mn	Cr	Mo	Ni	Nb	Co
1A	X10 CrNiNb 18/9	1.61	17.85	-	10.63	0.43	0.11
1B	" " "	1.53	18.03	-	10.53	0.62	0.18
1C	" " "	1.76	18.62	-	10.60	0.86	0.18
1D	" " "	1.53	18.15	0.11	10.10	0.33	0.05
2	Vessel basis	-	0.38	-	0.84	-	0.023
3A	Garigliano P.V.	Nucleco sample L17, (see text)					
3B	Garigliano P.V.	Nucleco sample T22, (see text)					
4A	AGR shield	0.70	0.05	0.02	0.10	-	0.011
4B	AGR fuel tie	<.20	15-18	2.5-4.0	42-45	-	<.25
5	AGR support strut	0.8-1.4	-	-	-	-	<.03
6	Reinforcement	0.75	0.01	-	0.06	-	-
7A	AISI 347	1.41	19.28	-	9.28	0.61	-
7B	AISI 347	1.17	17.99	-	9.12	0.38	-
8	CrNiNb 18/8	1.60	17.58	0.36	10.55	0.92	-

TABLE 1. Steel samples analysed, with concentrations of selected elements taken from specifications.

2.2 Molybdenum, cobalt and nickel.

These elements have been grouped together as conventional activation analysis proved satisfactory and yielded acceptable results for all samples.

Although ^{93}Mo has a weak decay scheme it has been included in the study due to its 3,500 year half-life. The most sensitive technique was found to be irradiation of samples in a reactor core tube for one week counting the 2.76 day activity of ^{99}Mo resulting from the ^{98}Mo (n,γ) reaction. The 739.5 kv gamma line was analysed with the aid of the anti-Compton detector.

Induced 5.27 year ^{60}Co is normally regarded as the dominant activity in determining gamma ray dose rates up to about 100 years decay of the reactor structure. Measurement is straightforward and for samples 3A and 3B agreement within experimental errors was obtained with the analyses carried out by the University of Pavia using neutron activation analysis and by AGIP Nucleare Laboratories using atomic absorption.

^{59}Ni decays by electron capture; the resulting atom emits a K X-ray of energy 7.5 keV followed by internal bremsstrahlung with an end point energy of 1.06 MeV. However, only a fraction 8.4×10^{-4} of electron captures lead to photon emissions. Nickel was determined using a one-week irradiation under cadmium to enhance fast neutrons yielding 71 day ^{58}Co via the (n,p) reaction. Where nominal specifications were available good agreement was obtained in all cases. For steels 3A and 3B, where analysis had been carried out by the University of Pavia, agreement was again obtained within the uncertainty of the measurements.

2.3 Silver

In the present work both activation and flameless atomic absorption have been carried out. Of the various silver activities which can be employed in activation analysis ^{110}Ag with a half-life of 20.6 seconds was found to be most sensitive when used in conjunction with the cyclic activation system. However, consideration of background activity levels lead to a detection limit of about 0.6 ppm and this was confirmed in practice where silver measurements were only achieved for samples 3A, 4A and 8. Atomic absorption spectrometry using graphite furnace atomisation, with calibration by standard additions, yielded results for all samples with a detection limit of approximately 0.1 ppm. Good agreement was obtained between atomic absorption results and those obtained using the cyclic activation system. Agreement was also obtained for steels 3A and 3B with the atomic absorption results obtained by AGIP Nucleare Laboratories. In all the steels studied the levels of silver lay in the neighbourhood of 1 ppm.

	Ni %	Co %	*Nb ppm	°Ag ppm	Eu ppb	Sm ppb
1A	10.32 ± .13	.107 ± .007	3535	2.4 ± .1	<33	<35
1B	10.84 ± .14	.184 ± .007	4968	1.7 ± .1	+	+
1C	11.08 ± .12	.379 ± .009	7544	1.1 ± .1	<28	<39
1D	9.38 ± .11	.053 ± .008	2819	0.3 ± .1	+	+
2	.762 ± .029	.021 ± .001	9.0	0.7 ± .1	<41	<31
3A	.752 ± .026	.0187 ± .0007	6.0	1.4 ± .1	15 ± 24	43 ± 42
3B	.732 ± .026	.0178 ± .0007	6.0	0.7 ± .1	+	+
4A	.269 ± .009	.0096 ± .0008	6.0	1.6 ± .1	<16	<7
4B	37.34 ± .34	.205 ± .017	41.0	0.7 ± .1	+	+
5	.287 ± .008	.0191 ± .0016	6.0	0.9 ± .1	<12	<11
6	.047 ± .015	.0253 ± .0042	5.0	0.3 ± .1	+	+
7A	8.92 ± .11	.0148 ± .007	5306	0.6 ± .1	+	+
7B	8.85 ± .10	.134 ± .006	3302	0.5 ± .1	+	+
8	10.01 ± .13	.159 ± .007	7710	4.5 ± .1	+	+

+ not measured ° Atomic absorption * ICP (det. limit ~ 5 ppm)

Note : Mo and Ho concentrations are not listed as they do not contribute significantly to gamma activities.(Ho because of low concentrations)

TABLE 2. Results of analyses for the most significant trace elements for gamma activity. Unless otherwise indicated, method used is neutron activation analysis.

2.4 Niobium

Niobium is of significance for decommissioning due to the activation product ^{94}Nb with a 20,300 year half-life which decays with the emission of 871 and 703 keV gamma rays. Activation analysis for niobium is difficult for various reasons. 6.3 minute ^{94m}Nb decays mainly by a highly internally converted isomeric transition with only 0.5% of decays going by a beta decay to an excited state with the emission of an 871 keV gamma ray. There is thus a choice between measuring the relatively intense K x-rays of energies 16.6 keV and 18.6 keV, where self absorption may occur, or the very weak gamma rays, where other stronger gamma emitters will compete. Simple attempts at prior chemical separation of niobium are rendered difficult due to the sharing of the niobium content of steel between that present "in solution" and that in particulate form as niobium carbide. The proportion present as niobium carbide particles increases with niobium content and with the ageing of the steel sample (Williams (9)).

Direct counting of the 16.6 keV x-ray was attempted using the cyclic activation system. There was clear evidence, for high niobium steels, of self-absorption depressing the results while for low niobium steels an inherent detection limit of approximately 200 ppm was evident. Inductively coupled plasma (ICP) spectrometry was carried out on all samples and found to give results which, in the main, were above the detection limit of about 5 ppm and which agreed reasonably well in the case of the high niobium steels with the manufacturers' specifications. Apart from the specific high niobium steels, the general level of niobium concentrations was lower than has been assumed in the past when setting nominal concentrations.

In view of the significant difference between the low levels of niobium detected in steels 2, 3A and B, 4A, 5 and 6, and the detection limit of activation analysis, a third method of analysis, secondary ion microprobe spectrometry (SIMS) has been used. In this technique a plasma is formed of oxygen ions which are then accelerated to 10.5 keV. The resulting beam is then rastered over a 200μ square of the sample surface and a few microns are eroded. Reactive sputtering takes place and the resulting ions are collected electrostatically and filtered magnetically for mass. To overcome the dependence of sensitivity on element, measurements for steels 3A and 3B were calibrated by measuring also two steels having known concentrations of Nb. Results were consistent with the ICP measurements as shown in Table 3.

2.5 Rare earth elements, europium, samarium and holmium.

In their review of trace elements, Stephens and Pohl reported a determination of europium in a steel sample of 0.41 ppm while samarium and holmium were present in concentrations of less than 0.01 ppm. They discounted the importance of rare earths in view of these small concentrations and noted further that, due to the large burn-up cross-sections, no ^{152}Eu or ^{154}Eu remain after 40 years of irradiation. Woollam (7) emphasised the potential importance of rare earth elements in steels and reported a determination of activity in a charge pan sample taken from a MAGNOX reactor and which was subjected to radiochemical separation prior to counting. The europium level determined was 2 x 10^{-5}ppm while the upper limit for holmium was 1 x 10^{-2} ppm. In practice, the activities resulting from both europium and samarium are complex functions of burn-up and, due to these complex relationships of decay and capture, irradiation of samarium leads principally to the formation of europium gamma emitting nuclides. Activation of europium can, potentially, lead to a 13.3 year ^{152}Eu, 8.8 year ^{154}Eu and 4.96 year ^{155}Eu. Activation of samarium leads to the production of ^{154}Eu and, to a lesser extent, ^{155}Eu and ^{152}Eu (the principal long-lived samarium isotope being ^{151}Sm which is a beta emitter and not considered further here.) Activation of holmium leads, with a much smaller cross-section, to the production of 1,200 year ^{166m}Ho.

In the present work an ion exchange separation technique was used to remove parent elements of potentially competing activities. Samples were dissolved in hydrochloric acid with the residues being retained for separate analysis. Following passage of the solution through the ion exchange material, and the eventual removal of the rare earth elements from the resin, samples were prepared for activation analysis. Typically, for the six samples which were analysed using this method, upper limit concentrations of about 30 ppb for samarium and europium were obtained and about 20 ppb for holmium. Analyses by the University of Pavia for sample 3a had yielded 4 ppb for samarium and 4 ppb for europium.

2.6 Discussion

Results of the analyses are summarised in Table 2. The most significant results are those for Nb and Ag. The relatively low levels of niobium observed in steels, other than those of specifically high niobium content, indicates it may be possible to achieve levels of niobium less than ~10 ppm by materials selection. Such a level of niobium would give external gamma dose rates comparable to that from ^{59}Ni in type 304 steel (see, for example, the data of Smith (4)). Levels of Ag clearly need careful control in view of the levels seen in this survey; 1 ppm of Ag can lead to external dose rates equivalent to that arising from 10 -100 ppm of niobium depending on reactor type (11).

3. QUANTITATIVE ESTIMATION OF TRACE ELEMENTS IN THE GARIGLIANO PRESSURE VESSEL.

3.1 Description

The Garigliano nuclear plant is a dual cycle BWR which began operating on 26.11.63 and was shut down completely on 8.8.78. The nominal reactor power was 506 megawatts thermal with an electrical power output of 160 megawatts. The plant is situated on the Garigliano river a few kilometres from the Tyrrhenean sea about 50 miles north of Naples. The pressure vessel is illustrated in Figures 1 and 2. It was formed by welding a series of rings together, each part identified with a section number as shown in Figure 1. The material used for the forging was a steel classified by ASTM A 336 ASME code 1236. It was clad internally with stainless steel by welding; the first welding run used AISI-309-LC electrodes; there were then two successive runs with AISI-308-LC electrodes. Of the 127 mm thick cylinder walls 6.35 mm are therefore stainless steel. The vessel comprises about 23 cubic metres of steel and weighs 180 tonnes. The thermal shield is a cylinder of AISI 304 steel, 3,651.5 mm high, with an external diameter of 3480 mm

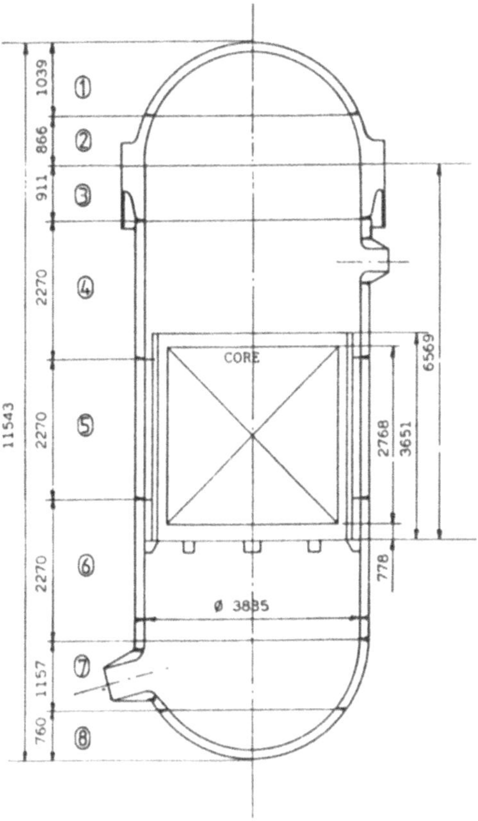

FIGURE 1. Elevation of Garigliano reactor vessel showing ring numbers and dimensions (mm).

and a wall thickness of 25.4 mm. The total weight of this component is 8,000 kg. The core consisted of 208 elements. The active section of the core was 2,768.6 mm high and began 5791 mm from the top join. Each fuel element consisted of 64 rods of UO_2 in an 8x 8 lattice; the initial fuel material was uranium oxide, while in the final cycles of the reactor some plutonium oxide was incorporated.

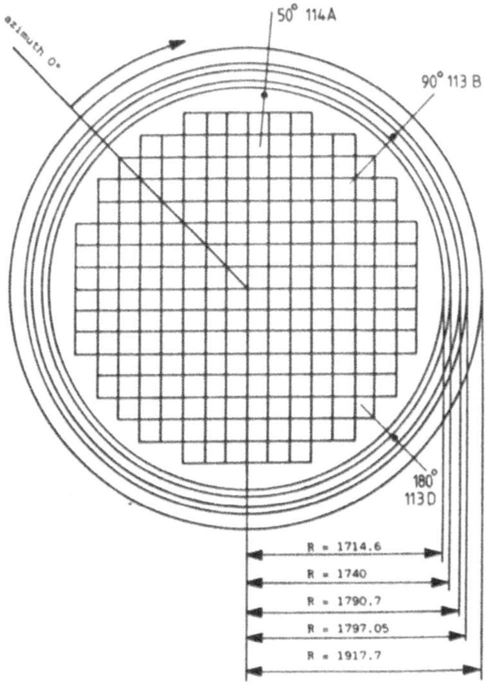

FIGURE 2. Plan of Garigliano reactor vessel, liner and thermal shield giving dimensions (mm) and the location of the 3 sample containers.

3.2 Chemical composition of the vessel steel

To determine the trace element composition of the vessel two non-radioactive samples, referred to as T22 and L17, of the type 336 steel were available for analysis. Neutron activation was carried out at the University of Pavia and, in selected instances, atomic absorption analysis was performed by the AGIP Nucleare Laboratories. These steel samples were also made available to Imperial College and the results of the analyses organised by both institutions are summarised in Table 3. Except in the case of niobium, where ICP and SIMS analysis indicates a substantially lower result than activation analysis, all results are within an acceptable degree of agreement.

	T22 (ring (5))		L17 (ring (4))	
	Nucleco	Imperial College	Nucleco	Imperial College
Ni	7853 ± 380	7320 ± 260	7795 ± 390	7520 ± 260
Co	166 ± 8	178 ± 7	175 ± 9	187 ± 7
	*180 ± 36	-	*180 ± 36	-
Ag	1.17 ± 0.12	< detection	1.1 ± 0.11	1.33 ± 0.11
	*0.7 ± 0.14	*0.7 ± 0.1	*1.3 ± 0.26	*1.4 ± 0.1
Mo	7200 ± 400	6970 ± 170	5800 ± 406	5770 ± 160
Nb	201 ± 32	<223	238 ± 47	<183
	-	+6.0	-	+6.0
		°<8		°<11
Eu	<0.004	-	<0.004	0.043 ± 42
Sm	<0.004	-	<0.004	0.015 ± 24

° SIMS + ICP * Atomic Absorption

TABLE 3. Comparison of analyses by Nucleco and Imperial College (ppm) of Garigliano reactor vessel steels T22 and L17 (steels 3B and 3A in Tables 1, 2). Activation analysis was used unless stated otherwise.

In the absence of analyses of the cladding and thermal shield, the following concentrations were adopted from reference (6):

Co	1500	ppm
Nb	160	ppm
Ag	0.4	ppm
Eu	1.0	ppm
Sm	0.01	ppm

3.3 Estimation of the time dependent neutron flux to which components were exposed.

The parameters which influenced the distribution of flux throughout the reactor life are many (power level, spatial consumption of fissile material, movement and extraction of control rods, etc). The problems of representing the spatial distribution of neutron flux are also significant in that a two dimensional transport theory calculation representing core and vessel used 6 hours of computing time on a powerful computer, in this case an IBM 370.

The principal simplification made was that the neutron flux at the vessel was directly proportional to thermal power using a coefficient of proportionality. Throughout the reactor life 8 cycles were identified by the numbers 1A, 1B, 1C, 2, 3, 4, 5, 6; cycle 6 only generated energy for 2 months. The proportionality coefficient between flux and power was assessed theoretically and was finally evaluated for the core in such a configuration that on the average the neutron flux was somewhat higher than would be indicated by the corresponding power. This was particularly true at the end of cycle 5.

The parameters at the end of cycle 5 which were evaluated were ;

- the distribution of water density in three dimensions
- the distribution of fissile nuclides including plutonium isotopes
- distribution of xenon and samarium poisons.

This assessment began at the start-up of the core and took into consideration all the actions carried out such as power production, control rod movements, research activities, shut-down etc. The presence of mixed oxide fuel was also taken into account. Before the project began the power distribution calculated was compared with the results of gamma scanning of irradiated elements. Results were excellent in most core areas (to within a few %) except in some outer elements where errors exceeded 25 - 30 %.

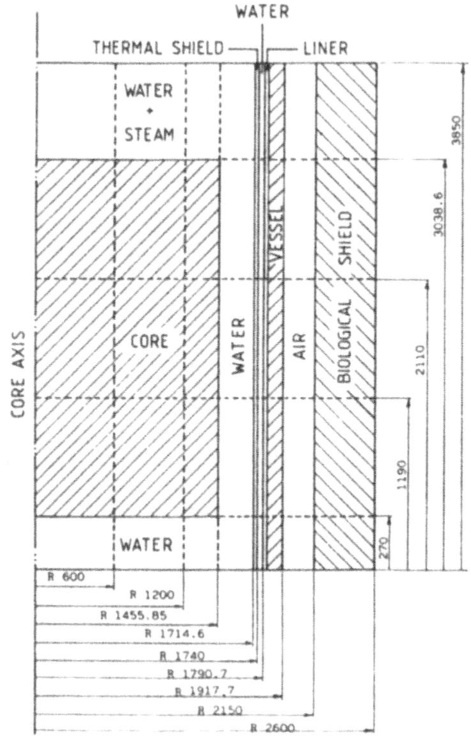

FIGURE 3. Representation of Garigliano reactor structure in DOT neutron transport calculations.

Since the reactor structure could be represented by a symmetrical cylinder, the two dimensional transport theory code DOT 3.5/2 (12) was used. The initial data set was the DLC100 G/3 (13) set which was condensed to 26 groups, with the one-dimensional transport theory code ANISN/4 (14) being used to condense the various region cross-sections in an appropriate manner. The reactor was represented in the calculation as shown in Figure 3; the thermal shield, cladding and vessel sections have been represented in detail with the core being assumed to be a cylinder with the same height and volume. A part of the biological shield was also represented to take account of those neutrons which might be reflected back to the vessel.

An assessment was made of the most cost-effective expansion of the scattering cross-sections ; these studies were carried out with the ANISN code. P_0 expansion was found not to be acceptable (errors in the flux at the vessel approached 400 %) while P_3 differed to a negligible extent with P_1 calculations. P_1 cross-sections were therefore used.

3.4 Assessment of activity

In order to make use of the spatial neutron flux information and the reactor operating history, a computer program EVA (Evaluation of Activity) was devised to follow the time dependent reactions of the isotopes of interest. EVA incorporated a simplified set of activation and decay data.

The original intention had been to compare predictions with the activity of various vessel material test samples which had been inserted at various points around the vessel and close to the core before the reactor began operation. However, due to the temporary delay in the various programmes devoted to the closing of the reactor these samples have not yet been extracted and were not therefore available. However, comparison was possible using various steel test samples which had been previously extracted together with information made available by ENEL concerning the measured ^{60}Co activity and the necessary temporal and geometric data.

The position of the samples is indicated in Figure 2. They were inserted into the reactor at the start of its life in three large containers, each with 5 capsules and each capsule having 12 steel samples. The containers, identified as 114A, 113D, and 113B, were extracted, in the order just given, after cycle 1C, after cycle 2 and after cycle 5. The samples were analysed using a Ge(Li) detector, directly, and in the case of the container 113D, after radiochemical separation to concentrate the activities ^{54}Mn, ^{60}Co and ^{59}Fe.

3.5 Comparison between calculated and experimental values of ^{60}Co activity.

In estimating cobalt activities in the samples account was taken of the variation in the calculated flux across the dimensions of the container and also the range of uncertainty in the ^{59}Co concentration. This yielded a band of uncertainty of about $\pm 15\%$ in the ^{60}Co predictions.

The agreement was very satisfactory in that for the three containers 114A, 113D and 113B, the percentages of the ^{60}Co activities of samples lying within this basic uncertainty band was 40%, 67%, and 40% with a significant proportion of the remainder of the samples lying within a further 20% range.

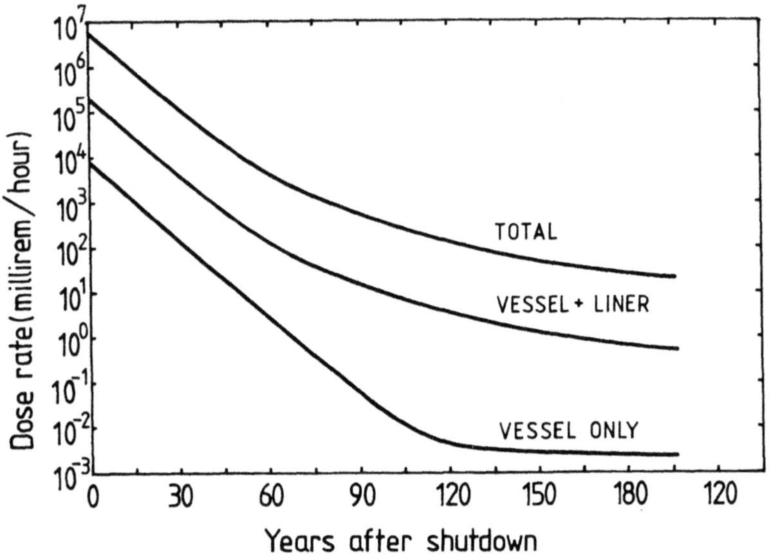

FIGURE 4. Variation of gamma dose rate at the centre of the vessel (core removed) with liner and thermal shield in place. Contributions of vessel and vessel + liner to total are shown.

4. INTENSITY OF GAMMA DOSE IN THE CENTRE OF THE VESSEL AS A FUNCTION OF TIME

During the eventual dismantling of the Garigliano reactor the gamma emitting activities will create a gamma flux that will require the use of a suitable shield to reduce the exposure to the operators and in many cases will call for remote dismantling procedures. As an indication of the consequent levels of dose rate as a function of time, the dose rate at the centre of the empty vessel has been estimated. This estimate was obtained by summing over: the various zones which could give exposure at the centre of the vessel, the number of nuclides selected (^{60}Co, ^{94}Nb, ^{108m}Ag, Eu isotopes) and the number of gamma energy groups, and took into account the flux of photons at the centre of the vessel produced by sources in the various regions. These photon fluxes were obtained from a series of ANISN calculations which treated the photon flux anisotropy and the scattering anisotropy in the S12 and P5 approximations. A number of approximations are implicit in this procedure but the dominant uncertainty is considered to be the trace element composition of the vessel cladding and of the thermal shield; the nominal composition listed in section 3.2 was used. The results of these calculations led to the following conclusions:

- the dose at the centre of the vessel is due largely to the presence of the thermal shield which even after 200 years of shut down produces and intensity of about 15 mR per hour

- if the thermal shield is removed the dose intensity would be sustained by the stainless steel cladding adjacent to the core. An intensity of about 15 mR per hour will exist 90 years after shutdown. The vessel alone would generate a comparable intensity after 45 years.

- the curve of dose intensity relative to the vessel alone shows a different form from the other two components. This is due mainly to the low percentage of europium present in the vessel steel based on the analyses which were carried out. For the liner and thermal shield the dose intensity is due almost entirely to ^{60}Co in the first 50 years; due to ^{152}Eu from 50 - 120 years and finally to ^{94}Nb and ^{108m}Ag. For the vessel, however, there is a rapid transition from a period dominated by ^{60}Co to when ^{94}Nb and ^{108m}Ag become important.

5. CONCLUSIONS

Methods for measuring trace elements exist which are of adequate sensitivity for the principal steel components of PWR and BWR reactors. Analysis of a range of samples indicates that material selection for low niobium content should be feasible but that special provision may be necessary to reduce silver content significantly below 1 ppm.
The extensive theoretical calculations of activity distributions accompanied by experimental analyses of inactive and active samples have demonstrated that activities in light water reactor vessels may be estimated satisfactorily. Predicted dose rates as a function of time for the Garigliano BWR reactor have demonstrated the importance of an understanding of the trace element content of all major vessel components and not simply the vessel base material.

7. REFERENCES

1. Stephens, J.J. Jr., and Pohl, R.D. Trace elements in reactor steels : implications for decommissioning, Nucl. Engn. and Design, 47, 125 - 134, 1978.

2. Woollam, P.B. A perspective on potential contributions to reactor decommissioning problems caused by nickel-59, CEGB Report RD/B/N4232. 1978.

3. Woollam, P.B. An assessment of the data for decommissioning calculations on Ag-108 metastable, CEGB report RD/B/N4393, 1978.

4. Smith, R.I. Technology, Safety and Costs for Decommissioning a Reference Pressurised Water Reactor NUREG CR 0130, US Nuclear Regulatory Commission, June, 1978.

5. Watzel, G.V.P. et al., Decommissioning of Nuclear Power Stations at the end of their service life. (CEGB Trans 7870) Progess Reports of the Association of German Engineers, (Fortschr.-Ber. VDI-Z) Series 15,no 18,1982

6. Oak, H.D. et al. Technology, Safety and Costs of Decommissioning a Reference Boiling Water Reactor Power Station. NUREG/CR-0672-V.1. US Nuclear Regulatory Commission, June 1980.

7. Woollam, P.B. and Pugh I.G. The radioactive inventory of a decommissioned MAGNOX power station. Part 2. A summary of neutron induced activation, waste disposal and dose equivalent rates for the reactor island structure. CEGB Rep. RD/B/R4350, 1978.

8. Bibby, D. and Goddard, A.J.H. Long-lived activity of structural material and its implications for containment. Proc. IAEA seminar Decommissioning of Nuclear Facilities, IAEA, Vienna, 1979.

9. Williams. T.M. Private communication, 1982

10. Peselli, M. Private communication, 1983.

11. Boothby, R.M. and Williams, T.M. The control of cobalt content in reactor grade steels. European Applied Physics Reports, 5, 2, 121 - 143, 1983.

12. Dot 3.5. Two dimensional discrete ordinates radiation transport code. OAK Ridge National Laboratories.

13. ZZ-DLC-100G. 100-Group neutron transport code cross-section data generated by SUPERTOG from ENDF/BS. NEA.

14. ANISN. One dimensional discrete ordinates transport code with anisotropic scattering. OAK Ridge National Laboratories.

ESTIMATION OF RADIOACTIVE WASTE QUANTITIES ARISING DURING DECOMMISSIONING

M. Lasch *, K.H. Schaller **, W. Stang *, G.V.P. Watzel ***

* Kernkraftwerke Gundremmingen Betriebsgesellschaft mbH

** Commission of the European Communities, Brussels

*** Rheinisch-Westfälisches Elektrizitätswerk AG, Essen

Summary

The quantities of radioactive waste arising from the decommissioning of nuclear power plants, which are in operation, under construction or ordered in the countries of the European Community, have been roughly estimated.

As compared to earlier evaluations, measurements and decontamination work at the 237 MWe Boiling-Water Reactor KRB-A Gundremmingen, which has been shut down since 1977 after 11 years of operation, allowed a more precise forecast in two fields, penetration of contamination in walls and floors, and decontamination of metallic components and piping at a large scale.

The penetration of contamination has been measured at selected locations. The surfaces of almost all of the restricted area are covered by a protective layer as in most of the European nuclear power plants. 90% of the activity has been found to be cumulated in the coatings. An extrapolation to the total plant surface showed, that only about 500 Mg of waste would be generated during dismantling in the process of decontaminating walls and floors to values for unrestricted release.

Electrochemical decontamination of metallic components to unrestricted release has been demonstrated successfully at KRB-A at a large scale. The treatment of the decontamination solutions allowed the estimation of the quantities of secondary waste produced in the event of a decontamination of the whole machine-house equipment.

Supposing that, in the average, reactors will be dismantled definitively after a decay-time of at least 30 years after final shut-down, the arising quantities of medium- and low-level radioactive waste have been estimated for the different types of reactors.

Cumulation of the quantities shows, that only after the year 2050 an annual quantity would be reached, which is of the same order of magnitude than the waste produced annually by normal reactor operation (including reprocessing of fuel elements) in the years 1990 to 2000.

1. Introduction

The quantity of radioactive waste arising from the decommissioning of nuclear power stations must be estimated in order to give an idea of the order of magnitude of the quantities of waste arising from the power reactors in the countries of the European Community.

The data from the measurements and tests on the Gundremmingen 250 MWe boiling water reactor, which was decommissioned in 1977, can be used to assess the amount of waste - hitherto regarded as negligible - arising from the decontamination of large components of facilities and the quantities arising from the removal of contaminated layers on walls and floors.

Since there are no generally recognized criteria for distinguishing between radioactive and non-radioactive waste these test results are specific to KRB-A. A pessimistic estimate has been made of the quantity of radioactive waste arising throughout Europe from decommissioning.

2. Large-scale decontamination in Block A of the Gundremmingen nuclear power-station

Since the incident in January 1977 the first German demonstration nuclear power station with a boiling water reactor - KRB Block A - has been shut down. For commercial reasons the two companies concerned, Rheinisch-Westfälisches Elektrizitätswerk AG and Bayernwerk AG, decided in 1980 to shut down the facility permanently. Before the plant could be completely demolished the activity was to be allowed to decay further over a "safe containment" period of approximately 25 years. However, the large amounts of lightly contaminated decommissioned systems and components in the turbine hall are safe to dismantle and decontaminate at an earlier stage.

This procedure appears to be advantageous for the following reasons :

it considerably reduces the size of the safe containment area (by restricting it to the reactor and treatment buildings) ;

personnel familiar with the operation of the plant are still available for planning and supervising the dismantling work ;

The main component of the contamination is still the nuclide ^{60}Co (accounting for more than 70%), which provides an indicator which is easy to measure. In 25 years time, when the ^{60}Co has decayed, the main radiation will come from low-level beta emitters, e.g. ^{63}Ni, which are very difficult to measure.

Since at present no store is available for final storage of the low-level radioactive waste arising from the decommissioning it is to be stored provisionally in the reactor building. Because of the space it occupies it is important that when decontamination is being carried out in the turbine hall the quantity of low-level radioactive waste should under no circumstances exceed 10% of the quantity of steel treated. A demonstration test has been carried out with approximately 100 tonnes of representative components of the facility to show that this is possible.

2.1 Choice of representative components

When choosing components with regard to the degree of difficulty of :

dismantling ;
decontamining and
measuring them for freedom from contamination

the widest possible selection should be made.
For this reason a variety of components from the primary steam, condensate and feedwater sections was selected with regard to these aspects.

Since at rough estimate approximatively 75% of the total mass of the turbine hall can be dismantled and decontaminated so that it can subsequently be released as being under a surface threshold limit of 0.37 Bq/cm^2, the 100 tonne trial was confined to these components. The remaining 25%, in the form of pipes of nominal bore under 150 mm, instruments,clusters of tubes and the like, were at first disregarded, since there is as yet no regulation governing the release of these materials without surface measurement. The relative quantities of radioactive waste should, however, remain within similar orders of magnitude.

Initial activity

The inner contamination of the components affected by water or steam was approximately 10-1 000 Bq/cm^2. Surfaces which had been in contact with other low-level reactive media, such as air, oil etc, were found to have maximum contamination levels of approximately 10 Bq/cm^2.

The possibility of radioactivity in the area of the turbine hall could be discounted.

2.3 Decontamination methods

Under a research programme supported by the EC for developing economic decontamination methods, electropolishing appeared to be a very useful decontamination method. This process works on the principle of reversing the galvanizing process, in which the surface is removed rather than applied. 70% phosphoric acid proved to be a very good electrolyte.

In the 100 tonne test this method was used to decontaminate approximately 65 tonnes of components which had been in contact mainly with primary steam and water. The remaining 35 tonnes with looser decontamination were cleaned using generally familiar methods such as steaming etc.

2.4. Quantities of waste

The quantity of radioactive secondary waste arising from the usual decontamination methods such as steaming depends on the thickness of the layer and how the waste water is treated. The unit quantity of waste arising when this method is used certainly remains equal to or less than the quantity arising from electropolishing.

It was found on the basis of the electropolished steel components in the 100 tonne test that when appropriate methods are used for regenerating and treating the phosphoric acid the radioactive secondary waste can be kept to under 2%.

Further quantities of waste arise when components of complex shapes are dismantled which could not be decontaminated economically, e.g. instrument housings. A total of approximately 2.5 tonnes of small pieces of this type of waste was accumulated.

The inevitable quantity of disposable materials such as plastic sheet, clothing, filter inserts etc. should not exceed 2 tonnes after incineration or compression. The processing of the washing water generated less than 1 tonne of solids. The total quantity of low-level radioactive waste is thus approximately 8.5 tonnes, i.e. 8.5% of the 100 tonnes of steel treated.

Making pessimistic assumptions, which also take account of the uncertainties regarding the small pieces which still remain to be treated, we calculate that the quantity of waste should be less than 10% of the components of average contamination in the turbine hall of KRB-A.

This gives a total figure of around 200 tonnes of waste for a total quantity of approximately 2000 tonnes of steel in the form of decommissioned plant components. This low-level radioactive waste can be packed, transported and put into final storage without being shielded.

3. Penetration of the decontamination in walls and floors as exemplified by KRB-A

The literature on this subject /5/ shows that the quantities of radioactive waste which arise when the floors and walls of buildings are decontaminated represent a high percentage of the total waste.

For these reasons tests were carried out in KRB-A on :

the amount of radioactivity ;
distribution of nuclides ;
degree of penetration

with the aim of determining the quantities of waste that may be expected to arise from the entire structure of the building.

In addition, these test results were confirmed within certain orders of magnitude when the former solid materials store was decontaminated.

3.1 Test to determine the amount of radioactivity, nuclide distribution and degree of penetration

3.1.1 Sampling

At the outset we assumed that the depth to which activity had penetrated into concrete was only a few mm. For this reason we needed to use a sampling process whereby the concrete could be removed in 1 mm layers.

The samples were taken using a hand drill fitted with a hollow bit. The cores extracted were 55 mm in diameter and some 30 mm long. Particular care was taken during sampling that no contamination from the surface of the sample should be transferred to the underlying layers of the concrete because of the sampling method.

A total of 17 concrete samples was taken (see Fig. 1). Samples 1-10 were taken from wall and floor areas which had been contaminated by leakages of water from the reactor. Samples 11-13 had been contaminated through the action of primary steam. Samples 14 and 15 had been affected by contamination transferred from various areas of the reactor building. Samples 16 and 17 were from the turbine hall.

Sample No	Source	Total activity (Bq)	Depth of penetration (mm)
1	Wall/primary flushing pump room	818	3
2	Wall/primary flushing pump room	5 741	4
3	Floor/circulation pump room	1 615	4
4	Wall/circulation pump room	96	3
5	Wall/zero-power cooling room	48	3
6	Wall/room containing collecting tank for radioactive material	41	3
7	Floor/scram facility container room	5 502	4
8	Floor/scram facility container room	6 942	3
9	Floor/control rod drive room	1 011	3
10	Floor/rod control rod drive room	2 831	3
11	Floor/pipe conduit to turbine hall	9 272	3
12	Floor/pipe conduit to turbine hall	810	5
13	Wall/pipe conduit to turbine hall	15 507	3
14	Floor/in front of personnel air lock of reactor building	91	3
15	Floor/control room 31 m	395	3
16	Floor/turbine room	40	3
17	Floor/7.15 m turbine hall	4 961	3

Fig. 1 : Sources of samples and concrete contamination figures

3.1.2 Measurements

The values measured are listed in Fig. 1 by source. Most samples were covered with a coloured coating which it was possible to decontaminate. This layer was removed from the drill core and subjected to separate measurements before the grinding operation was carried out.

It should be noted that the depth of penetration of the radioactivity bore no relation to the surface contamination. It is not true that a high degree of contamination causes particularly deep penetration of the concrete. When the samples were tested for the presence of specific nuclides more ^{137}Cs was found in the deeper layers than ^{60}Co.

Sample no 4 (see Fig. 2) may be regarded as representative as regards the contamination distribution in concrete. More than 90% of the total activity in this sample was confined to the protective coating, so that the concrete below contains only traces of contamination.

Sample no 12 (see Fig. 2) is a typical example of the penetration pattern in concrete when there is no protective coating. This demonstrates how important it is to provide the concrete with a protective coating.

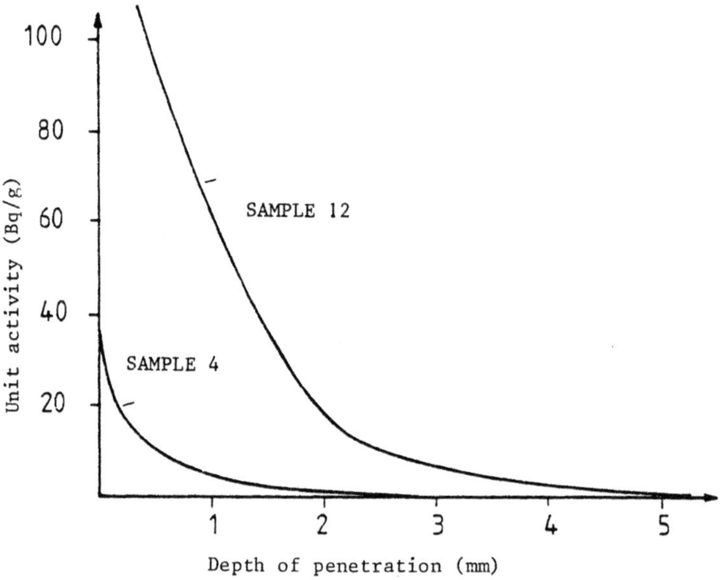

Fig. 2 : Typical graph of radioactivity in contaminated concrete

3.1.3 Extrapolation of measurement values to the entire facility

Taking account of the above activity profiles, where sample 4 should be taken as representative for KRB-A, a rough estimate can be made of the surface area of the building to be decontaminated and the volumes to be disposed of.

Sample 4 shows clearly that the limit value of 0.37 Bq/cm^2 for surface contamination can be achieved theoretically with the removal of a layer of concrete approximately 3 mm thick.

In the case of a building with an inner surface area (walls and floor) of approximately 50 000 m^2, removal of a layer averaging 3 mm gives a total volume of waste of approximately 150 m^3 of concrete.

3.2 Decontamination trial in the former solid waste store of KRB-A

3.2.1 Test method

The former solid waste store was used for packaging and conditioning solid radioactive waste. This produced an above-average degree of contamination on the floors and the lower areas of the walls. The protective coating was renewed on a number of occasions so that most of the contamination was confined to the coating.

Following a series of preliminary tests, which included flame scarfing, the favoured process was to remove the surface of all floors with a concrete milling machine while simultaneously removing the dust by suction. In areas which were difficult of access, e.g. corners, stairs etc. commercially available electric chisels proved effective. The walls were decontaminated mainly by washing or applying a corrosive product.

The KRB radiation protection service showed the building to be contamination-free, i.e. with radioactivity below 0.37 Bq/cm^2, by checking the whole with a large-area counter.

3.2.2 Quantities of waste

At the end of this decontamination process 12 200 litre containers had been filled. Approximately half of these were filled with concrete dust removed by milling, and the other half were filled with concrete chips which had ben removed by chisel.

On the basis of a mechanically-treated surface of approximately 200 m^2, this means that an average thickness of 6 mm was removed.

Experience has shown that even when the depth of penetration of the contamination is only 3 mm surface removal must be carried out with a margin of 100% because of the removal method selected.

In addition particular features of the building such as sumps, staircases etc. can affect the quantity of waste considerably, since it is not possible to remove a thin layer from such features.

If the total quantity of waste is calculated on the basis of the mechanically scarified surface area of approximately 200 m^2, the thickness of radioactive concrete to be removed arising from the former solid waste store in KRB-A is theoretically approximately 12 mm.

3.3. Comparison of the analyses of core samples and the tests in the solid materials store

In those areas of the facility in which it is necessary and possible to remove thin layers it should be assumed for the purposes of calculating the quantity of waste that concrete must be removed to twice the depth to which contamination has penetrated.

Since in calculating the figure of 50 000 m^2 for the area of contaminated surfaces inside the building all the walls were included, while in the solid materials store it turned out that only a small portion of the walls had to be decontaminated mechanically, the figure of 150 m^3 for the total quantity of radioactive concrete removed can be confirmed as realistic.

The only guide to quantities of waste from heavily irradiated and poorly protected areas of the facility is the experience with the solid materials store, which indicates that the above quantity of concrete should be approximately doubled.

4. Assessment of quantities of radioactive waste arising by type and capacity of reactor

The radioactive inventory has been calculated in a large number of studies for many reactors. When a large power reactor is finally closed down after reaching the end of its design life the inventory lies somewhere between 10^{17} and 10^{18} Bq. After periods of 30-100 years isotopes with long decay times become crucial. The remaining inventory, which consists almost exclusively of low-energy emitters, lies between 10^{14} and 10^{15} Bq. The greatest part of the inventory is concentrated on a relatively small quantity of material, i.e. the fixtures near the core.

4.1 Working hypotheses

This assessment relates only to power reactors, which are assumed, with the exception of prototypes, to produce electricity for approximately 40 years from when they were commissioned (which corresponds to an equivalent full-power operating time of 30 years). In the period of up to 10 years following shutdown all fuel elements and all liquid radioactive material are normally removed. From this time on the power station becomes a temporary store for the remaining waste. The possibility of sealing off the reactor block and allowing it to decay indefinitively is not at present under consideration in Community countries.

The length of time which must elapse before the facility is finally demolished depends on many factors. Demolition within 10 years of shutdown means that experienced personnel can be used and releases the building site, but produces higher accumulated doses of radiation and greater quantities of radioactive waste. After a period of 30-50 years the gamma emitter ^{60}Co is less important than more long-lived isotopes in the radioactivity inventory and there is no advantage in allowing more time to elapse before dismantling the more highly reactive components. After about 100 years any further delay in demolition brings practically no reduction in the radiation doses to personnel or in the quantity of waste. The disadvantages of the waiting period are of course the mounting costs of supervising and maintaining the decommissioned facility and the related radiation doses.

For this estimation a waiting period of 30 years before demolition has been assumed, in the full knowledge that for the purposes of demonstration or for other reasons many facilities are dismantled well within this period. The policy on dismantling differs from one Member State to another. It is also influenced by the availability of final storage facilities.

A further assumption must be made in order to classify the waste with a view to final storage, although there are no uniform and generally recognized categories for waste arising from decommissioning, which normally contains no or at most only occasional traces of alpha emitters. For simplicity the unit activity of typical combinations of nuclides in decommissioning waste has been used.

Limits on the quantities of these nuclides in material for final storage are however dependent on a large number of secondary considerations, such as the radiation at the surface of the storage site, the intensity of radiation at the surface of the containers, the total inventory of nuclides per final storage site and the maximum permitted concentration of nucides in water and in air.

A large part of the waste is of low activity and shalow land burial or ocean bed disposal are the most economical options. The upper limit for this type of waste, which we shall term "low activity" (LA) waste is, for instance, 2.2 x 10^9 Bq/m^3 in England (ref. /1/). In France the limits are related to the permitted values for individual nuclides in water and the atmosphere. The values for the important nuclides ^{90}Sr and ^{137}Cs for instance are 1.5 x 10^{11} Bq/m^3 and 3.7 x 10^{12} Bq/m^3 respectively, although it is laid down that special handling methods must be used in the final store in La Hague. For this estimate a value of 740 Bq/g (2 x 10^{-8} Ci/g) has been taken.

The radioactive waste can be divided into two groups :

"Medium-activity waste" (MA), with an average radioactivity per component or individual lot of more than 740 Bq/g, for which the following final storage options are available :

. deep burial (mostly in former salt, potassium or mineral strata) ;
. shallow burial with special precautions ;
. ocean bed disposal (if permitted by existing legislation).
N. B. This type of waste also contains a proportion of material (e.g. fixtures from near the core) which can only be burried at great depths because of its radiological properties. In the case of water-cooled power reactors these quantities are approximately 50-100 Mg. For graphite waste with ^{14}C as almost the only emitter, incineration and conditioning of the residues is also possible.

"Low-activity waste" (LA) with an average radioactivity of 3.7-740 Bq/g can typically be finally stored by :

. shallow burial ;
. ocean bed disposal.

Limit values and release criteria for material below an "exemption limit" have not yet been standardized. When it is a case of releasing for unmonitored re-use or disposal on ordinary refuse tips or at the reactor site, the values under discussion range from 0.37 Bq/g to 74 Bq/g (ref. /2/). In this study a value of 3.7 Bq/g (10^{-10} Ci/g) has been assumed.

Quantities of simple building débris and scrap, which are clearly neither contaminated nor activated, and waste with a radioactivity content which is less than 3.7 Bq/g and therefore negligible, have not been included in the calculation.

A further stipulation should be that classification by radioactivity level is only approximate (e.g. dividing the biological shield into cylindrical layers with thicknesses of the order of a metre). Contaminated layers which it would be uneconomic to remove should be regarded as integral parts of the contaminated material.

4.2. Activated materials

Once limit values have been defined and the neutron flow distributions established neutron flow limits can be calculated for the individual materials. In terms of thermal flow these upper and lower limits for low-activity waste (see also ref. /3/) are as follows :

stainless steel	5×10^7	LA	10^5 n/cm^2 sec.
ordinary steel and concrete	5×10^8	LA	10^6 n/cm^2 sec.

The ^{59}Co content is the decisive factor. In the activity specifications permitted ranges are generally given for alloy elements and contamination, so that in relation to these values the recorded activity fluctuates within a corresponding range.

Thus the quantities for large water-coled reactors (900 MW - 1300 MW) are (see also ref. /2/ and /4/):

MA :	Steel components	700 Mg
	concrete (with reinforcement)	400 Mg
LA :		400 Mg

For gas graphite reactors (GGR) similar to the "Magnox" type, according to ref. /5/ :

MA :	Steel components	1 900 Mg
	graphite	2 200 Mg
LA :	Steel components	800 Mg
	concrete	5 200 Mg

For advanced gas-cooled reactors (AGR) the values are :

MA :	Steel components	2 800 Mg
	graphite	1 800 Mg
LA :	Steel components	5 000 Mg
	concrete	5 000 Mg

There are not so many reactors of other types. For small prototype reactors of the water-cooled type approximately half these quantities should be taken.

4.3 Contaminated steel

Calculations should include only steel which cannot economically be decontaminated for unrestricted release. It is assumed here that as a result of advances in decontamination techniques the marginal cases mentioned in ref. /2/ - such as turbines in boiling water reactors - will be classified as capable of being decontaminated. The remaining contaminated quantities can be classified as "low-activity waste" when mixed with other material or melted down.

Pressurized water reactors over 800 MWe	5 300 Mg
Boiling water reactors over 800 MWe	4 700 Mg
Gas graphite reactors	2 000 Mg

For medium-sized water-cooled reactors (400-800 MWe) 75% of these quantities should be assumed, and for small reactors 50%.

4.4 Contaminated concrete

For a large reactor, with no protective coating on the walls and floors, penetration depths (see ref. /5/) of up to 15 cm should be assumed. This means that the quantity of concrete in the LA category is approximately 14 000 Mg. In reactors in the Community protective coatings are used extensively.

Extrapolating the values obtained in KRB-A for a 1 200 MW facility gave a quantity of waste of approximately 500 Mg (concrete and protective coating).

Since no check has been made on whether the values for KRB-A can be applied to other facilities a quantity of 3 000 Mg should be assumed per reactor.

4.5. Secondary waste from decontamination

The approximate quantities of waste in the LA category are as follows :

Large pressurized water reactor	1 800 Mg
Large boiling water reactors	2 000 Mg
Gas graphite reactors	1 000 Mg

For smaller water-cooled reactors the quantities should be adjusted as in 4.3.

4.6 Other waste

The calculations do not include particular types of waste which arise in unforeseable quantities which require special treatment.

These include :

asbestos from thermal insulation ;
sodium from liquid metal cooled reactors.

4.7 Totals by type and capacity of reactor

For the final total the values given in table 1 are used.

No	Reactor type	MA (Mg)	LA (Mg)	Total (Mg)
1	Large water-cooled reactor (over 800 MWe)	1 100	10 900	12 000
2	Medium-sized water-cooled reactor	1 100	8 900	10 000
3	Small water-cooled reactor (under 400 MWe)	600	7 400	8 000
4	Gas-cooled reactors (GGR)	4 100	11 900	16 000
5	Advanced gas cooled reactors (AGR)	4 600	17 400	22 000
6	Prototype low-power reactors	600	7 400	8 000

Table 1 : Quantities by type of reactor

5. Estimation of quantities of waste arising in the Community countries from the decommissioning of power reactors

Table 2 summarizes the expected quantities. These figures cover power stations which have already been decommissioned, those in operation or under construction and those for which a definite order has been placed (see ref. /6/).

Among the decommissioned reactors in the nuclear power plant for the ship "Otto Hahn" has been excluded. The KKN power station in Niederaichbach, which was operational for only a short time, has been entered as a special case (S) with a quantity of 2 700 Mg.

Year of commissioning	No of reactors	Type No	Total MA Med. Activity Mg	Total LA Low Activity Mg	Total quantity Mg
Decommissioned before 1984 or commissioned before 1960	1	8	-	2 700	2 700
	3	3	1 800	22 200	24 000
	10	4	41 000	119 000	160 000
	3	6	1 800	22 200	24 000
	17		44 600	166 100	210 700
Commissioned 1961-1970	5	3	3 000	37 000	40 000
	20	4	82 000	238 000	320 000
	5	6	3 000	37 000	40 000
	30		88 000	312 000	400 000
Commissioned 1971-1980	16	1	17 000	174 400	192 000
	4	2	4 400	35 600	40 000
	3	3	1 800	22 200	24 000
	5	4	20 500	59 500	80 000
	4	5	18 400	69 600	88 000
	1	6	600	7 400	8 000
	33		63 300	368 700	432 000
Commissioned later than 1980 or planned to be commissioned before 1990	53	1	58 300	577 700	636 000
	8	5	36 800	139 200	176 000
	1	6	600	7 400	8 000
	62		95 700	724 300	820 000
Totals	142		291 600	1 571 100	1 862 700

Table 2 : Total quantities of waste

Despite all conservative assumptions, it follows from Table 2 that in the year 2030 approximately 20 000 Mg of radioactive waste will be produced each year. Since research reactors and other facilities are due to be dismantled before this date, the annual quantities, which will still be very small in the year 2000, should rise slowly to this figure. By the year 2050 an annual figure of 40 000 Mg can be expected and subsequently this value will rise to 80 000 Mg.

For comparison the anticipated quantities arising from the operation of reactors in the Community and from the re-processing of the spent fuel elements should be taken into consideration (according to ref. /7/). However, Table 3 gives only the quantities of treated low- and medium-activity waste, and does not include the highly radioactive waste containing alpha emitters which arises from reprocessing.

Period	1981-85	1986-90	1991-95	1996-2000
Low activity (Mg)	591 000	659 000	738 000	790 000
Medium activity (Mg)	95 000	127 000	189 000	239 000
Total (Mg)	686 000	786 000	927 000	1 029 000

Table 3 : Low- and medium-activity waste arising from the operation of nuclear power plants

Accordingly, after 1990 annual production of low- and medium- level waste will be 180 000 Mg, rising to 200 000 Mg by the year 2000.

Very roughly, the annual arisings from the decommissioning of plants after 2050 will be equal to the annual amount of waste arising from the normal operation of the reactors in the European Community during the period 1990-2000.

References

/1/ Woollam, Pugh : Paper IAEA-SM-234/10, International Symposium on the Decommissioning of Nuclear Facilities, Vienna, 13-17) Nov. 1978.

/2/ Lörcher, Piel : "Dekontamination von Komponenten stillgelegter Kernkraftwerke für die freie Beseitigung", EUR 8704.

/3/ Francioni et al : "Anfall von aktiviertem Material aus dem Bereich des Reaktorkerns bei der Entsorgung eines stillgelegten Druckwasserreaktors", EIR-Bericht Nr. 368.

/4/ "Decommissioning of Nuclear Facilities", Technical Document IAEA-179.

/5/ Moore : "Facilitation of Decommissioning Light Water Reactors", NUREG/CR-0569.

/6/ Kempken : "Verzeichnis der Kernkraftwerke der Welt", Atomwirtschaft, Nov. 83.

/7/ "Analysis of the present situation and prospects in the field of radioactive waste management in the Community". COM (83) 262 of 16 May 1983.

METHODOLOGY FOR EVALUATING RADIOLOGICAL CONSEQUENCES OF THE MANAGEMENT OF VERY LOW-LEVEL SOLID WASTE

M.J. CLARK, G.M. SMITH and C.R. HEMMING
National Radiological Protection Board, UK

A-M. CHAPUIS
Commissariat a l'Energie Atomique, France

Summary

A radiological assessment has been made of the management options for waste materials containing very low levels of radioactivity arising in the decommissioning of a pressurised water reactor (PWR). Estimates have been made of the masses of materials activated and contaminated to within three limits of specific activity, 0.37, 3.7 and 37 Bq g^{-1}, at three times after reactor shutdown, 5a, 25a and 100a. The disposal options considered are shallow land burial, at a municipal landfill site or at the reactor site, and disposal on the seabed in coastal waters or the deep ocean. It is assumed that there is no special treatment to reduce or eliminate the potential radiological hazard. Consideration has also been given to the recycle of contaminated steel.

The radiological impacts of all the management options for the wastes considered were found to be small. It may therefore be acceptable to allow such very low-level wastes to be disposed of without special restrictions provided that the level and type of activity of the wastes can be verified.

1. INTRODUCTION

The aim of the study is to develop a methodology for evaluation of individual doses and collective dose commitments arising as a result of various management modes for very low-level solid radioactive waste arising from the dismantling of nuclear power plants.

The management modes studied include disposal at a normal municipal refuse tip, burial at the reactor site, disposal on the seabed, and recycling of materials. In each case it has been assumed that no special measures have been taken relating to the potential radiological hazard. Assessments have been made of disposal of wastes of less than 0.37, 3.7 and 37 Bq g^{-1} specific activity. The three specific activity limits for unrestricted release have been chosen to investigate their suitability, or otherwise, for the kinds of waste under consideration. For comparison, the level of 0.37 Bq g^{-1} is the current limit implied by legislation (1) for unrestricted disposal for many types of solid wastes in the United Kingdom.

Within the study responsibility for the definition of the wastes, in terms of materials and radionuclide inventory, rests with the Commissariat a l'Energie Atomique (CEA), as does assessment of the recycle management mode. Assessment of the other management modes is the responsibility of the National Radiological Protection Board (NRPB). Full details of the study will be given in a Final Report, which is to be published by the Commission of the European Communities.

2. WASTE COMPOSITION

The quantities and activities of steel and concrete activated and contaminated at levels near to the limits given above have been estimated. Information has been used from several reactors including for example PALUEL, a 1300 MWe pressurised water reactor (PWR) (2), and BUGEY, a 540 MWe graphite moderated reactor (3). Inventories have been estimated at three times after reactor shutdown, 5a, 25a and 100a. Results presented here are representative of those for a PWR.

2.1 Contamination

The investigation of various radioactive wastes and their classification indicates that the mass of contaminated wastes decreases by a factor of two when their specific activity increases by a factor of ten. This relationship has been assumed for wastes arising from dismantling of nuclear power plants. Contamination arises on surfaces rather than throughout the body of a waste. Therefore, to relate levels of contamination to specific activity, assumptions must be made about the surface area associated with a given mass. For the purpose of the study it is supposed that the concrete is cut into blocks having a contaminated surface area of 1 m^2 per tonne. Primary circuit steel is expected to be too contaminated for consideration here and would remain as radioactive waste requiring special disposal. Steel near the limits for unrestricted release would come from auxiliary circuits, effluent circuits, pools, treatment and storage tanks, runways, gratings and stairs. A contaminated surface of 10 m^2 per tonne of steel has been assumed.

Estimates of masses and activities of contaminated concrete and steel for a PWR falling within the assumed limits are given in Table I. Considering surface contamination in relation to a specific activity limit of, eg 37 Bq g^{-1}, the level of contamination averaged over 400t gives rise to an average specific activity somewhat below the limit, ie. 10 Bq g^{-1}. Similarly, for the other limits, the level of contamination averaged over a large mass is about ¼ of the limit. The radionuclides in the waste have been determined from studies of contaminants in primary coolant, and liquid and gaseous effluents, and are given in Table II.

2.2 Activation

Estimates have been made of the level of thermal and epithermal neutron activation of the reinforced concrete of a PWR biological shield. Neutron fluxes within layers of the shield are given in Table III. Reinforcement is assumed to include 50 kg of iron per m^3 of concrete and the elemental compositions of concrete and iron have been derived from references (4-7) and (8-9) respectively. Waste masses and activities falling within the assumed limits are given in Table IV. The radionuclides in activation products are taken to be the same as those given in Table V for the specific activity in the fourth layer of concrete. The level of activity in each successive outer layer is about an order of magnitude lower than that of the layer inside it.

Some steels within the reactor could be activated near to the levels under consideration but the masses and corresponding neutron fluxes are very difficult to analyse. No estimates of levels of activation of steel are presented here.

2.3 Source terms for disposal calculations

Any particular quantity of waste may contain radioactivity arising from contamination and from activation. To determine masses and activities of material falling within any given limit, it has been assumed that the most activated material is also the most contaminated, unless the combination itself leads to the limit being exceeded. Because of uncertainties about steel, only reinforced concrete has been considered in disposal assessments. The specific activity limit of 37 Bq g^{-1} implies that approximately the inner 1 m of the biological shield of a PWR, about 400t, would be too active for consideration here.

3. DISPOSAL IN SHALLOW LAND BURIAL FACILITIES

The methodology for calculating the radiological impact of shallow land burial of wastes is described in reference (10). In the case of a municipal landfill facility the waste is assumed to be placed in a simple trench which is located about 10 m above the water table and sited in glacial sand overlying a bedrock of sandstone. The landfill is therefore surrounded by permeable material and is well drained. Infiltrating rainwater will tend to leach radioactivity from the waste and carry it through an unsaturated zone to the water table whence it is assumed to move downslope to a stream. In the case of disposal at the reactor site the trench is assumed to lie within the water table and groundwater passing through the waste is assumed to drain into a river or into coastal waters. Leach rates assumed for the various radionuclides fall within the range from 10^{-2} y^{-1} to 2 10^{-5} y^{-1}.

Estimates have been made of the doses arising from use of the water in the stream, farming of the land adjacent to the sites, ingestion of seafoods, and from direct intrusion into the waste during the course of building construction work. These include maximum annual doses to members of critical groups and collective dose commitments, truncated at various times. Results for the disposal of reinforced concrete from a PWR for the given range of specific activity limits and delays following reactor shutdown are presented in Table VI.

4. DISPOSAL ON THE SEABED

Consideration has been given to disposal in coastal waters and disposal on the deep ocean bed. In neither case was encapsulation allowed for and release of activity was assumed to commence on disposal at a fixed fractional release rate of 10^{-2} y^{-1}. In the former case a compartment model (11) for the dispersion of radionuclides in coastal waters of Northern Europe has been used to determine radionuclide concentrations in seawater, and hence in seafoods and on beach sediments. Doses to man arising from consumption of seafoods, and beach occupancy, have been calculated following disposal of the reinforced concrete from a PWR in UK coastal waters. Results are presented in Table VII. Corresponding doses arising from deep ocean bed disposal have been estimated using a model for the dispersion of radioactivity in the deep waters of the Atlantic Ocean (12). In this latter case predicted doses are very much lower.

5. RECYCLING OF STEEL

Consideration has been given to doses arising during the refabrication of contaminated steel and to those arising subsequently from use of the steel in furniture, buildings and vehicles. Steel contaminated within each of the assumed limits for unrestricted release is assumed to provide 10% of the annual feed to a small steel works, and the contamination is therefore diluted ten times. Although a proportion of radionuclides may be associated with slag, in determining the results presented here it was assumed that all activity would be retained in steel product.

In the smelting and tooling of the steel the highest doses are likely to arise through the inhalation of contaminated dust. The maximum annual dose to workers has been calculated assuming a dust level of 10 mg m^{-3} in the smelting plant, occupancy of 1800 h, and a breathing rate of 1.2 m^3 h^{-1}. This leads to inhalation of 21.6 g a^{-1} of dust assumed to have the same radionuclide composition as the contaminated steel. Collective doses have been calculated on a similar basis assuming that the number of workers involved is in the same proportion to the total workforce as the contaminated steel is to total steel production.

The major source of exposure to members of the public arising from the articles made from the steel is the gamma radiation from ^{60}Co and ^{137}Cs. The proportions of steel product going to uses leading to the most significant exposures are 7.4% to automobile manufacture, 1.7% in light industry and 3% in the building industry. Doses have been calculated using simplifying assumptions about exposure times and the geometry of manufactured articles; results are presented in Table VIII.

6. DISCUSSION

The results presented here have been obtained on the basis of very simple assumptions. Furthermore, there are considerable uncertainties associated with many of the parameter values adopted to define the wastes and to assess the radiological impact of management modes. Nevertheless the following observations can be made.

6.1 Waste composition

The quantity of material below any of the specific activity limits for unrestricted release is not significantly affected by the particular choice of limit. This is because most of the many thousands of tonnes of waste is contaminated or activated at below the lowest limit considered. On the other hand, the quantity of material above the limit could vary considerably, from about 400t of reinforced concrete for a specific activity limit of 37 Bq g^{-1} to about 1600t for a limit of 0.37 Bq g^{-1}.

The definition of a specific activity limit should include the volume or mass over which activity is to be averaged. The specific activity of iron in reinforced concrete arising as a result of neutron activation may be somewhat higher than that of the concrete as a whole, but not drastically so. Otherwise activation of concrete is likely to be uniform. High levels of contamination may arise in very small areas, but such spots should be relatively easily identified and dealt with under whatever procedures are in hand for decontamination. In the waste composition given here it is implied that it is acceptable for the average specific activity in about 100t of material to be about 1/3 of the limit.

Most of the activity in very low-level wastes is likely to arise through contamination rather than activation since the level of activation drops off very rapidly with distance from the core, whereas contamination can be more widely spread. In terms of abundance the most notable radionuclides are different for the different dismantling times and specific activity limits. At 5a several short-lived fission products are important as well as tritium, ^{60}Co, ^{55}Fe, ^{152}Eu (arising from activation) and several actinides. At 100a the short-lived radionuclides are no longer relevant, their place being taken by ^{41}Ca, ^{63}Ni and ^{14}C, and a higher proportion of actinides. ^{137}Cs and ^{90}Sr are significant at all times.

Several of these radionuclides are not easily detected particularly at the very low levels under consideration.

It is also of interest to note that the natural level of activity to be expected in concrete would be in the range from 0.4 to 1.0 Bq g^{-1}. This level would arise principally from ^{40}K, and uranium and thorium and their daughters.

6.2 Radiological impact of management modes

Disposal of very low-level waste reinforced concrete is predicted to give rise to very small doses for all the options considered. The highest annual individual dose is predicted for disposal at the reactor site following intrusion into the waste trench for building purposes in the year following disposal. Even given these unlikely circumstances the predicted dose is only 3 10^{-5} Sv for the highest specific activity limit. This may be compared with the annual limit for members of the public recommended by the ICRP of 5 10^{-3} Sv (13). Individual doses predicted for all other pathways for all options are considerably lower. Predicted collective dose commitments are also very small, the largest being 3 10^{-2} man Sv. No single radionuclide dominates the results but ^{60}Co is particularly important for the 5a dismantling time, whereas ^{137}Cs, ^{152}Eu and the actinides are important for the 100a option.

The recycle of very low-level contaminated steel is also predicted to have a small radiological impact. The maximum annual individual dose, 3 10^{-6} Sv, is predicted to arise during refabrication of waste steel, and the largest collective dose commitment arising from all pathways is 3 man Sv. ^{60}Co, ^{137}Cs and the actinides are again the important radionuclides.

For all the management modes, for all the wastes considered here, individual doses and collective dose commitments are very low. On the basis of suggestions given in reference (14) the detriment cost associated with the radiological impact of each management option for the wastes considered is less than a few thousand pounds. Therefore consideration of radiological impact will play little part in the choice of management options for these wastes.

However care is needed in drawing conclusions from these results. Only one type of reactor has been considered. Furthermore, the operating history of any particular reactor, eg with regard to fuel failure, may be critical in determining the quantity and quality of wastes. None of the three specific activity limits appears unreasonable as a basis for defining wastes which can be disposed of without special restrictions. The adoption of the highest limit, 37 Bq g^{-1}, would have the advantage that the quantity of waste requiring special measures for disposal would be minimised. However demonstrating compliance with any given limit may not be straightforward because of the difficulty of measuring such low levels of radioactivity.

References

1. Radioactive Substances Act, 1960. HMSO (1960).
2. Rapport de Surete - PALUEL - tranche 1.
3. Rapport de Surete - BUGEY - tranche 1.
4. Lunning, W.H. Decommissioning of WAGR, Technical Survey, UKAEA N.D.M. 90(R) (1977).
5. Regan, J.D. et al. Catalogue of design features facilitating the decommissioning of AGR's. CEC contract report DE-G-001-81 UK(H) (1983).

6. May, S. Determination Analytique d'Elements Traces dans des Echantillons de Bêtons Utilisés dans des Réacteurs Nucléaires de la Communauté Européanne. CEC contract report DE-FF-005-F(SD) (1983).
7. Pascal, P. Nouveau Traité de Chimie Minérale, Ed Masson and Cie (1960).
8. Acière ALPA, Gargenville, France.
9. Harbecke, W. et al. Die Aktivierung der Biologischen Schilds im Stillgelegten Kernkraftwerke Lingen. CEC contract report DE-F-001-D (1982).
10. Pinner, A.V., Hemming, C.R. and Hill, M.D. An assessment of the radiological protection aspects of shallow land burial of radioactive wastes. NRPB-R161 National Radiological Protection Board, Chilton (to be published).
11. Clark, M.J., Grimwood, P.D. and Camplin, W.C. A model to calculate exposures from radioactive discharges into coastal waters of Northern Europe. NRPB-R109 National Radiological Protection Board, Harwell (1980).
12. Mobbs, S.F. and Hill, M.D. Assessment of the radiological impact of intermediate level waste disposed on the seabed. Report for the Department of Environment, contract no. PECD7/9/03. National Radiological Protection Board, Chilton (1982).
13. International Commission on Radiological Protection. Recommendations of the ICRP. Oxford, Pergamon Press. ICRP Publication 26. Ann. ICRP, 1 No 3 (1977).
14. ASP4. Cost benefit analysis in optimising the radiological protection of the public : a provisional framework. National Radiological Protection Board. Chilton (1981).

Table I Masses and activities of very low-level contaminated reinforced concrete and steel from a PWR

		Time from shutdown of reactor, a					
		5		25		100	
Surface activity Bq cm^{-2}	Average specific activity Bq g^{-1}	mass t	total activity Bq	mass t	total activity Bq	mass t	total activity Bq
Concrete							
370 - 3700	10	400	$4\ 10^{9}$	220	$2.2\ 10^{9}$	120	$1.2\ 10^{9}$
37 - 370	1	800	$8\ 10^{8}$	440	$4.4\ 10^{8}$	240	$2.4\ 10^{8}$
3.7 - 37	0.1	1600	$1.6\ 10^{8}$	880	$8.8\ 10^{7}$	480	$4.8\ 10^{7}$
0.37- 3.7	0.01	3200	$3.2\ 10^{7}$	1760	$1.8\ 10^{7}$	960	$9.6\ 10^{6}$
Steel							
37 - 370	10	800	$8\ 10^{9}$	440	$4.4\ 10^{9}$	240	$2.4\ 10^{9}$
3.7 - 37	1	1600	$1.6\ 10^{9}$	880	$8.8\ 10^{8}$	480	$4.8\ 10^{8}$
0.37- 3.7	0.1	3200	$3.2\ 10^{8}$	1760	$1.8\ 10^{8}$	960	$9.6\ 10^{7}$

Table II The relative proportions of radionuclides in contamination

Radionuclide	Activity at different times after reactor shutdown (arbitrary units)		
	5a	25a	100a
^{55}Fe	10^{1}	$7.8\ 10^{-2}$	0
^{60}Co	10^{2}	7.2	$3.8\ 10^{-4}$
^{59}Ni	10^{-2}	10^{-2}	10^{-2}
^{63}Ni	2	1.7	1
^{90}Sr	5	3.1	$4.8\ 10^{-1}$
^{93}Zr	10^{-4}	10^{-4}	10^{-4}
^{94}Nb	10^{-3}	10^{-3}	10^{-3}
^{99}Tc	10^{-3}	10^{-3}	10^{-3}
^{125}Sb	10^{1}	$7\ 10^{-2}$	0
^{134}Cs	4	$3.9\ 10^{-3}$	0
^{135}Cs	$3\ 10^{-5}$	$3\ 10^{-5}$	$3\ 10^{-5}$
^{137}Cs	10^{1}	6.3	1.1
^{147}Pm	10^{1}	$5\ 10^{-2}$	0
^{151}Sm	10^{-2}	$8.6\ 10^{-3}$	$4.8\ 10^{-3}$
^{238}Pu	$7.5\ 10^{-2}$	$6.4\ 10^{-2}$	$3.5\ 10^{-2}$
^{239}Pu	$9\ 10^{-3}$	$9\ 10^{-3}$	$9\ 10^{-3}$
^{240}Pu	$1.2\ 10^{-2}$	$1.2\ 10^{-2}$	$1.2\ 10^{-2}$
^{241}Pu	2.1	$7.9\ 10^{-1}$	$2.1\ 10^{-2}$
^{241}Am	$2.5\ 10^{-2}$	$6.6\ 10^{-2}$	$8.3\ 10^{-2}$
^{244}Cm	$3.5\ 10^{-2}$	$1.5\ 10^{-2}$	$8.4\ 10^{-4}$
Total	$1.5\ 10^{2}$	$1.9\ 10^{1}$	2.8

Table III Neutron fluxes within the biological shield of a PWR

layer	thickness cm	concrete mass t	iron mass t	neutron fluxes, n cm^{-2} s^{-1}	
				thermal	epithermal
1	25	90	1.9	3.5 10^9	4.5 10^9
2	"	100	2.1	1 10^9	2 10^8
3	"	105	2.2	1 10^8	6 10^6
4	"	113	2.4	1 10^7	2 10^5
5	"	120	2.55	1 10^6	1 10^4
6	"	130	2.7	1 10^5	4 10^2
7	"	135	2.9	1 10^4	2 10^1
8	"	143	3	1 10^3	1 10^0

Table IV Masses and activities of activated reinforced concrete from a PWR

specific activity Bq g^{-1}	time from reactor shutdown					
	5a		25a		100a	
	mass t	activity Bq	mass t	activity Bq	mass t	activity Bq
3.7 - 37	120	1.7 10^9	115	1.7 10^9	105	1.5 10^9
0.37 - 3.7	130	1.9 10^8	120	1.7 10^8	115	1.7 10^8
0.037- 0.37	140	2 10^7	130	1.9 10^7	120	1.7 10^7

Table V Specific activity arising through neutron activation of the fourth layer of reinforced concrete in a PWR

Radionuclide	Specific activity at different times after reactor shutdown, Bq g^{-1} 5 a	25 a	100 a
^{3}H	$1.0\ 10^{2}$	$3.4\ 10^{1}$	$5.0\ 10^{-1}$
^{14}C	$5.0\ 10^{-2}$	$5.0\ 10^{-2}$	$5.0\ 10^{-2}$
^{36}Cl	$3.2\ 10^{-2}$	$3.2\ 10^{-2}$	$3.2\ 10^{-2}$
^{41}Ca	$1.5\ 10^{0}$	$1.5\ 10^{0}$	$1.5\ 10^{0}$
^{55}Fe	$2.65\ 10^{2}$	$1.3\ 10^{0}$	$2.6\ 10^{-9}$
^{59}Ni	$1.1\ 10^{-3}$	$1.1\ 10^{-3}$	$1.1\ 10^{-3}$
^{63}Ni	$1\ 10^{-1}$	$9\ 10^{-2}$	$5.8\ 10^{-2}$
^{60}Co	$7.9\ 10^{0}$	$5.7\ 10^{-1}$	$3.0\ 10^{-5}$
^{79}Se	$9.3\ 10^{-6}$	$9.3\ 10^{-6}$	$9.3\ 10^{-6}$
^{93}Zr	$4.7\ 10^{-6}$	$4.7\ 10^{-6}$	$4.7\ 10^{-6}$
^{94}Nb	$1.6\ 10^{-3}$	$1.6\ 10^{-3}$	$1.6\ 10^{-3}$
^{93}Mo	$1.1\ 10^{-5}$	$1.1\ 10^{-5}$	$1.1\ 10^{-5}$
^{99}Tc	$1.3\ 10^{-6}$	$1.3\ 10^{-6}$	$1.3\ 10^{-6}$
^{108m}Ag	$1.6\ 10^{-2}$	$1.4\ 10^{-2}$	$9.3\ 10^{-3}$
^{109}Cd	$6.0\ 10^{-3}$	$8.4\ 10^{-8}$	0
^{113m}Cd	$4.8\ 10^{-2}$	$1.7\ 10^{-2}$	$3.8\ 10^{-4}$
^{121}Sn	$7.6\ 10^{-6}$	$6.3\ 10^{-6}$	$3.3\ 10^{-6}$
^{125}Sb	$3.5\ 10^{-4}$	$2.1\ 10^{-6}$	0
^{133}Ba	$4.2\ 10^{-2}$	$7.0\ 10^{-3}$	$9.0\ 10^{-6}$
^{134}Cs	$5.0\ 10^{-1}$	$9.1\ 10^{-4}$	0
^{137}La	$1.7\ 10^{-6}$	$1.7\ 10^{-6}$	$1.7\ 10^{-6}$
^{151}Sm	$2.0\ 10^{-1}$	$1.7\ 10^{-1}$	$9.2\ 10^{-2}$
^{147}Pm	$3.3\ 10^{-2}$	$1.7\ 10^{-4}$	0
^{152}Eu	$4.9\ 10^{1}$	$1.5\ 10^{1}$	$2.0\ 10^{-1}$
^{154}Eu	$2.1\ 10^{0}$	$8.9\ 10^{-1}$	$3.5\ 10^{-2}$
^{155}Eu	$7.9\ 10^{-2}$	$4.8\ 10^{-3}$	$1.4\ 10^{-7}$
^{166m}Ho	$1.5\ 10^{-3}$	$1.5\ 10^{-3}$	$1.4\ 10^{-3}$
^{171}Tm	$1.2\ 10^{-2}$	$7.9\ 10^{-6}$	0
Total	426	54	2.5

Table VI Radiological impact of shallow land burial of very low-level waste reinforced concrete arising from the decommissioning of a PWR

a Maximum annual individual dose, Sv

	Dismantling time after reactor shutdown, a								
	5			25			100		
	Activity limit, Bq g^{-1}			Activity limit, Bq g^{-1}			Activity limit, Bq g^{-1}		
	0.37	3.7	37	0.37	3.7	37	0.37	3.7	37
Municipal landfill									
Building	1 10^{-7}	7 10^{-7}	4 10^{-6}	3 10^{-8}	3 10^{-7}	2 10^{-6}	1 10^{-8}	2 10^{-8}	2 10^{-7}
Highest other pathway	3 10^{-14}	4 10^{-15}	3 10^{-10}	3 10^{-12}	2 10^{-10}	2 10^{-9}	5 10^{-11}	5 10^{-10}	5 10^{-8}
Reactor site, river									
Building	2 10^{-6}	5 10^{-6}	3 10^{-5}	5 10^{-7}	2 10^{-6}	1 10^{-5}	6 10^{-8}	1 10^{-7}	4 10^{-6}
Highest other pathway	5 10^{-13}	5 10^{-12}	2 10^{-9}	3 10^{-11}	2 10^{-9}	2 10^{-8}	2 10^{-10}	2 10^{-9}	2 10^{-7}
Reactor site, coastal									
Building	2 10^{-6}	5 10^{-6}	3 10^{-5}	5 10^{-7}	2 10^{-6}	1 10^{-5}	6 10^{-8}	1 10^{-7}	4 10^{-6}
Highest other pathway	1 10^{-14}	8 10^{-14}	2 10^{-12}	8 10^{-14}	2 10^{-12}	2 10^{-11}	2 10^{-13}	2 10^{-12}	2 10^{-10}

b Collective dose commitment from all pathways, man Sv, truncated at given times

	Dismantling time after reactor shutdown, a								
	5			25			100		
	Activity limit, Bq g^{-1}			Activity limit, Bq g^{-1}			Activity limit, Bq g^{-1}		
	0.37	3.7	37	0.37	3.7	37	0.37	3.7	37
Municipal landfill									
10^2 a	$5\ 10^{-9}$	$3\ 10^{-8}$	$2\ 10^{-6}$	$4\ 10^{-8}$	$2\ 10^{-6}$	$1\ 10^{-5}$	$2\ 10^{-7}$	$1\ 10^{-6}$	$3\ 10^{-4}$
10^4 a	$1\ 10^{-7}$	$1\ 10^{-6}$	$2\ 10^{-4}$	$3\ 10^{-6}$	$2\ 10^{-4}$	$2\ 10^{-3}$	$3\ 10^{-5}$	$2\ 10^{-4}$	$3\ 10^{-2}$
10^6 a	$2\ 10^{-6}$	$1\ 10^{-5}$	$3\ 10^{-4}$	$2\ 10^{-5}$	$3\ 10^{-4}$	$2\ 10^{-3}$	$2\ 10^{-4}$	$4\ 10^{-4}$	$3\ 10^{-2}$
infinity	$2\ 10^{-6}$	$1\ 10^{-5}$	$3\ 10^{-4}$	$2\ 10^{-5}$	$3\ 10^{-4}$	$2\ 10^{-3}$	$2\ 10^{-4}$	$4\ 10^{-4}$	$3\ 10^{-2}$
Reactor site, river									
10^2 a	$1\ 10^{-8}$	$8\ 10^{-8}$	$9\ 10^{-6}$	$2\ 10^{-7}$	$7\ 10^{-6}$	$6\ 10^{-5}$	$8\ 10^{-7}$	$6\ 10^{-6}$	$1\ 10^{-3}$
10^4 a	$8\ 10^{-7}$	$6\ 10^{-6}$	$2\ 10^{-4}$	$9\ 10^{-6}$	$2\ 10^{-4}$	$2\ 10^{-3}$	$1\ 10^{-4}$	$3\ 10^{-4}$	$3\ 10^{-2}$
10^6 a	$9\ 10^{-7}$	$7\ 10^{-6}$	$2\ 10^{-4}$	$1\ 10^{-5}$	$2\ 10^{-4}$	$2\ 10^{-3}$	$1\ 10^{-4}$	$3\ 10^{-4}$	$3\ 10^{-2}$
infinity	$9\ 10^{-7}$	$7\ 10^{-6}$	$2\ 10^{-4}$	$1\ 10^{-5}$	$2\ 10^{-4}$	$2\ 10^{-3}$	$1\ 10^{-4}$	$3\ 10^{-4}$	$3\ 10^{-2}$
Reactor site, coastal									
10^2 a	$3\ 10^{-12}$	$3\ 10^{-11}$	$7\ 10^{-9}$	$1\ 10^{-10}$	$7\ 10^{-9}$	$6\ 10^{-8}$	$7\ 10^{-10}$	$6\ 10^{-9}$	$1\ 10^{-6}$
10^4 a	$1\ 10^{-8}$	$9\ 10^{-8}$	$2\ 10^{-5}$	$3\ 10^{-7}$	$2\ 10^{-5}$	$2\ 10^{-4}$	$3\ 10^{-6}$	$2\ 10^{-5}$	$2\ 10^{-3}$
10^6 a	$3\ 10^{-6}$	$2\ 10^{-5}$	$1\ 10^{-4}$	$2\ 10^{-5}$	$2\ 10^{-4}$	$6\ 10^{-4}$	$3\ 10^{-4}$	$4\ 10^{-4}$	$4\ 10^{-3}$
infinity	$3\ 10^{-6}$	$2\ 10^{-5}$	$1\ 10^{-4}$	$2\ 10^{-5}$	$2\ 10^{-4}$	$6\ 10^{-4}$	$3\ 10^{-4}$	$4\ 10^{-4}$	$4\ 10^{-3}$

Table VII Radiological impact of the seabed disposal in UK coastal waters of very low-level waste reinforced concrete arising from the decommissioning of a PWR

	Dismantling time after reactor shutdown, a								
	5			25			100		
	Activity limit, Bq g^{-1}			Activity limit, Bq g^{-1}			Activity limit, Bq g^{-1}		
	0.37	3.7	37	0.37	3.7	37	0.37	3.7	37
a Maximum annual individual dose, Sv									
Ingestion of:									
fish	3 10^{-12}	2 10^{-11}	8 10^{-11}	2 10^{-12}	1 10^{-11}	4 10^{-11}	7 10^{-13}	1 10^{-12}	5 10^{-12}
crustacea	2 10^{-12}	1 10^{-11}	7 10^{-11}	3 10^{-10}	1 10^{-11}	5 10^{-11}	6 10^{-13}	8 10^{-13}	4 10^{-12}
molluscs	4 10^{-12}	2 10^{-11}	7 10^{-11}	6 10^{-12}	3 10^{-11}	1 10^{-10}	1 10^{-11}	1 10^{-11}	7 10^{-11}
seaweed	4 10^{-13}	1 10^{-12}	6 10^{-10}	1 10^{-11}	6 10^{-11}	2 10^{-10}	2 10^{-12}	3 10^{-12}	2 10^{-11}
External irradiation on beach	3 10^{-10}	2 10^{-9}	8 10^{-9}	1 10^{-10}	7 10^{-10}	2 10^{-9}	1 10^{-12}	2 10^{-12}	2 10^{-11}
b Collective dose commitment from all pathways, man Sv, truncated at various times									
10^2 a	4 10^{-6}	3 10^{-5}	1 10^{-4}	4 10^{-6}	2 10^{-5}	7 10^{-5}	6 10^{-6}	8 10^{-6}	4 10^{-5}
10^4 a	4 10^{-6}	3 10^{-5}	1 10^{-4}	5 10^{-6}	2 10^{-5}	8 10^{-5}	7 10^{-6}	9 10^{-6}	1 10^{-4}
10^6 a	4 10^{-6}	3 10^{-5}	1 10^{-4}	5 10^{-6}	2 10^{-5}	8 10^{-5}	7 10^{-6}	9 10^{-6}	1 10^{-4}
infinity	4 10^{-6}	3 10^{-5}	1 10^{-4}	5 10^{-6}	2 10^{-5}	8 10^{-5}	7 10^{-6}	9 10^{-6}	1 10^{-4}

Table VIII Doses arising from the recycle of very low-level contaminated steel from the decommissioning of a PWR

	Dismantling time after reactor shutdown, a								
	5			25			100		
	Activity limit, Bq g^{-1}			Activity limit, Bq g^{-1}			Activity limit, Bq g^{-1}		
	0.37	3.7	37	0.37	3.7	37	0.37	3.7	37
a Maximum annual individual dose, Sv									
Refabrication	$8\ 10^{-9}$	$5\ 10^{-8}$	$3\ 10^{-7}$	$3\ 10^{-8}$	$2\ 10^{-7}$	$9\ 10^{-7}$	$8\ 10^{-8}$	$5\ 10^{-7}$	$3\ 10^{-6}$
Exposure from									
furniture	$5\ 10^{-8}$	$3\ 10^{-7}$	$1\ 10^{-6}$	$2\ 10^{-8}$	$1\ 10^{-7}$	$6\ 10^{-7}$	$2\ 10^{-9}$	$1\ 10^{-8}$	$7\ 10^{-8}$
buildings	$2\ 10^{-10}$	$2\ 10^{-9}$	$8\ 10^{-9}$	$9\ 10^{-11}$	$6\ 10^{-10}$	$3\ 10^{-9}$	$8\ 10^{-12}$	$5\ 10^{-11}$	$2\ 10^{-10}$
vehicles	$8\ 10^{-8}$	$5\ 10^{-7}$	$2\ 10^{-6}$	$3\ 10^{-8}$	$2\ 10^{-7}$	$1\ 10^{-6}$	$4\ 10^{-9}$	$2\ 10^{-8}$	$1\ 10^{-7}$
b Collective dose commitment, man Sv									
Refabrication	$2\ 10^{-5}$	$5\ 10^{-5}$	$1\ 10^{-4}$	$1\ 10^{-4}$	$3\ 10^{-4}$	$8\ 10^{-4}$	$6\ 10^{-4}$	$2\ 10^{-3}$	$4\ 10^{-3}$
Exposure from									
furniture	$4\ 10^{-2}$	$2\ 10^{-1}$	1	$2\ 10^{-2}$	$1\ 10^{-1}$	$5\ 10^{-1}$	$4\ 10^{-3}$	$2\ 10^{-2}$	$1\ 10^{-1}$
buildings	$3\ 10^{-3}$	$2\ 10^{-2}$	$8\ 10^{-2}$	$1\ 10^{-3}$	$8\ 10^{-3}$	$4\ 10^{-2}$	$3\ 10^{-4}$	$2\ 10^{-3}$	$1\ 10^{-2}$
vehicles	$5\ 10^{-2}$	$3\ 10^{-1}$	2	$2\ 10^{-2}$	$1\ 10^{-1}$	$6\ 10^{-1}$	$3\ 10^{-3}$	$2\ 10^{-2}$	$1\ 10^{-1}$
Total all pathways	$1\ 10^{-1}$	$5\ 10^{-1}$	3	$4\ 10^{-2}$	$2\ 10^{-1}$	1	$7\ 10^{-3}$	$4\ 10^{-2}$	$2\ 10^{-1}$

SOME EXAMPLES OF RELEASE OF MATERIALS AND SITES, INCLUDING MEASUREMENT TECHNIQUES

A.M. CHAPUIS and M. HULOT
Commissariat à l'Energie Atomique
W. JAGER
GKSS Forschungszentrum, Geesthacht
R. PANTER
United Kingdom Atomic Energy Authority, London

Summary

Many test reactors and some power reactors have already been decommissioned and partially or entirely dismantled. Although some materials are classified as radioactive waste, others may be considered as ordinary waste or reutilized for nuclear or non-nuclear work. There is no international recommendation on activity limits for the unconditional reutilization of material. The levels applied in the various countries differ according to the materials and their use or destination. In Germany the limits applied for dismantling the Otto Hahn reactor were 10^{-4} μCi.g^{-1} and 10^{-5} μCi.cm^{-2}. In France there are no statutory limits for release for public use. Each case is submitted for authorization to the Service Central de Protection contre les Rayonnements Ionisants (Central authority for protection against ionizing rays), a department of the Ministry of Health, which decides on the basis of the characteristics of the waste in question. We give examples of release in Germany (GKSS), France (CEA) and the United Kingdom (UKAEA).
In Germany, the reactor on the ship NS Otto Hahn was dismantled between 1980 and 1982, after 10 years' operation. The secondary circuit, with low contamination, was decontaminated. Monitoring of the surface activity before unconditional release required more than 1 million measurements. In September 1982 the ship was resold and in the following months was equipped with a diesel engine.
We also describe the dismantling of some French reactors, in particular PEGASE and GL. In the case of the latter, the building and the interior equipment, such as the gantry, were reused for non-nuclear work. Some materials such as steel and copper pipes can be considered as uncontaminated. In the PEGASE reactor, the vessel and tanks were reused after decontamination, for the temporary storage of irradiated fuels. The steel loops were sent, after decontamination, to the smelting works.
At the Harwell site a series of low-power reactors built between 1947 and 1965 have since been decommissioned and the buildings reused for non-nuclear work. This was the case, in particular, with BEPO, a 6.5 MW air-cooled-graphite reactor, and with LIDO, a 300 kW swimming pool reactor.

1. INTRODUCTION

Among the numerous test or low-power reactors which have been decommissioned in Europe in recent years, we give some examples of operations which led to the unconditional reutilization of buildings or the reutilization of decontaminated materials in the public sector.

This procedure must be based on statutory regulations laying down limits for mass and/or surface activity below which the materials and their utilization no longer require to be monitored. There are as yet no international regulations or recommendations on this. The procedures and regulation therefore vary from one country to another. To ensure compliance with the standards a great number of measurements, sometimes difficult measurement, need to be taken, which then have to be verified by an independent supervisory body.

2. REACTOR OF THE GERMAN SHIP NS OTTO HAHN

One of the most typical cases of dismantling was that of the first European marine propulsion reactor in the ship NS Otto Hahn. This reactor was shut down in 1979 after 10 1/2 years in operation (1). Following a study carried out in liaison with NIS (Nuklear-Ingenieur-Service) on the various options open, it was decided to dismantle the reactor and the nuclear part of the plant completely (2,3).

About 1 200 tonnes of material were removed, including 480 tonnes for the vessel and its protection, and 300 tonnes of decontaminated material which was transported to the GKSS site and placed in a special room for further study. The interior structures of the reactor, representing 1.5% of the mass of dismantled material, contained almost all the activity, mainly arising from cobalt 60.

Public release and the unconditional reutilization of the structures of the ship and part of the dismantled componenets (370 t) was fixed, in accordance with the Regulation on protection against ionizing radiation of 13 October 1976, at a surface activity limit of 10^{-5} μCi.cm^{-2} (β emitters).

An independent expert (PTB)* verified conformity with this regulation by sampling and measuring with the aid of a highly sensitive dose rate meter developed by PTB. The walls and ceiling of the nuclear zone were made of steel. The external protection of the containment vessels and the protection of the fuel tank were in reinforced concrete. These installations of relatively simple geometric shape, involving relatively low non-ingrained contamination, were decontaminated. This work was monitored by direct measurement with a Minicont detector. This procedure was monitored by the authorities, who laid down that, for the final monitoring measurements, the universal probe should be replaced by:

- a sensitive α - β probe for the components of surface activity, with slightly higher level background
- a sensitive β - γ probe for painted or rusted components, with low level background.

The dismantable components, amounting to 3 600 m^2, were measured in a low level background zone. The fixed parts were measured on the spot, after reducing the ambient background by removing all the radioactive sources; 14 400 m^2 were measured in this way.

* (P.T.B. Braunschweig)

28% of this surface area underwent a second decontamination
5.5% underwent three decontaminations
3% underwent four decontaminations.

There were a total of 1 440 000 measurements, including 1 350 000 using the Minicont with the universal probe, requiring 11 500 hours of work. 90 000 final monitoring measurements carried out by the authorities (Service for Protection against Radiation and the independent expert) took 1 250 hours.

The activity of components of complicated geometric shape was measured indirectly on samples whose surface activity was determined after γ-spectrometry analysis.

Decontamination procedures were monitored by means of electrolytic sampling and γ-spectrometry measurements on 60 samples of slightly higher activity, about 10^{-4} $\mu Ci.cm^{-2}$ (5), and decontamination was assessed with the aid of the Minicont detector.

Many measurements were carried out on the more active parts of the plant, in particular using Ge-Li spectrometry, in order to obtain a good knowledge of the contaminations and their distribution and to apply this to the strategy of unconditional release.

Ambient and contact dose rate measurements carried out by the independent PTB expert confirmed the care with which the decontaminations were executed and the accuracy of the measurements. Where the average surface activity obtained after decontamination ($\cong 6.10^{-6}$ $\mu Ci.cm^{-2}$) was exceeded slightly this was discovered quickly and surely with the selected dose rate meter.

3. THE FRENCH EXPERIMENTAL REACTOR PEGASE

PEGASE was a 35 MW thermal swimming pool reactor situated at Cadarache and designed to test on a true scale, in 18 gas loops, the fuel elements of natural uranium-graphite-gas or heavy water-gas reactors. These tests were extended to the fuel elements of HTR and fast reactors. PEGASE was in operation from 1963 to 1975. In May 1984 there was a fuel can breakage with perforation and partial fusion of several slabs. In the loops, studies and long-term tests were carried out on the development of ruptures of cans of UNGG fuels. The internal parts of the plant were thus contaminated by the fission products and by the activates. The loops include strong tubes intended to receive the fuel elements for testing, which were immersed and therefore activated. They could be detached from the rest of the loop in order to be repaired or replaced by tubes of different diameter or materials, or stored in the open air. The non-immersed circuit in the loops included a cooling circuit, a pump and an exchanger controlled by a valve. Depending on the tests, these loops were made of stainless steel or AG3 aluminium.

The dismantling of the plant has been carried out in several stages between 1976, when the fuels (core and experimental fuels) were removed, and 1984. A few checks and some dismantling remain to be carried out on materials which were transported outside of the plant, in particular on the external parts of the loops.

No building was pulled down. The outlying buildings are being reused for non-nuclear work. The reactor building, which includes the pool, the storage tank and the stripping tank (disassembly of loops), is being utilized for the temporary storage of irradiated fuels in tanks.

The tanks are coated in 18/8 stainless steel, and some are painted. Contamination of the pool, measured by smears shortly after shutdown of the reactor, was estimated at 0.2 μCi.cm^{-2}. The spectrum of contaminants was determined on dusts suspended in the pool water. The main contaminants were as follows : Ce 144 + Pr 144: 20%, Cr 51: 16%, Ce 141: 11%, 60 Co : 10%, Ru 106 + Rh 106 : 9%. Decontamination tests were carried out on samples taken from the material in the pool. Following decontamination the surface activity was of the order of 10^{-5} μCi.cm^{-2}, the contaminants of the aluminium samples being Sb 124 : 96%, Co 60 : 4% and those of the steel samples Cr 51 : 91%, Co 60 : 9%.

The pool underwent decontamination by water jets under pressure and by codine. The residual surface activity was between 1.8×10^{-6} and 6.3×10^{-5} μCi.cm^{-2}, allowing access to workers and then reutilization of the pool.

In the building itself the floors and walls were monitored by smear tests, and the highest surface activity was between 10^{-6} and 10^{-7} μCi.-cm^{-2}.

All the materials in the plant were dismantled and classified in different categories according to their activity :

1. **High activity** ; dose rate higher than 200 mrad.h^{-1}; 1200 kg 3270 Ci. These parts were cut up under water in the stripping tank, while activity was measured with the aid of an ionization chamber by removing the parts from the water. The preponderant radioactive element was cobalt 60. 800 kg were sent to the Centre de la Manche for permanent surface storage. 400 kg, with too high activity, were placed in temporary storage.
2. **Medium activity** - waste which cannot be decontaminated; 25 000 kg, 168 mCi.
   This involves either parts activated in the pool or parts which cannot be decontaminated because of their shape. They have been stored as radioactive waste.
3. **Low activity** - waste which can be decontaminated; 24 700 kg of stainless steel and 2 900 of aluminium. Following decontamination these materials could be classified in four categories :
   3 280 kg containing 1.5 mCi were stored as radioactive waste.
   10 965 kg of stainless steel and 1 785 kg of aluminium containing 40 and 80 μCi respectively, as well as the deactivation torus of 4 700 kg containing about 0.5 mCi, were earmarked for resmelting. The corresponding limits of specific activity were 10^{-4} μCi.g^{-1} and 10^{-5} μCi.cm^{-2}.
   6 780 kg which contained no detectable activity were sent to the scrap heap.
   The decontamination treatment produced 23 m^3 of liquid effluents containing 4.6 mCi.
4. **Non-active materials** : 75 700 kg.
   These materials are considered to be neither irradiated nor contaminated as a result of their previous utilization. After verifying the absence of activity, measured with a highly sensitive apparatus used at a site selected for the low level of background, they were sent to the scrap heap.

Methods of measurement

A substantial proportion of the materials consisted of tubes, pipes and ventilation shafts and therefore presented problems in monitoring the surface activity. On the deactivation torus of 97 m^2 in area, analyses were carried out by spectrometry on 20 samples taken from inside and outside, followed by the measurement of surface activity inside (684 measurement points) and outside (650 points). On the loops, of smaller diameter, it was not possible to measure interior surface contamination directly. The spectrum of contaminants, the decontamination factor and the interior and exterior surface activity of the pipes were determined on the basis of samples taken before and after decontamination. After decontamination, cobalt 60 is the principal radioactive nuclide. Interior surface contamination was estimated by means of dose rate measurements carried out with the aid of a highly sensitive detector.

Tests carried out on pipes and tubes - smear tests, aerosol samples taken from the circulating air following repeated shocks and vibrations on the circuits - showed that the contamination remaining after decontamination was fixed. There was no evidence of any sweating-out phenomenon.

4. OTHER FRENCH REACTORS

Among the other dismantled French reactors there is Zoé, where the building has been reused for non-radioactive work, Minverve, where the reactor has been transferred from the Fontenay-aux-Roses centre to the Cadarache centre and put into operation again, the buildings being reused for other non-nuclear work.

The case of the G1 reactor at Marcoule - a natural uranium, graphite and air-cooled reactor - resembles that of Pégase most closely. This 40 MW reactor was in operation from 1956 to 1968. The premises and the interior plant (gantry etc.) are being reutilized as a testing shop. It is planned to recover the steel of the piping and the copper of the slightly contaminated can-rupture detection piping for resmelting. In the first instance, measurements are carried out by scanning with the aid of a sensitive β detector, since the interior of the pipes is accessible. In the second instance, measurements are carried out by γ-detection on batches of about 20 kg. In all cases it is essential to take samples for spectrometric analysis. In this reactor as well, cobalt 60 is the principal radioactive nuclide.

5. CONCLUSION

The dismantling of power reactors, like the dismantling of test reactors, produces a large quantity of waste, owing to their situation and the equipment they contain, the buildings can generally be reused for nuclear or non-nuclear work. Among the dismantled materials some have very low radioactivity, which makes it possible to consider them as non-radioactive and, depending on the case to consign them to the rubbish heap or to recycle them. Such a solution requires strict monitoring of the residual activity. It is always necessary to take samples for spectrometric analysis and determination of the radioactive nuclides to be measured. In all cases studied, cobalt 60 is the principal radioactive nuclide, which facilitates the checks. For parts of complicated shape appropriate calibrations must be made to ensure that the statutory limits are respected.

REFERENCES

1. GKSS - Forschungszentrum Geesthacht GmbH (Hrsg): NS OTTO HAHN. Erstes deutsches Kernenergieschiff. GKSS 81/E/20. Geesthacht 1981.

2. KROEGER, J., LETTNIN, H.K.J., BIRKHOLD, U., OBST, J. Die Stillegung des Kernenergiehandelsschiffes Otto Hahn. GKSS 83/4/55.

3. LETTNIN, H.K.J., SCHMIDT, K., SCHAFSTALL, H.G. Stillegung des NS Otto Hahn. GKSS Jahresbericht, 1982, pp. 8 - 19.

4. Regulation on the protection against damage from ionizing rays (Irradiation Protection Law - StrlSchV) of 13 October 1976. Bundesgesetzblatt Teil I, p. 2905

5. HANULIK, J., SCHENKER, E., ULLRICH, G. Die Bestimmung der Kontamination und Dekontamination des Brennelementlagerbeckens des NS Otto Hahn. EIR 432, Würenlingen, Switzerland, December 1981.

6. Rapport de sûreté de mise à l'arrêt du réacteur Pégase, Technicatome.

SESSION I : SUMMARY OF DISCUSSION

Paper : "Activation of biological shields"

A Swiss participant asks, as Lithium is missing in the paper's tables, if the research for Lithium (leading to Tritium) has been omitted. Mr. May answers that Li is not included simply because the reaction n,gamma does not yield any significant radio-isotope which would be measurable. There are nevertheless reactions leading to ^3H, and an analysis by radio-chemical separation and analysis for Li and ^3H separately is feasible, but difficult to perform.

A French participant thinks that traces of Uranium in concrete will not be of importance, but believes that the analysis of Transuranian elements is necessary to eliminate any doubt. Mr. May explains that Uranium has been analyzed and found to be present in very small quantities; no other alpha emitters have been searched.

A representative of CEGB Berkeley adds that his organization has analyzed the Li content of concrete, not by neutron activation but by ICP techniques. The Li content has been found to be very low but if, for radiological considerations in waste disposal, ^3H will become an important isotope, the Li content also becomes important. Another comment is that it has not been underlined why trace element contents, for elements like Europium, are so similar over the whole of Europe, varying only slightly from Scotland to Italy. Mr. May, who had not been able to present all the results obtained, agrees that, given time and funding, some more elements could have been analyzed.

Paper : "Estimation of radioactive waste quantities arising during decommissioning"

An FRG participant would like to have more information on cooperation with licensing authorities giving permission for melting of contaminated items. One of the co-authors, Mr. Stang (KRB, Gundremmingen) confirms that the licensing authorities have given agreement to the melting of 6 tons of small scrap in a conventional furnace; this is meant to be a demonstration test and the measurements are intended to be the base for an authorization allowing the melting of the total quantity of small scrap from the machine house.

A French participant asks for the type of paint applied at the KRB-A reactor which has allowed to limit contamination of concrete to values 10 times lower than those on unpainted surfaces. Mr. Lasch indicates that the paints usually employed in German reactors, principally resin- and epoxy-paints, had been used depending on the surface to be protected.

An FRG participant asks if calculations or considerations exist to evaluate the quantity of contaminated material which will still be at higher levels than an 0.37 Bq/cm^2 free release limit after a decay time of 30 to 50 years. Mr. Lasch, confirming that such considerations have been made, points out that the objective of the test has been, for the actual situation, to know whether the amount of residual waste for present day decontamination would be in the order of 5 to 10% or much different. A detailed extrapolation for a decay time of 30 years has not yet been performed, but a problem will certainly be that after such a long time other nuclides than the easily measured ^{60}Co will be the lead nuclide.

A Belgian participant is interested in activity levels, conditioning and disposal of the iron-phosphate sludges generated in neutralizing the electrolyte. For Mr. Lasch, this is an essential point, in view of the destination of the secondary waste. As the contamination in the machine house is very low, the activity of the solutions is also very low. The sludges reach, depending on items which have been decontaminated, only 10^3 to 10^6 Bq/g. The sludges are stored actually in a half wet form and conditioning is in the preparation phase; the procedure is to form a dry powder.

A U.K. participant thinks that an important downstream activity will be the assaying of waste for disposal as a quality assurance procedure. The treatment of these problems for medium active waste raises his interest. The Chairman proposes to delay discussion of these problems, as Session IV will discuss in detail waste management and disposal.

Paper : "Methodology to evaluate radiological consequences of the management - of very low level solid waste"

An Irish participant was surprised to see that the collective dose commitment figures were so low. He asks for the size of the population taken into consideration and for confirmation that the calculations went to infinity in time. Mr. Clark answers that the full global population was used and dose commitment going on for all time has been assumed.

An FRG participant asks for clarification on the modifications with decay time in the nuclide compositions considered, and questions particularly the assumed actinide contents. Mrs. Chapuis, co-author of the paper, explains first that at the time of shut-down, an actinide content of 1‰ has been assumed in accordance with existing measurements on this type of waste. With decay time, the relative content of actinides will, of course, increase as shorter lived beta and gamma emitters disappear. Equally, as by decay a class of higher radioactive waste quantities is down-graded into the class of very low level waste - masses which are added to the already existing quantities - there is an apparent increase in activity.

A participant representing an international organization asks, considering a specific activity limit of 37 Bq/g and the inconsequent nature of the doses which are received, assuming that these levels can be measured on these large volumes of waste, how far could this activity limit be taken, without leading to consequential levels of the doses. Mr. Clark means that an answer would be difficult, as for a higher classification there would be more radionuclides which have a higher radiological significance; consequently, there is no linear relationship and a study would have to be made. The participant thinks that for concrete such a linear relationship could be established. Mrs. Chapuis comments that in the present study only the doses after effective dismantling have been considered, not taking into account the doses received by dismantling workers. For higher activity limits, it may be that the doses received by the workers will be the limiting values.

A French participant agrees with Mr. Clark insofar as it is necessary to reach an agreement on a specific activity limit now, but asks if decisions should be taken on a case by case basis or if, on the contrary, general recommendations should be proposed. For Mr. Clark, the point he was making was not to agonize too much over a particular limit chosen. It is necessary to chose one and the effect of the radiological impact of that choice is very small. The actual impact on people having to do the dismantling of the concrete could be, on the contrary, enormous. In Mr. Clark's strictly personal view (not speaking for NRPB or the UK) general recommendations on such limits could be set up now for this type of waste.

An FRG participant asks for the reason of giving average values of 10 Bq/g for waste classified as being between 37 and 3.7 Bq/g. Mrs. Chapuis explains that the lower the specific activity is, the larger the corresponding waste quantities are.

Paper : "Some examples of release of materials and sites including measurement techniques"

A French participant, after recalling that the amount of work needed for control and measurement has been given with a high degree of precision, asks if the total quantity of work necessary for dismantling these installations has been evaluated. One of the co-authors, Mr. Jager, does not have present the complete figures for decommissioning the nuclear part of NS Otto Hahn. In the average, 26 workers were occupied with this task, per month; work has been performed between June 1980 and May 1982. The number of working hours has been 72 611 in total. Dismantling has been performed by Noell, Würzburg, for GKSS at a cost of about 20 mill. DM. Mr. Crégut contributes figures for dismantling of the Pégase reactor; 8 workers have been occupied in the average during 14 months; these figures include work done to arrival of waste at the conditioning installation.

A representative of KRB Gundremmingen expresses his astonishment on the exposure value of 30% of total exposure attributed to the radiation measurement and surveillance personnel; similar work for measurements to unrestricted release at KRB-A has resulted in very low doses for this group of workers. The Chairman thinks that the 30% figure does not only include measurements after treatment, but also the doses received during measurements before decontamination.

Mr. Jager contributes the total exposure, 30 rem, and gives the average exposure of 600 mrem/year (with a highest value of 1300 mrem/year) which has to be compared to the 5000 mrem/year limit authorized for nuclear workers.

SESSION II

DECONTAMINATION FOR DECOMMISSIONING PURPOSES

Chairman : T. VAN DER PLAS, KEMA

Secretary : K.H. SCHALLER, CEC, Directorate-General Science, Research and Development

Introduction by the Session Chairman

Composition of contamination layers

Chemical and electrochemical decontamination

Decontamination of concrete, with particular reference to flame scarfing

Specific decontamination methods : water lance, erosion by cavitation, application of gel-based decontaminants

Factors to be considered in deciding whether to decontaminate for unrestricted release

Particular aspects of post-accident plant decontamination

Summary of discussion

INTRODUCTION BY THE SESSION CHAIRMAN

Th. VAN DER PLAS
KEMA, The Netherlands

This session deals with decontamination, i. e. removal of radioactivity from installations and components by chemical, electro-chemical and mechanical means. The material removed is transported elsewhere, in order to be treated for disposal. In this way the handling and treatment of the installations and components is made easier, at the price of treating part or all of the radioactivity in another place.

Accordingly, a decision to decontaminate is based on a compromise between the advantages and disadvantages of separating the radioactivity or dealing with the fully active installation or component. The decontamination processes might be carried out just to ease the handling of materials or to bring the radioactivity down to a level where unrestricted release is possible. The latter objective presupposes the existence of such a level, agreed upon with the competent authorities; a question which is dealt with extensively in one of the papers. The same paper gives a comparison of costs for full decontamination as compared with the disposal of the untreated components as well as a successful example.

Structure, morphology, chemical composition and adherence to the metal are important features of the oxide layer which determine its resistance to chemical and mechanical attack. Investigations of the oxide layer are reported in several papers, which show the influence of the operational history of the power plant on the properties of the layer. Results of the chemical and electrochemical treatments are given.

It would seem desirable that future investigations of this kind be grouped together in a more coordinated fashion in order to arrive rapidly at an optimal method of treatment for each oxide layer. Such studies could be usefully supplemented with investigations on the influence of plant operation on the structure of the oxide layers.

However, it might well be that a full removal of the oxide layers is not sufficient. In one of the papers strong indications can be found that radioactive nuclides have penetrated into the metal itself.

The chemical decontamination methods give rise to large volumes of radioactive, toxic and aggressive liquids. The efficient treatment of these secondary wastes forms an important step in the sequences that should be optimised. Some indications are given in the papers.

The mechanical methods for the removal of the oxide layer, viz. the high pressure water lance and the method of erosion by cavitation use much water but this is pure which is an important advantage for waste management. Further development is needed.

The removal of superficial layers from concrete is addressed in one paper which gives a review of the available methods and goes more deeply into the method of flame scarfing.

Finally, it should be noted that a detailed description of the decontamination procedures carried out at the TMI-2 site is given in the last paper.

COMPOSITION OF CONTAMINATION LAYERS

J.P. GAUCHON (CEA), F. BREGANI (ENEL)
W. AHLFAENGER (KWL), M. LASCH (KGB)

Summary

In the framework of the development of decontamination methods, the first studies to be made concern the oxide layers. This paper presents the research carried out in several European countries, on thickness, distribution and characterization of the contaminated layers and penetration depth of each radionuclide.

K.W.Lingen performed studies on composition of contamination layers. These investigations have been performed on samples taken out of the primary circuit of the Lingen reactor (250 MWe BWR shut down in 1977 after nine years of operation).

Activity depth profiles and characteristics of the oxide layers have been determined on the pipe samples :

- after removal of 86 μm by elecrtrolytic erosion in the primary water pipe samples, the ^{60}Co content is equal to the content of the outer part of the pipe;
- the same result is achieved after removal of about 80-90 μm on the condensate pipe.

ENEL CRTN - Milan : developed vigorous decontamination techniques using contaminated samples from the Garigliano reactor (160 MW BWR shut down in 1978 after 15 years of operation. A characterization of the oxide layers (OM, SEM, EMPA) showed the presence of magnetite, nickel ferrite and copper oxide. Radiometric measurement on some specimens of various primary tubing showed surface contamination of about 10^7 nCi/m^2 of gamma activity mainly due to ^{60}Co and about 10^4 nCi/m^2 due to alpha activity.

KRB Gundremmingen studied the development of economic decontamination in the KRB-A nuclar power plant (a 237 MWe BWR shut down in 1977 after 11 years of operation). Even in special areas of very high contamination and frequent damages of the protective paint during plant operation, the penetration did not exceed 7 mm.

CEA/CEN-Cadarache performed studies on metal decontamination by chemical methods. Several experiments have been made with contaminated oxide from electromagnetic filter connected to the primary circuit of the PWR (CAP Cadarache). The best result is obtained with a fluonitric bath after a pre-treatment. The tests mentioned have been carried out on samples from the German BWR ISAR. The specimens have been taken of a non activated ferritic steel component which had been operating for 5 years in the primary cooling circuit. Among the reagents which are available for decontamination, we have chosen sulfuric acid, halogenous and formid acid. An eletrolytic decontamination in two steps (reduction followed by oxidation) is used.

OM : Optical microscopy; SEM : Scanning electron microscopy;
EMPA : Electron microprobe analysis.

1. INTRODUCTION

The main criteria to be borne in mind in setting up a decontamination procedure are :

- the nature of the contaminated metal or alloy;
- the characteristics of the surface coating on the metal (e.g. the nature of the oxide, anodization, phosphating, metallic deposits, corrosion products, etc.);
- the nature of the contaminants and, especially, the nature of the bonds between the contaminants and the metal support.

In the absence of neutron activation of the metal (only the core being involved) most of the retained radiochemical activity present on light-water reactor pipes is in the oxide layers.

The aim of the research work summarized in this paper is to establish the nature and embrittlement and/or dissolution conditions of the oxide layers. The oxide deposits observed on the internal parts of water reactors originate from the corrosion of the material of which the reactor is made up. In practical terms, the creation and deposition of activated corrosion products in the primary circuits of water reactors depend on a number of factors, the most important of which (Fig. 1) are as follows :

- the effectiveness of the purification circuits or, to be more precise, conservation of the physical and chemical characteristics of the reactor water;
- variations in certain physical parameters along the hydraulic circuits (e.g. hydrodynamic factors, temperature gradients, magnetic factors, adsorption phenomena causing oxide deposition or removal, etc.);
- chemical and isotopic exchanges;
- in more general terms, the "history" of the reactor up to the time of decommissioning (e.g. the loading-unloading cycle, decontamination, operational incidents, unforeseeable corrosion of certain elements, clad rupture, etc.).

It is generally agreed that at least two oxide layers form on the pipes of water reactors (4), viz.

- an external layer of varying thickness and porosity, made up of coarse, flake-type particles;
- a homogeneous internal layer of greater density and without mechanical properties similar to metal.

Of the oxide layers deposited in a reduction medium in the primary circuits of PWRs, the internal layer usually comprises (5) chromium ferrite-rich magnetite, while the external layer generally comprises magnetite and nickel ferrite.

In BWRs, the oxide layer is formed in an oxidizing medium (radiolysis oxygen), and the deposits (4) are usually made up very largely of hematite.

The radionuclides present in the oxide are produced by neutron activation of the corrosion product deposits in the casing (Table 1). Activity transfer takes the form of displacement of the oxide and - to a lesser degree - by isotopic exchange with the radioactive species in solution (Fig. 1). Oxide deposits are most liable to form in the steam generators and in the parts of the circuits with the lowest water flow rates.

TABLE 1 - Main activation products formed in light-water reactors

Nuclear reaction	Period (days)	Source	Activation cross section (barn)
^{50}Cr (n, γ) - ^{51}Cr	28	Chromium/steel and water	12,1
^{54}Fe (n, p) - ^{54}Mn	313	Iron/steel and water	0,068
^{58}Fe (n, γ) - ^{59}Fe	45	Iron/steel and water	1,01
^{58}Ni (n, p) - ^{58}Co	71	Nickel/steel	0,098
^{59}Co (n, γ) - ^{60}Co	1912	Cobalt/steel	27,2
^{60}Ni (n, p) - ^{60}Co	1912	Nickel/steel	
^{181}Ta (n, γ) - ^{182}Ta	115	Tantalum/steel	
^{45}Sc (n, γ) - ^{46}Sc	85		
^{64}Zn (n, γ) - ^{65}Zn	245	Corrosion products	
^{94}Zr (n, γ) - ^{95}Zr	65	Corrosion products	0,05
^{135}Sb (n, γ) - ^{124}Sb	60	Impurities in the steel	2,5
^{46}Ti (n, p) - ^{46}Sc	85	Corrosion products	

The other radioactive elements deposited on the pipes in the course of declassification originate essentially from fuel leaks caused by can bursts. The elements concerned are, in descending order :

- fission products ^{137}Cs, ^{144}Ce, ^{106}Ru, ^{90}Sr
- alpha emitters (^{240}Pu, ^{239}Pu, ^{238}Pu, ^{241}Am, ^{244}Cm, etc.)

2. WORK CARRIED OUT BY ENEL-CRTN

2.1 Research aim

The aim of the research was to characterize the oxide contamination layers in the reactor circuits of the Italian boiling water reactor in Garigliano, a first-generation, 150 MW(e) General Electric BWR (6-7) (Fig. 2) shut down in 1978 after operating for 15 years. Characterization of the contaminated layers enabled vigorous decontamination methods to be developed, and these were subsequently tested on a DECO loop, using samples taken from the reactor.

2.2 Nature of the samples. Radiochemical and metallographic characteristics

The samples of stainless AISI 304 steel on which the tests were carried out were taken from three points on the reactor, i.e.

- a 24-inch recirculating tube between the secondary steam generator and the reactor;
- a 2-inch sampling tube on the above pipe;
- on the secondary steam generator separation plate.

The results of the radiochemical analyses carried out four years after reactor shut-down are set out in Table 2 below.

TABLE 2 - Radiochemical characteristics of samples taken from the reactor

Sample	^{60}Co (KBq/cm^2)	Other beta emitters (Bq/cm^2)	Alpha emitters (Bq/cm^2)	^{90}Sr (Bq/cm^2)
2-inch tube	11.5	-	5.8	0.26 - 220
Secondary steam generation separation plate	44.5	337.9	50.4	0.74

These samples indicate a preponderance of ^{60}Co over the other gamma emitters, which are (in descending order) $^{106}Ru/^{106}Rh$, ^{144}Ce, ^{54}Mn, ^{58}Co and ^{22}Na. There was a marked difference in radioactivity between the two samples.

Metallographic analysis was carried out in the JRC laboratories in Ispra (LMA), using optical microscopy (OM), scanning electron microscopy (SEM) and electron micro-probe analysis (EMPA), with an X-ray spectrophotometer (8) working on either the energy dispersion (XES) or the wave length dispersion (XWS) principle. Chemical analysis was also carried out by atomic absorption of oxides separated from the base metal by cathode pickling.

a) **24-inch pipe**

SEM analysis of the inner surface of the 24-inch pipe showed highly irregular particles with dimensions varying from 1 to 10 μm. The thickness of the oxide layer, estimated with the help of the optical microscope, was 15 μm. The cross sections of the oxide layer (Figs. 4 and 5) and the quantitative results obtained by EMPA-XES and EMPA-XWS are set out in Tables 3 and 4. Fig. 6 sets out the qualitative concentration profiles of the main elements.

TABLE 3 - EMPA-XES analysis at the points set out in Fig. 4 (% weight)

Elements	Points			
	1	2	3	4
Fe	44.6	41.8	40.6	61.8
Ni	26.1	25.2	24.7	8.6
Cr	1.4	2.8	2.9	16.3
Mn	0.7	0.8	0.8	1.3
O	27.3	29.5	31.1	-

TABLE 4 - EMPA-XWS analysis at the points set out in Fig. 5 (% atomic concentration and weight)

Elements	Atomic concentration (%)			% weight		
	1	2	3	1	2	3
Fe	20.36	22.18	60.78	38.21	40.90	67.40
Ni	11.63	11.51	7.66	22.95	22.32	8.93
Cr	0.68	1.16	17.76	1.18	1.99	18.34
Cu	0.42	0.15	0.17	0.89	0.31	0.21
Mn	0.61	0.11	1.01	1.13	0.19	1.10
O	66.3	64.9	-	35.63	34.28	-
Fe/Ni	1.75	1.93	-	-	-	-

The results obtained prove that the chemical composition of the oxide does not vary, and is similar to that of nickel ferrite (NiO, Fe_2O_3). The Ni and Fe atoms appear in some cases to be replaced, in small quantities, by Mn, Cr and Cu atoms wiwth a distribution which varies widely in the thickness of the oxide itself.

b) **2-inch pipe**

The optical microscope revealed a roughly 2 μm-thick layer of oxide, and SEM analysis (Fig. 7) showed that the surface was made up of particles which were homogeneously distributed and with peaks and troughs of varying widths. The particles were between 1 and 2 μm in width, and appeared to be bonded by an amorphous oxide.

c) **Plate separating the water tank from the secondary steam generator**

SEM analysis (Fig. 8) showed a number of crystals of between 0.5 and 2 μm surrounded by small particles of a width of between 0.1 and 0.2 m. Quantitative (XES) analysis at the top of a crystal showed that the oxide was a ferrite of nickel with a trace quantity of Cr.

The results of atomic absorption analysis are set out in Table 4, indicating substantial quantities of Fe, with levels of chromium and nickel below those in AISI 304. The presence of Cu cas also identified.

There was a marked degree of scatter in the results, which may have been due to the different origins of the samples analysed.

2.3 Characteristics of the samples after decontamination

a) **24-inch pipe**

After decontamination by a strong mineral acid (9), the oxide was completely eliminated, as the SEM surface analysis proves (Fig. 9). The peaks are probably due to the attack mechanism of the chemical solution used. XES qualitative analysis appeared to indicate a higher level of Cr in the outer metal surface.

b) **2-inch pipe**

Decontamination by a strong mineral acid (Fig. 10) again eliminated the oxide layer. XES qualitative analysis appeared to show traces of Cu on the surface.

Atomic absorption measurements gave percentage readings similar to those of the matrix (AISI 304).

2.4 Discussion of the results

The morphology and the thickness of the contaminated oxide layer varies widely in the samples observed. However, the composition of the oxide was fairly uniform, and corresponded to that of nickel ferrite, in which the Fe and Ni atoms can be substituted by Mn, Cr and Cu. In particular, the variations in the concentrations of Cu (high in the more internal layers) can apparently be explained by the installation of monel and cupro-nickel preheaters used from the time the power station started operations until 1968 (10).

Apart from the variations in Cu and the other low-level elements present, no definite evidence was found of the presence of a Cr-enriched oxide layer. A possible reason is that in the power station's first years of operation a large quantity of suspended solids may have produced an oxide layer which prevented any accretion of the deposit by corrosion of the base metal.

3. WORK CARRIED OUT AT CEN-CADARACHE BY DRDD/SEDFMA

3.1 Research aim

Under the terms of the French nuclear programme, SEDFMA at Cadarache is responsible for carrying out study and research on low- and medium-level waste.

From the point of view of decommissioning, the aim is to formulate feasible decontamination methods to :

- improve decontamination effectiveness;
- reduce the amount of secondary waste products;
- devise methods for purifying effluents and treating secondary waste.

The work summarized in this chapter is concerned with studying embrittlement or dissolution conditions either chemically or electrochemically of oxide layers deposited in the circuits of light-water reactors.

3.2 Nature of the samples. Radiochemical characteristics

For our tests, we used representative samples from the primary circuit of a PWR taken on an electromagnetic filter which had been in use for 20 months connected to the primary circuit of the prototype advanced boiler (CAP) at Cadarache. Tests were also carried out on samples form the emergency water cooling circuit in the German Isar BWR. The characteristics of the samples are summarized in Table 5 below.

TABLE 5 - Characteristics of the samples used by SEDFMA/Cadarache

Nature of the samples Characteristics	Samples taken on the CAP electromagnetic filter	Samples taken on the German Isar reactor
Nature of the material	Stainless 316 L steel	Ferritic steel DIN 17 MnMo V 64
Period of operation	20 months	5 years (emergency circuit)
Neutron activation	NA	
Geometric characteristics of samples	Tube diameter 2.5 cm oxide layer approx. 16 cm^2	30 pipes diameter 50cm oxide layer approx. 16 cm^2
Radiochemical activity	18 - 144 KBq/cm^2	2 - 37 KBq/cm^2
^{60}Co	99 %	approx. 60 %
^{54}Mn	-	approx. 40 %

3.3 Embrittlement of the oxide layer in the pipes of PWRs

Currently the most widespread method of decontaminating PWR components during operation is the APAC process, which comprises two stages (11) :

- oxidation in an alkaline medium (ph = 14) using permanganate (t = 60°C)
- decontamination using a bath filled with chelating acids (i.e. citric acid and oxalic acid) and a corrosion inhibitor.

It is generally acknowledged that the first stage has an embrittling effect on the thickest and toughest layer of oxide by oxidizing the Cr III in the ferrites into soluble Cr IV (12-13-14).

In our tests, we took the APAC process as a reference, and sought to establish the optimum oxide embrittlement and dissolution conditions and the effect of these factors on decontamination effectiveness.

Three oxidation pretreatment methods were tested on samples cut from a tube linking the electromagnetic filter and the reactor :

- oxidation treatment using permanganate in an alkaline medium;
- anodic oxidation treatment of samples in an alkaline medium;
- oxidation treatment using 2.5 % oxygen in demineralized water.

The latter reaction is used in the Canadian CAN-DECON process (15) for the soft contamination of water reactors.

The ozonized oxygen tests were carried out in various media (e.g. alkaline, acid, buffer, ph = 6.5). The best results were obtained by sparging the ozonized oxygen in demineralized water.

The reference pretreatment method (i.e. soda-permanganate) (Fig. 11) permits removal of 165 $\mu g/cm^2$ of chromium from the sample at 80°C, only slightly more than the amount removed (162 $\mu g/cm^2$) at 60°C; in both cases, the maximum amount was obtained after a contact period of one hour.

Anodic oxidation of the sample (ia = 100 mA/cm^2) yielded a comparable result (160 $\mu g/cm^2$) in 1 hour). In the best instance, this latter figure was only obtained after 7 hours immersion in ozonized oxygen-saturated demineralized water (Fig. 12). Unlike the alkaline treatment methods, part (approximately 50 %) of the ^{60}Co activity of the sample was removed, a phenomenon which can be put down to the mechanical effect of sparging on the oxide layer.

Table 6 compares the relative effectiveness of decontamination of the samples with and without pretreatment. Without pretreatment, only anodic oxidation in oxalic acid yielded significant results.

The samples used had an initial activity level of between 5 and 8 x 10^4 cps (counts per second). After oxidation by the permanganate, residual activity levels lower than 10 cps could be obtained only by the application of a hot fluonitric mixture (0.5 ml/l/HF) and anodic oxidation of the sample in an oxalic medium.

If the samples are pretreated with ozonized oxygen, residual activity is between 3 and 5 times greater at the end of the acid treatment than in the above case.

Prior electrolytic oxidation of the samples in an alkaline environment produces (in the best possible instance) DFs ten times lower than after permanganate oxidation.

These latter results prove that the amount of chromium removed from the samples is not always an adequate criterion for assessing the effectiveness of the decontamination process, and that more detailed analysis of the oxide layers must be carried out after embrittlement.

3.4 Conditions favourable to the dissolution of the oxide layer present on BWR components

The oxide layer present on BWR ferritic steel pipes is usually made up largely of hematite. In the course of our tests, we came across three

TABLE 6 - Decontamination of representative samples from the primary circuit of a PWR. Effect of pretreatment on the DFs.

Characteristics of the acid baths				Absence of pre-treatment	Nature of pretreatment		
Reagents (mol/l)		t (°C)	Time (h)		Permanganate in alkaline medium	Ozonized oxygen in demineralized water	Electrolytic oxidation in alkaline medium
HNO_3	2	25	2		1 500	630	
HF	0.5		4		2 000	860	
HNO_3	2	85	1	10	10 000	2 000	
HF	0.5						
HNO_3	2	25	2		530	140	60
HF	0.05		4		1 000	240	110
			8		2 000	400	170
HNO_3	2	85	1	6.4	900	230	
HF	0.05						
Anodic oxidation		25	0.033		20	50	
Oxalic acid	0.16		0.15		1 100	400	
$i_A = 100$ mA/cm^2			0.50	1 040	5 500	2 000	

characteristic types of deposit :

- Type 1 : The metal is covered by a thin, uniform, bright red layer, and the elements thus covered emit very little radiation.
- Type 2 : Uniform layer thicker than in type 1, slightly violet and blue tints.
- Type 3 : A thick, irregular layer, duller than type 1.

The highest levels of radiation (up to 300 mrad/h for a pipe of diameter 40 cm) were associated with types 2 and 3.

The decontamination tests carried out on these samples indicated which were the most favourable media and conditions (16). Generally speaking, we succeeded in reducing the residual activity of the samples to fairly low levels ($\simeq$ 3.7 Bq/cm^2) by concentrating on the length of treatment (up to 16 hours for the sulphuric medium).

The speed of decontamination of type 2 samples in a sulphuric medium of 2 mol/l was increased by a factor of 2 to 3 when the temperature varied between 25 and 40°C (Fig. 13) without any change to the erosion rates of the oxide and the metal (Fig. 14). (At 60°C the rate of erosion is multiplied by 10). By pretreating this type of oxide with permanganate in an alkaline medium, it is possible to greatly reduce the decontamination period (by a factor of 2). In this case, it would appear that the rate of oxide erosion is reduced and is more regular (Fig. 14).

Electrolytic decontamination in sulphuric, phosphoric or formic media of a type 3 oxide is shown in Fig. 15. By using sulphuric acid, it is possible to obtain DFs comparable to phosphoric acid DFs 6 times weaker and with a current density of 100 mA/cm^2 (compared with 360 mA/cm^2 for the phosphoric test).

The rate of decontamination can be increased 3 times by 2-phase electrolysis :

- reduction (embrittling the oxide layer)
- oxidation (removing the oxide layer and inducing erosion of the metal support).

Two-phase decontamination is less effective with a type 2 oxide (Fig. 16), where all that is needed is to initiate decontamination by means of an anodic sequence in an alkaline medium, the residual activity in the sample falling as quickly as with a type 3 sample.

3.5 Discussion of the results

To be really effective, acidic decontamination of PWR components **must** be preceded by oxidation, permanganate oxidation in an alkaline medium being the reference process yielding the highest DFs. Oxidation with ozonized oxygen, and electrolysis in an alkaline medium, permit the chromium in the oxide layer to be removed as effectively as with the reference treatment.

However, the action of fluonitric solutions subsequent to sensitization of the oxide layer by these processes yields DFs 10 times lower for electrolysis and 4 to 5 times lower for ozone.

Advance oxidation of the oxides present on PWR components can, in certain cases (e.g. type 2 oxides), reduce the time needed for decontamination. In this case - as for PWR oxides - the process of oxidation can be carried out with permanganate in an alkaline medium or by electrolysis. For other types of oxides, better results are obtained by 2-phase electrolytic decontamination (i.e. reduction followed by oxidation) in a sulphuric medium (residual activity 3.7 Bq/cm^2).

4. WORK CARRIED OUT BY LINGEN POWER STATION

4.1 Research aim

The aim of the research work was to determine the nature of the contamination layers formed during the life of the reactor, and the distribution of the specific radiochemical activity as a function of the depth of erosion of the metal. The tests were carried out on samples taken from the 250 MWe Lingen BWR, which was shut down in 1977 after a service life of 9 years.

4.2 Nature of the samples. Chemical and radiochemical characteristics

The samples used for these tests were taken from three points on the primary circuit, viz. (Fig. 3 and Fig. 17) :

- a tube linking the reactor to the steam generator (composition : Fe 68.1 - Cr 17.7 - Ni 10.9 - Co 0.11);
- a pipe located at the exit of the steam generator upstream of the preheater (composition : Fe 70.8 - Cr 17.5 - Ni 10.1 - Co 0.09);
- an element of piping located on the reactor water purification circuit upstream of the ion exchanger (composition : Fe 68.3 - Cr 18 - Cr 11 - Co 0.1).

The results of chemical and radiochemical analysis of the oxide taken at these three points are set out in Tables 7 and 8.

TABLE 7 - Radiochemical characteristics of the oxide sample taken from the primary circuit of the Lingen reactor (1 september 1983).

Nuclide	Sample upstream of steam generator		Sample downstream of steam generator		Purification circuit	
	Bq/g	Bq/m^2	Bq/g	Bq/m^2	Bq/g	Bq/m^2
^{60}Co	2.4×10^6	1.9×10^7	1.5×10^7	1.6×10^8	7.9×10^8	3.6×10^{10}
^{55}Fe	9.5×10^3	7.6×10^4	3.8×10^4	3.9×10^5	3.9×10^6	1.8×10^8
^{106}Ru	3.2×10^2	2.5×10^3	3.9×10^2	4.1×10^3	1.2×10^5	5.6×10^6
^{136}Cs	1.4×10^2	1.1×10^3	1.4×10^3	1.5×10^4	3.7×10^4	1.7×10^6
^{134}Cs			2.8×10^1	2.9×10^2		

TABLE 8 - Chemical composition of the oxide samples taken from the primary circuit of the Lingen reactor

Nature of the oxide	Sample upstream of steam generator		Sample downstream of steam generator		Purification circuit	
	g/m^2	%	g/m^2	%	g/m^2	%
Fe_2O_3	6.1	76.6	7.76	74.6	35.0	76.4
Cr_2O_3	1.29	16.2	1.91	18.4	7.2	15.7
NiO	0.52	6.5	0.70	6.7	3.4	7.4
Co O	0.054	0.7	0.03	0.3	0.24	0.5
	7.96		10.4		45.8	

Analysis brings out a predominance of ^{60}Co (the reactor had been shut down for 6 years), and a high level of corrosion products in the sample taken on the primary purification circuit.

The main form of oxide was hematite mixed with spinels, and in all cases, photomicroscopic analysis of the samples showed an intercrystalline form of corrosion, a phenomenon which was particularly marked on the samples taken from the primary purification circuit and upstream of the steam generator.

Electron microprobe analysis (EMPA) and chemical analysis showed a higher level of cobalt in the oxide layer compared with that in the material making up the components (average alloy level 0.1 %), the enrichment factor being 7.1 upstream of the steam generator, 3.8 downstream and 5.5 in the sample taken from the purification circuit.

The oxide layer was studied by secondary ion mass spectrometry (SIMS), and the results (Fig. 18) show that the deep layers had a higher level of ^{52}Cr, and the surface layer a higher level of ^{90}Zr. It should be noted that, in the primary circuit, zirconium is present only in the fuel element cladding, which means that it was probably the steam which was responsible for the appearance of the zirconium on the oxide surface.

4.3 Distribution of radiochemical activity in the metal

The distribution of radiochemical activity was determined by an electrolysis method developed in the Kernforschungsanlage (KFA) in Jülich.

A sample was first of all stripped of its oxide layer by a two-stage treatment process (APOx) :

- alkaline permanganate (5 % NaOH, 1.5 % $KMnO_4$, t = 95°C)
- oxalic acid (2 %, t = 95°C).

After two cycles of treatment had had virtually no effect on its surface (17), the sample was subjected to electrolysis (41 % H_2SO_4 - 53 % H_3PO_4 - 6 % water) monitored at approximately 1 μm erosion stages. Measurements were taken at each stage of :

- thickness loss on the part of the sample (using a micrometric system)
- loss of weight and levels of iron, chromium and nickel in the electrolyte.

Throughout the tests, correlation between the various measuring methods was found to be good.

The specific radiochemical activity (Bq/mg) corresponding, for each stage, to the ratio displaced activity of the sample : mass of dissolved metal was monitored (Fig. 19). On the basis of the results obtained from the various samples, it is possible to make the following points.

- The specific activity of the sample taken from the primary purification circuit fell rapidly, attaining (after 26 μm erosion) a level of activity similar to that present on the outside of the piping (2×10^{-4} Bq/mg).
- The sample taken downstream of the steam generator exhibited a slower rate of activity loss, a layer of 90 μm having to be dissolved before the result was comparable with that above.
- The specific activity of the sample taken in the hot part of the reactor (between the reactor and the steam generator) tended towards a higher limit (a factor of 300), which, remained high even at erosion depths of up to 100 μm.

  The latter behaviour could only be explained by neutron activation, although the distance between the sample and the reactor core would appear to rule this out.

4.4 Discussion of the results

The components of the Lingen BWR are covered by an oxide layer enriched in cobalt, the content of which varies depending on the function of the component in question. The oxide layer can be removed very easily by APOx treatment (i.e. by soda, permanganate and oxalic acid).

The metal had undergone intercrystalline corrosion, which was more marked on the samples taken upstream of the steam generator and on the purification circuit.

Electrolytic erosion of this sample resulted in a level of residual activity comparable to that of the external section of the piping after 26 μm erosion. The residual activity in the sample taken upstream of the steam generator approached a limit 300 times higher after 100 μm erosion.

5. WORK CARRIED OUT BY KGB – GUNDREMMINGEN

5.1 Research aim

The aim of the research undertaken by KGB was to identify the oxide layers present in the Gundremmingen KRB-A reactor piping with a view to developing optimum decontamination procedures for the dismantling of a BWR.

These processes were developed on a pilot scale, as a result of which declassification of this reactor will be used as a reference point for the declassification of a 1 200 MWe BWR. The KRB-A in Gundremmingen is a first-generation 237 MWe BWR (Fig. 2), which was shut down in 1977 after a service life of 11 years.

5.2 Nature of the samples. Chemical, radiochemical and crystallographic characteristics

The main characteristics of the KRB-A reactor components are as follows :

- **Primary recirculating circuit** : The oxide layer formed on austenitic steel at high temperature, producing one of the highest levels of contamination.
- **Primary steam circuit** : Contamination of the primary steam circuit was 100 to 1 000 times less than that of the primary recirculating circuit. The mass of oxide was approximately 1.6 mg/cm^2, and the level of contamination 17 KBq/cm^2, and over most of this component the level of contamination was highest in the lower part of the circuit (60 KBq/cm^2, 22 mg/cm^2 oxide). ^{60}Co accounted for 99 % of the contamination of the ferritic steel component.
- **Condensate circuit** : The oxide layer was formed at a lower temperature than in the steam circuit and – at 10 mg/cm^2 – was made up of the following elements : 70 % iron, 0.6 % copper and nickel, 1.8 % manganese and 0.2 % chromium and cobalt. The level of contamination increased from the upper (0.23 KBq/cm^2) to the lower (0.73 KBq/cm^2) part of the circuit (16 – 55 % ^{60}Co, 3 – 6 % ^{134}Cs, 41 – 77 % ^{137}Cs).
- **Water supply circuit** : Contamination similar to that of the primary steam circuit, but with a lower level of activity, undoubtedly due to purification of the condensates. The oxide present on the components was subjected to nuclear gamma resonance (Moessbauer spectroscopy). The results obtained are set out in Table 9, and can be interpreted as follows.

TABLE 9 - Crystallographic characteristics of samples taken from components of the KRB-A Gundremmingen reactor - (Moessbauer spectroscopy analysis)

Nature of the component	Composition of the samples (%)			
	Magnetite	Hematite	Goethite	Fe OOH
Primary recirculating circuit : sections				
- upper	29.1	54.7	6.2	7
- upper	23	66.4	4	3.6
- intermediate	18.6	75.1	3.3	2.5
- intermediate	21.5	68.9	3.4	5.7
- lower	17.7	77.1	3	1.6
- lower	27.6	59.9	2.1	7.7
Primary steam circuit : sections				
- upper	11.3	73.9	6.2	7.7
- upper	12	69.8	2.8	14.2
- intermediate	2.3	90.2	2	9.5
- intermediate	24.3	64.2	2	9.5
- lower	12.5	82.9	3	2.3
- lower	10.7	82.9	3	2.3
Condenser circuit :				
- upper section	82.8	0	13.1	1.7

- The predominant form of oxide on all the components was hematite, apart from the upper part of the condenser (82.8 % magnetite).
- Apart from the condenser, the part of the reactor with the highest level of magnetite (18 - 29 %) in the oxide was the primary recirculating circuit.

Generally speaking, the oxide deposit comprised two distinct layers (18) :

- a highly active, rust-coloured external layer, which could be removed with some difficulty with a brush and water;
- a thinner, grey underlying layer, which resisted dissolution even in the strongest mineral acids.

When the reactor was shut down (August 1977), the outer layer contained 20 % ^{54}Mn and less than 80 % ^{60}Co, whereas the inner layer contained more than 99 % ^{60}Co, which would appear to confirm the age of the inner layer and the secondary role played by the isotopic exchange process.

5.3 Discussion of the results

Hematite is the predominant form of oxide on all KRB-A reactor components apart from the condenser, which is covered largely by magnetite, an element which accounts for between 20 and 30 % of deposits taken from the primary recirculating circuit (the component featuring the thickest and most active layer of oxide).

6. CONCLUSIONS

The composition of deposits of oxides present on the components of the primary circuit of light-water reactors depends on the nature of the component and the type of reactor.

The main form of oxide in PWRs is magnetite associated with ferrites of nickel and chromium. This form of oxide requires a preliminary embrittlement phase before the acid decontamination phase proper.

Oxidation by permanganate in an alkaline medium yields the lowest levels of residual activity ($\simeq$ 3.7 Bq/cm^2) after the action of a mixture of strong mineral acid (fluonitric).

Oxidation may be either by ozonized oxygen or by electrolysis, the effectiveness of fluonitric baths then being reduced by a factor of between 5 and 10.

Hematite is the normally predominant form of oxide on the components of a BWR primary circuit. Deposits with more than 80 % magnetite were found only as a result of nuclear gamma resonance analysis on the condenser of Gundremmingen KRB-A. It should be noted, however, that some of the results can be explained by the presence of an oxide containing varying amounts of magnetite, as is the case with :

- a sample taken from the Lingen reactor condenser, the residual activity of which did not fall below 5.9 x 10^{-2} Bq/mg despite 100 μm electrolytic erosion;
- some of the samples taken from the Isar feed water circuit, decontamination of which benefited from prior oxidation using permanganate in an alkaline environment.

If the component is made of austenitic steel covered by hematite, the oxide layer can be removed chemically by strong acids (e.g. fluonitric and chloronitric acids) as in Garigliano, or by electrolysis in a sulphophosphoric medium, as at Lingen.

Removal of the oxide layer on a component made of ferritic steel requires less vigorous media (e.g. sulphuric, halogenous and formic acids), resulting in longer decontamination periods.

The effectiveness of the tests carried out on the samples from the Isar feed water circuit was improved by temperature selection and eletrolytic decontamination in two phases :

- cathode electrolysis to embrittle the oxide
- anode electrolysis to decompose the metal.

REFERENCES

1. Corrosion, NACE, 1979, Vol. 35
2. Chimie et radioactivation des circuits primaires des centrales électronucléaires à eau sous pression, R. DARRAS - CEA/DCAEA/SCEF 1980
3. Contamination mechanisms and decontamination techniques in light water reactors, 1979, EUR 6422 EN, CEC, Joint Research Centre
4. Nature of deposits on BWR and PWR primary system surfaces : relation to decontamination. A.B. JOHNSON, J.B. GRIGGS and F.M. KUSTAS, BNES, 1980, paper 54
5. Investigations on oxide formed in high temperature water on austenitic steel. Journal of nuclear materials 92(1980) 334-344, North Holland Publishing Company. R. BAUCH, M. CAMBINI, P. WEISBERGER, C. DE ASMUNDIS, C. PLOG
6. ENEL Garigliano plant. Internal communication. International decommissioning Seattle, 10-14 Oct.
7. Operational experience in KWU boiling water reactors, R. RIESS, N. HENZEL and K. FROHLICH - Kraftwerk, AG Erlangen BNES, London, 1978
8. F. BREGANI, P. BIANCHI (ENEL), P. BOTTELIER, G. BUSCAGLIA, M. COMBINI (CCR Ispra); Metallographic characterization contaminated oxide films of the BWR Garigliano nuclear power plant : relation to decontamination, Ispra-JRC, Tech. Note n° 1.06/00.83/101, Sept. 83
9. R. PASCALI, F. BREGANI, CC. RONCHETTI, R. RIZZI, Hard chemical decontamination of parts of the primary loop of serviced nuclear plants, 3rd BNES conference on water chemistry of nuclear reactors systems, Bournemouth, 17-21 Oct. 1983
10. A. AGOSTINELLI, F. BREGANI, A. GAROFALO, R. PASCALI, Chimica dell'-acqua della centrale elettronucleare del Garigliano, RNRL-DPT-SN e CRTN N 2/82/2, Luglio 82
11. Decontamination of nuclear reactor and equipment. J.A. AIRES, G.P. PANZEN, J.L. ZEGGER, US Pat. 3.013 909, 1960-61

13-14. R.M. WATKINS, YAEC 90.44 (1958) - YAEC 117 (1959)

15. Procédé de décontamination des réacteurs nucléaires. Atomic Energy of Canada, Limited, 9.4.1980, n° 80080-10
16. R. PASCALI, F. BREGANI (ENEL), W. AHLFÄNGER (KWL), M. LASCH (KGB) J.P. GAUCHON (CEA) - Chemical and electrochemical decontamination Decommissioning Conference, Luxembourg, May 1984
17. P.H. EFFERTZ, U.H. MEISEL - Verzunderung warmfester Stähle in Hochdruckdampf nach langen Betriebszeiten
18. In situ Dekontamination von Teilen des KRB Primärkreises - N. EICKPASSCH and M. LASCH, 1979

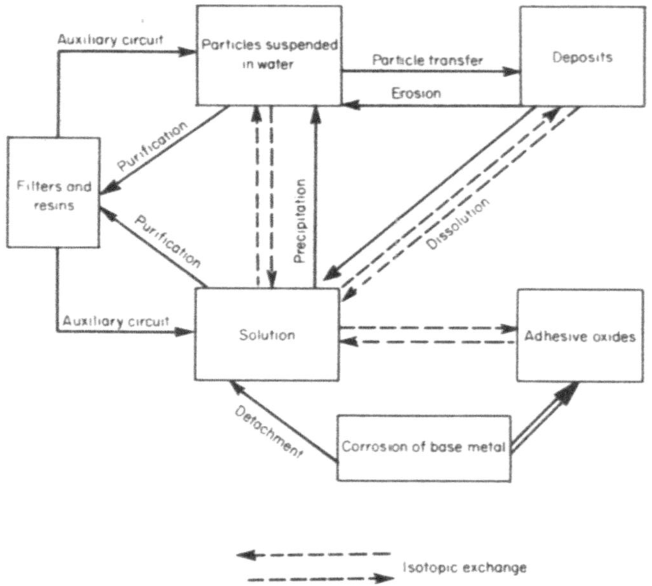

FIG. 1 - General diagram of the movement of corrosion in a light-water reactor circuit

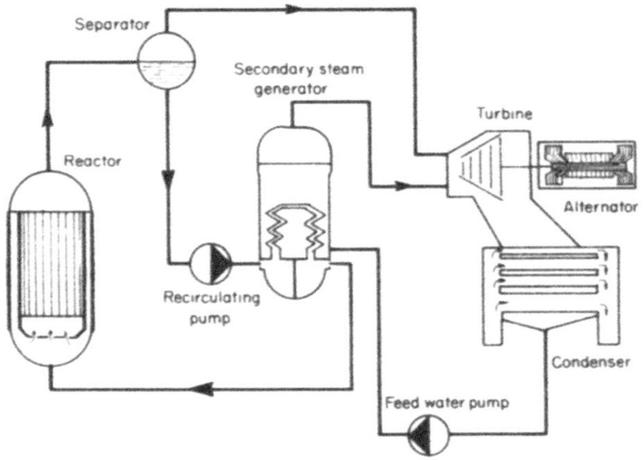

FIG. 2 - Basic design of a first-generation BWR (Garigliano - Gundremmingen)

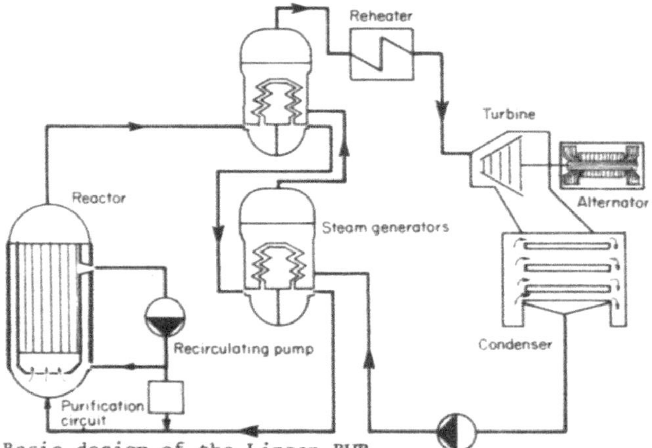

FIG. 3 Basic design of the Lingen BWR

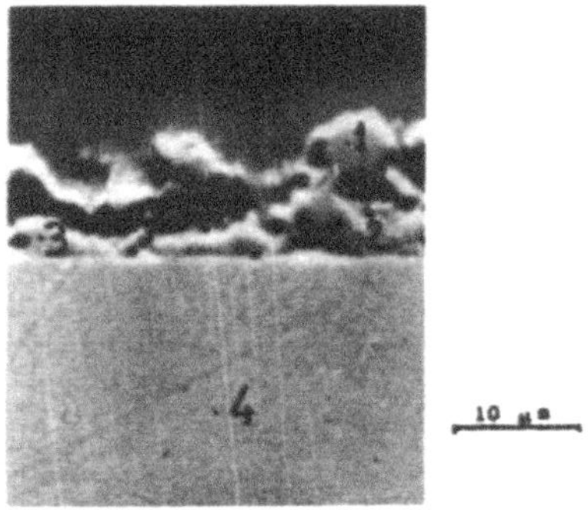

FIG. 4 - SEM analysis of the surface of the 24-inch pipe (The quantitative analyses are set out in Table 3)

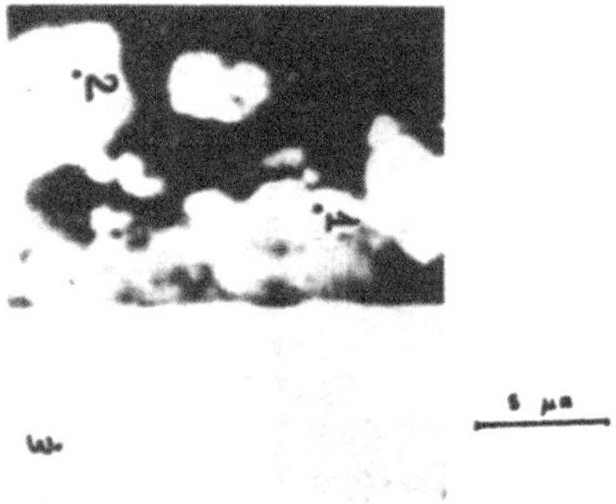

FIG. 5 - SEM analysis of the surface of the 24-inch pipe (The quantitative analyses are set out in Table 4)

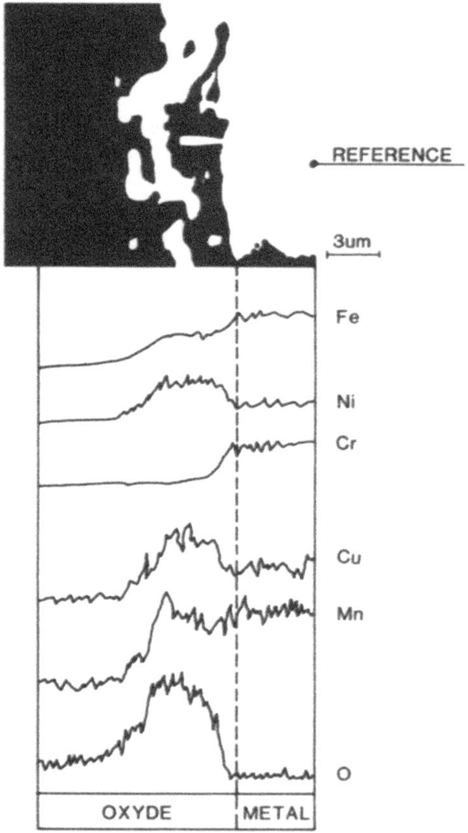

FIG. 6 - SEM analysis of the oxide deposit present on the 24-inch pipe

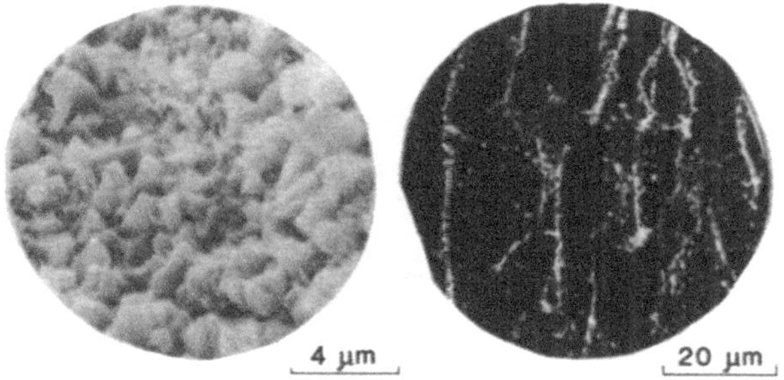

FIG. 7 - SEM analysis of the inner surface of the 2-inche pipe

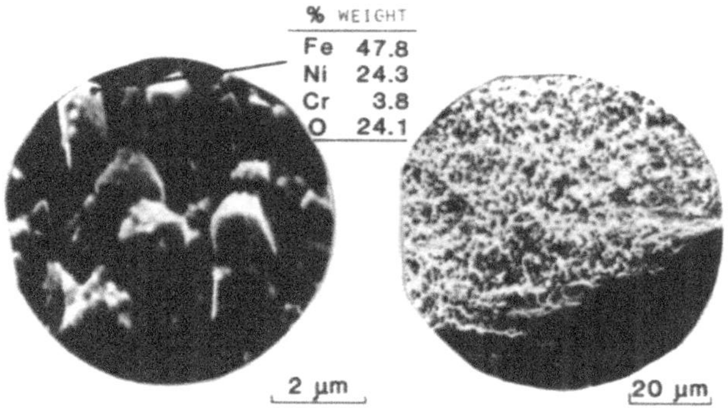

FIG. 8 - SEM analysis of the steam generator separation plate

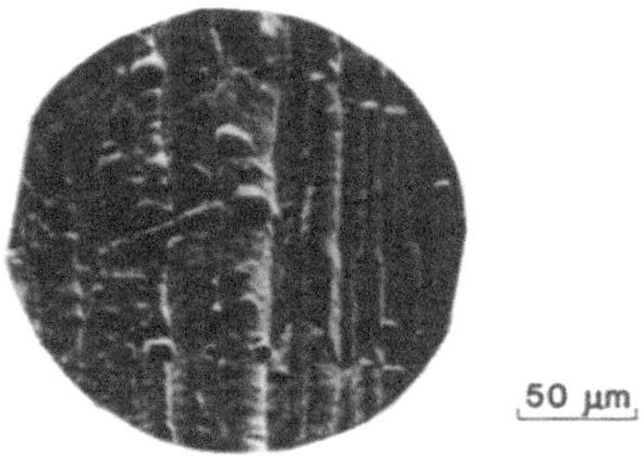

FIG. 9 - SEM analysis of the inner surface of the 24-inch pipe after decontamination

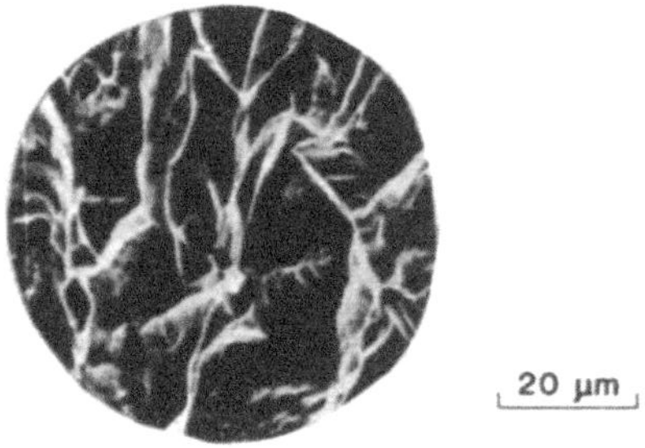

FIG. 10 - SEM analysis of the inner surface of the 2-inch pipe after decontamination

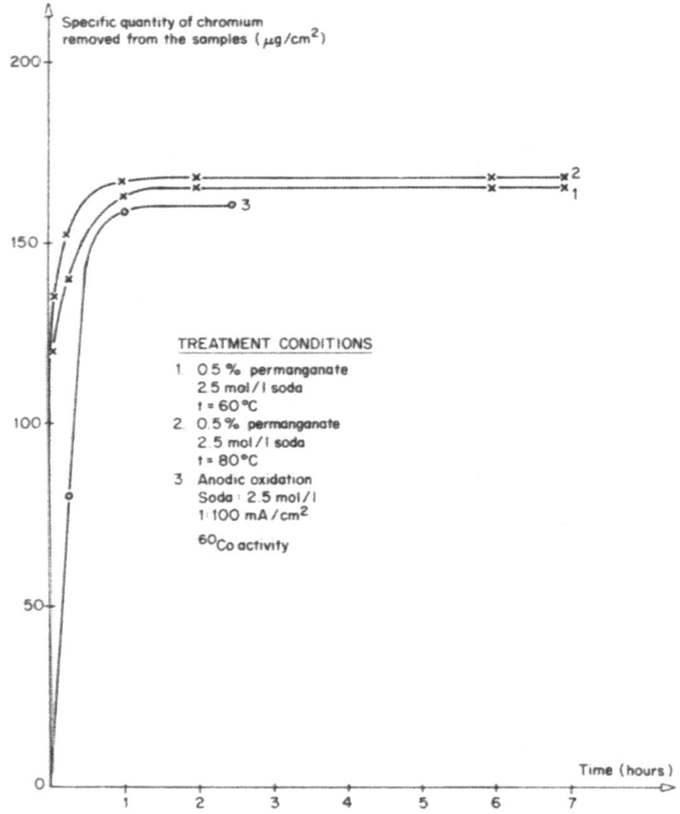

FIG. 11 – Solubilisation of the chromium by

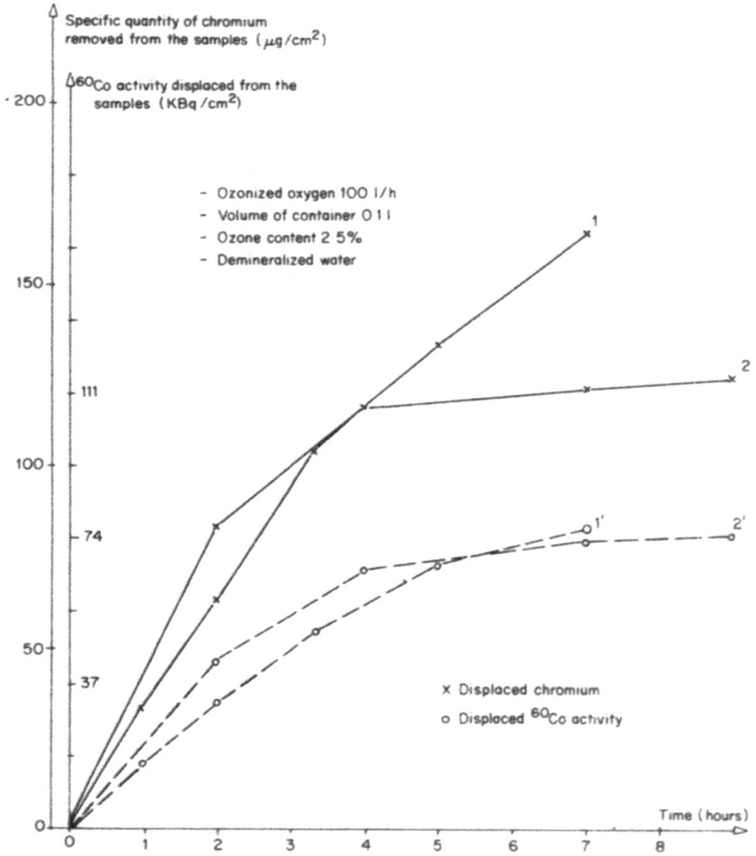

FIG. 12 – Displacements of chromium and ^{60}Co activity

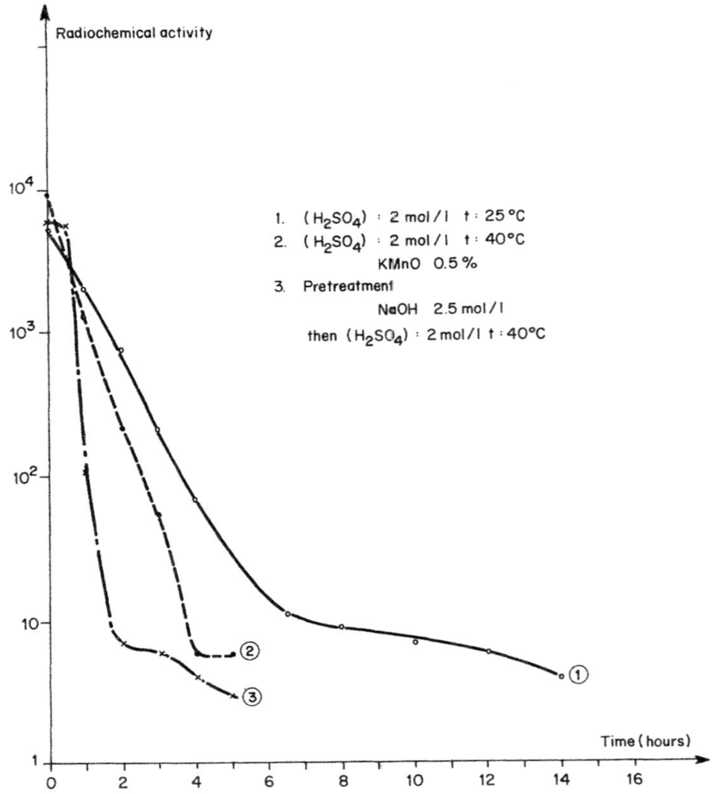

FIG. 13 - Type 2 oxide

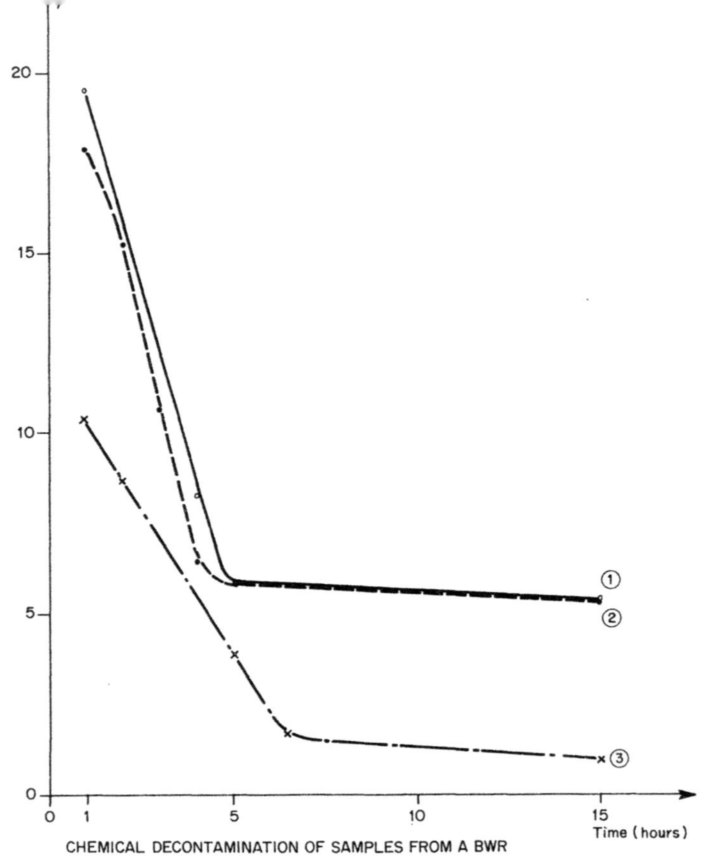

FIG. 14 Rate of erosion of decontaminated samples in a sulphuric medium

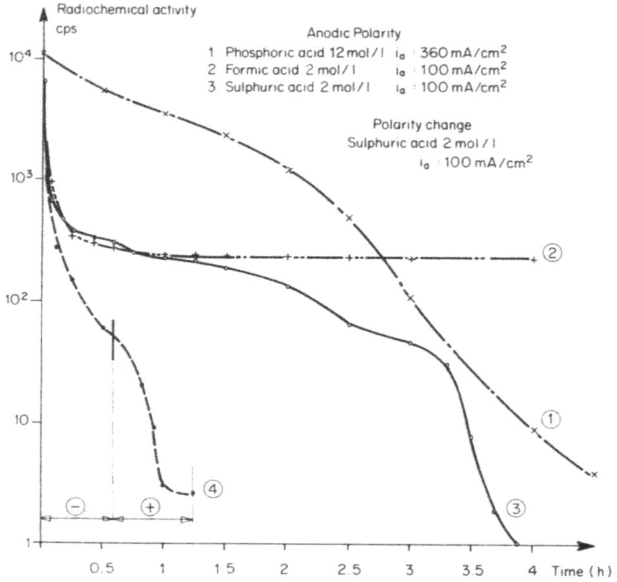

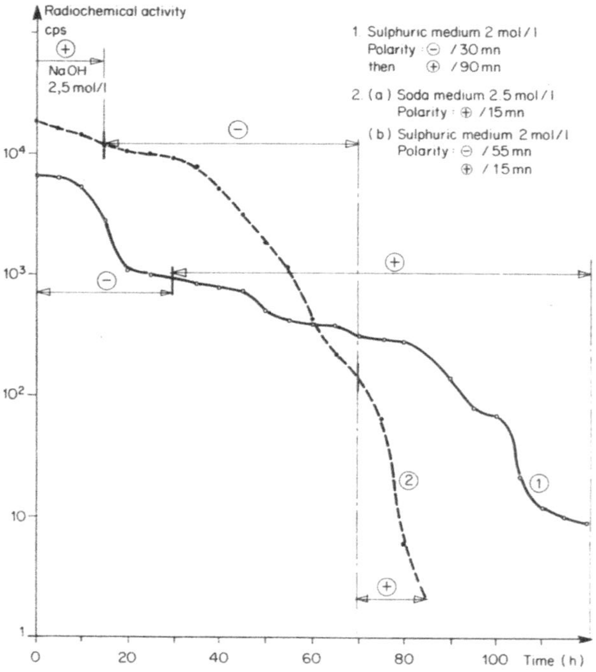

ELECTROCHEMICAL DECONTAMINATION OF SAMPLES FROM A BWR

FIG. 15 - Type 3 oxide

FIG. 16 Type 2 oxide

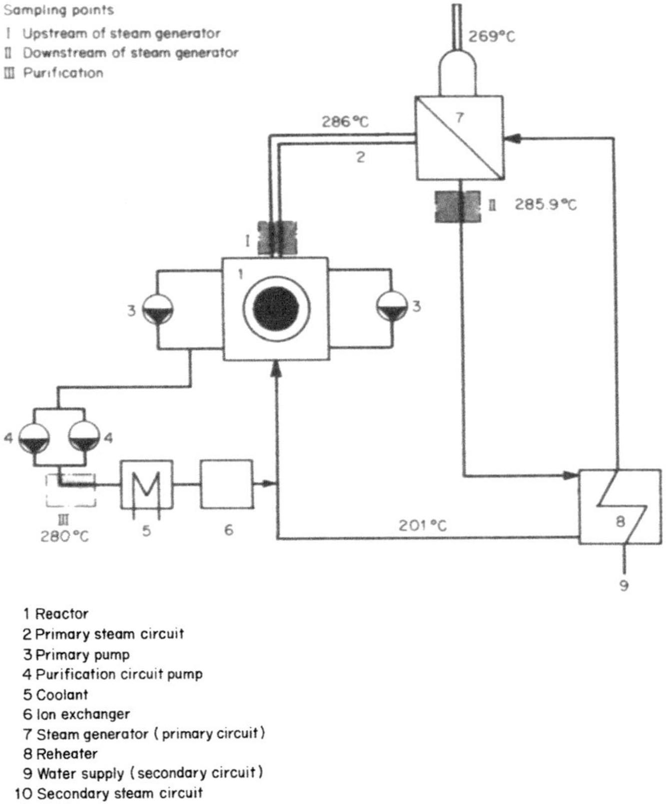

FIG. 17 – Lingen BWR – Primary circuit

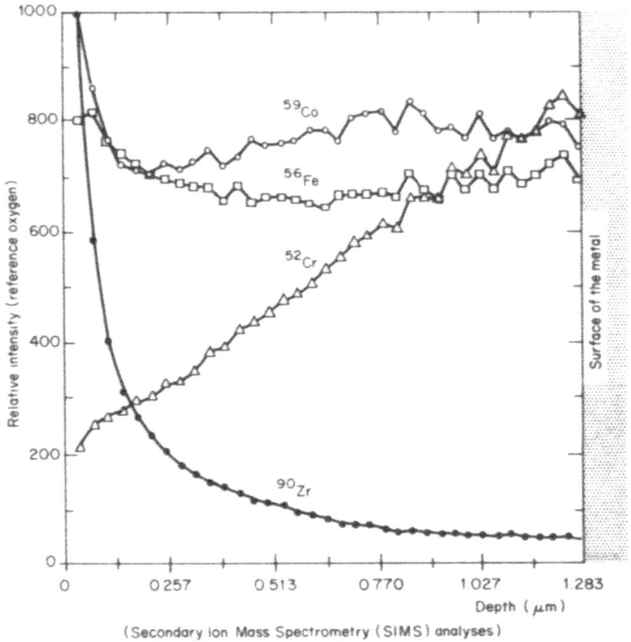

18 – Distribution of the various elements in the oxide layer present between reactor and steam generator

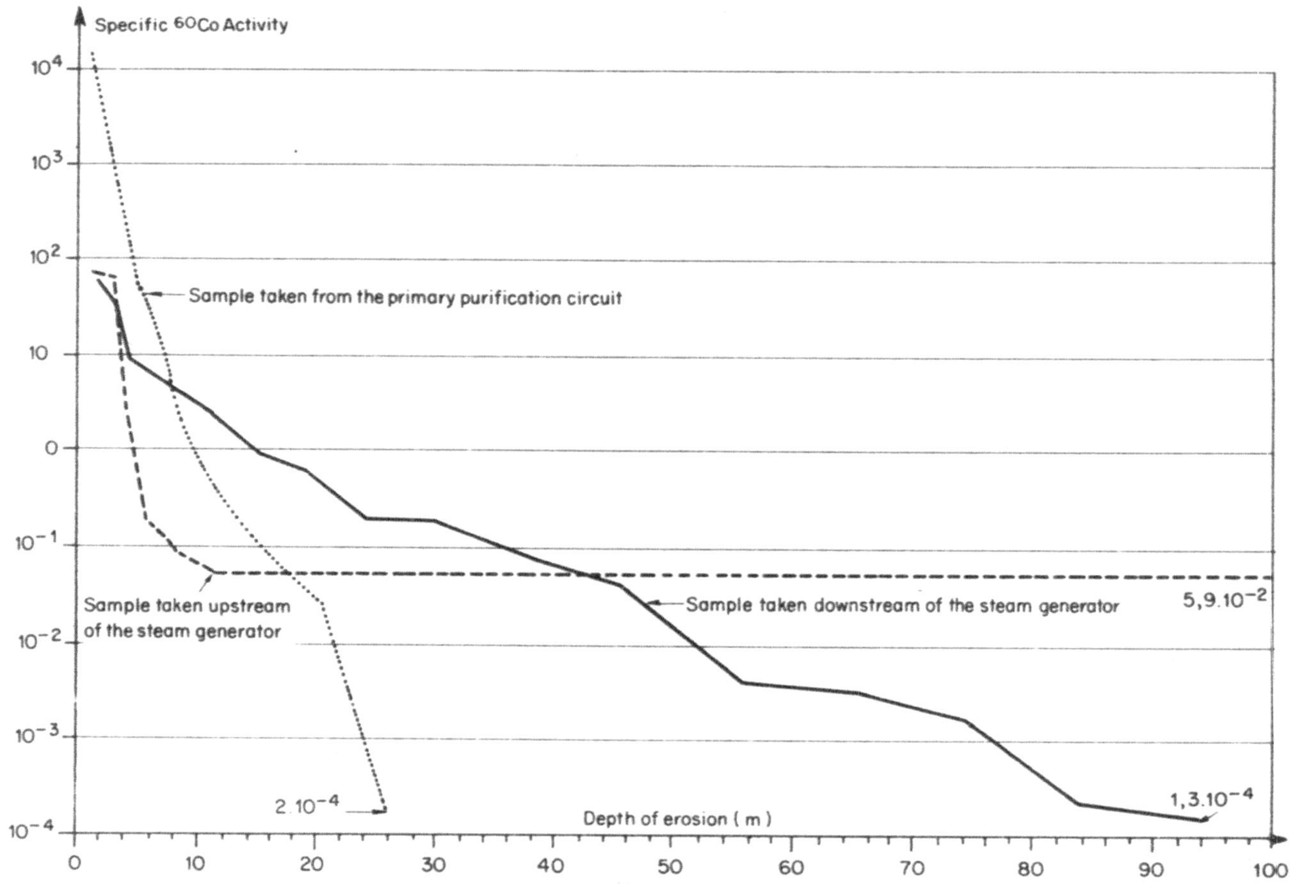

FIG. 19 - Development of specific ^{60}Co as a function of sample erosion depth
Samples taken from the Lingen reactor

CHEMICAL AND ELECTROCHEMICAL DECONTAMINATION

R. PASCALI (ENEL), F. BREGANI (ENEL), W. AHLFÄNGER (KWL),
M. LASCH (KGB), J.P. GAUGHON (CEA)

Summary

Among operative decommissioning problems the decontamination procedures, by chemical and electrochemical methods, of contaminated metallic surfaces of nuclear power plants have been studied.
The aim of these researches is to develop vigorous decontamination techniques, taking into account the cost of waste treatment, to possibly achieve unrestricted release of decontaminated materials, and to obtain know-how for in-situ applications.
In particular four addressed researches have reported. Studies on composition of contamination layers and on the efficiency and mechanism of decontamination procedure performed by KW Lingen. Vigorous decontamination tests of steel samples coming from the BWR Garigliano plant carried out by ENEL-CRTN Milan. The development of economic decontamination procedures, in particular electrochemical decontamination, studied by KRB Gundremmingen. Tests on metal decontamination by chemical and electrochemical methods performed by CEA-CEN Cadarache.

1. INTRODUCTION

The increase in the number of nuclear plants and the growing sensitivity of public opinion towards nuclear activities have led to great interest in decommissioning problems. In this context, R&D programme on decontamination, i.e. for unrestricted release of materials, are considered a priority stage. Particularly good decontamination of out-of-core surfaces can be highly beneficial. In fact, contamination activity, although a small part of the total activity inventory, is spread over very large surfaces. Therefore, decontamination is necessary both for safety, feasibility of dismantling and for unrestricted release of materials (1,2).

Radionuclide contamination on surfaces can occur by chemical absorption, ion exchange with free surface ions and by physical absorption or adhesion followed by nuclide migration towards material cracks and crevices (3). Contamination features are closely connected with the nature of the surfaces and the fluid carrier. In metallic surfaces, a chemical similarity is often present between the active ions and the metallic substratus, so contaminated oxide can grow in the substratum matter itself. In this case, total surface decontamination can be obtained only by complete removal of the metallic layer which includes active ions.

A nuclear plant is contaminated from the beginning of operation. Primary loop materials corrode and then activated corrosion products deposit on primary surfaces and tend to concentrate in crevices and in zones of weak flow circulation. Fission product and actinides contamination, arising from leakage of fuel elements pins, may also be present. The cleaning process is therefore closely connected both with the physico-chemical features of surfaces and the dissolution mechanisms of oxide layers (4).

Decontamination for decommissioning purposes differs from the usual in-service decontamination procedures. In fact,it allows even heavy damages of the material or of a group of materials, since they are not being returned to service. Moreover it entails the problem that it is not sufficient merely to achieve a level of decontamination consistent with limited access: the decontamination should preferably allow unrestricted release. Therefore, experiences acquired during in-service soft decontamination are not usually generalizable for decommissioning purposes.

In chemical and electrochemical cleaning, the choice of reagents must take into account the amount and nature of spent liquid radwaste: an exact definition of the treatment and disposal must be given because decommissioning requires both the reduction of radioactivity to safe dose levels and the radioactivity-displacement to another place. Waste treatment systems in nuclear power plants, at present, do not appear to be able to easily receive wastes resulting from decommissioning activities either in terms of volume or of technology processes (5). Finally, the assessment of unrestricted release contamination levels is a very complex matter both from the legal-philosophical and the technical points of view. In particular, a careful evaluation of each isotope encountered is required. It is well known that the decision of whether to release materials as industrial waste or not ought to be based on a cost-benefit analysis (with the ALARA principle).

The researches reported in this paper have been carried out in the framework of the European Communities Commission programme of the Decommissioning of Nuclear Power Plants, within the task n° 2: Decontamination for Decommissioning Purposes.

2. RESEARCHES CARRIED OUT BY KWL

The KWL (Kernkraftwerk Lingen GmbH) is the operator of the nuclear power station Lingen. The Lingen plant is a BWR (252 MWe)with two closed water steam circuits: the primary circuit with reactor and steam transformers and the secondary circuit with steam transformers, gas-fired superheater and hot steam turbine. Approximately, the plant has been running for nine years from August 1968 to January 1977.

The main purpose of the research activities carried out by KWL, has been the characterization of the contamination layers formed in long-term reactor operation. Nevertheless specific chemical decontamination procedures were also studied and tested.

2.1 Test materials

All the contaminated test materials were taken from the Lingen reactor plant. In particular pieces of tubes have been cut from the steam, condenser and reactor water loops.

The measurements of the chemical compositions and the 60-Co content of the oxide layers in the three types of samples have shown that the chemical composition is fairly comparable, but the oxide quantity and 60-Co activities differ: primary water-pipe 45.8 g/m^2 and $9.8\cdot 10^8$ Bq/g; steam pipe: 8.0 g/m^2 and $3\cdot 10^6$ Bq/g and condensate pipe: 10.4 g/m^2 and $1.8\cdot 10^7$ Bq/g, respectively.

The main evidenced oxide species have been Fe_2O_3, Cr_2O_3 and NiO. On the steam pipe sample the 60-Co radioactivity, after decontamination, remains constant at $6\cdot 10^{-2}$ Bq/mg over the wall thickness of 3.5 mm as shown in Figure 1. As this sample has been taken only at 1 m distance above the reactor pressure vessel, activation by delayed neutrons has probably taken place.

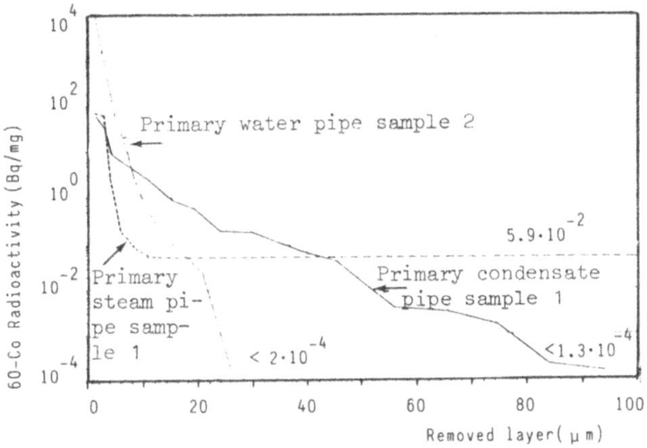

Figure 1 - Course of specific 60-Co radioactivity in removed layers.

2.2 Soft chemical decontamination

Test with soft chemical decontamination procedures have been performed in order to allow to remove only the oxide layer without any metal base attack. Initially, tests with typical two step procedure have been checked (50 g/l NaOH plus 15 g/l $KMnO_4$ followed by 20 g/l oxalic acid). The bath temperature was >95°C for all tests. A treatment cycle was the summary of the following partial treatments which also used in this order: alkaline oxidation, water flushing, acid treatment and water flushing. After these treatments the metal surface was visually free of oxidic deposits, nevertheless the residual contamination levels are still significantly high and correlable with the initial dose rate.

LOMI-reagents (low oxidation state metal ions) were experimented too. By this procedure, the ion Fe^{3+} is reduced to the ion Fe^{2+} in the oxides;

this process destroys the lattice structure of the oxides and leads to decomposition (6).

One stated that a complete removal of the oxide layers could not be achieved alone through the LOMI-treatment. The oxide layer was not completely removed even after 3 LOMI-treatments. They were completely removed only through a subsequent oxidic treatment with potassium permanganate and a further treatment with oxalic acid.

The decontamination of the primary water pipe sample was carried out with three treatment cycles (Table I). Contrarily to the condensate and

TABLE I - Results of LOMI decontamination tests carried out by KWL on Lingen samples.

Sample	N°	Surface (cm^2)	Initial dose (mrem/h)	Decontamination treatments (*)	Final dose (mrem/h)
Steam pipe	1	67	10	b-c-a-a-a	2-3
Steam pipe	2	60	6	a-b-c	<1.5
Condensate pipe	1	77	50-55	a-b-c	<1.5
Condensate pipe	2	104	25-30	b-a-b-c	<1.5
Water pipe		117	3000	a-b-c-a-b-c-a-b-c	10

(*) Decontamination treatments:

a- LOMI reagent : 1-1.8%, 70-120 min; >90°C, 4.5 pH;

b- oxidation : 0.4% HNO_3, 2h, 95°C, 2.5 pH;

c- oxalic acid : 0.4%, 3-5 min, 25°C

steam pipe samples, the oxide layer was completely removed after 3 cycles only; this could well be attributed to the essentially higher layer quantity for the primary water pipe (45.8 g/m^2) compared with the primary condensate pipe (10.8 g/m^2). It is noteworthy that the largest part of the activity was removed by the oxalic acid treatment but only a very small part of it was removed by the oxidation treatment.

2.3 Strong chemical decontamination

In order to reach a complete decontamination, a certain layer thickness has to be also removed from the base material surface of the primary circuit pipe samples. As no allowance should be made for the re-use of the components with respect to plants being shut-down, one could apply relatively concentrated mineral acids to solve the problems.

Pre-tests with corresponding acid concentrations, however, showed at the existence of an adhering oxide layer only an uncomplete attack with a strong formation of perforation. Due to these results the total decontamination course is basically divided into the stages: removal of the oxide layers (with the soft processes previously described) and then, removal of a determinated material layer thickness from the surface of the base material. After the first treatment stage, the material surface is only allowed to show insignificantly thin layer residues so that no perforation occurs with respect to the subsequent application of the acid mixture and a material removal as even as possible is achieved.

Preliminary tests on inactive samples with strong inorganic acids, showed that a marked surface removal of the austenitic metal plates only occured in two acid mixtures (HCl/HNO_3 and HCl with the addition H_2O_2).

The tests with the mordant 10% HCl + 1% H_2O_2 have not been continued in detail. The cause was the quick decay of the H_2O_2 and the very strong tendency to local attack. So perforation depths up to 150 μm already resulted after a pickling duration of 0.5 h.

A nitrochloric acid mixture (4.7 g/l HNO_3 + 13 g/l HCl) was finally chosen. In spite of the low acid concentration, treatment duration of 4 to 8 hours were sufficient to remove material thickness of 10–30 μm. Regarding the first studies of pipe samples, it was observed how a greyish green layer was formed on the metallic surface during the acid treatment. A complete removal of the layers was reached by an alkaline oxidation with a subsequent oxalic acid treatment.

Concerning the primary water pipe sample, it was necessary to apply a second total treatment course due to the essentially higher starting activity. The result of the condensate pipe sample shows that considerable quantities of activity are displaced by the directly successive acid treatment. The result of the primary steam pipe sample shows that one treatment cycle is sufficient for a not too high starting activity (Table II).

TABLE II - Results of strong decontamination tests carried out by KWL on Lingen samples already free of oxide.

Sample	Decontamination treatments (*)	Removed Material (μm)	Dissolved activity (Bq)	(Bq/mg)
Steam pipe	a-b-c	13.5	$3.4\cdot10^3$	
	Electrolytic	4.6	11	$7.9\cdot10^{-2}$
Condensate pipe	a	33.0	$2.7\cdot10^3$	
	a-b-c	9.0	137	
	Electrolytic	5.5	1.26	$1.4\cdot10^{-2}$
Water pipe	a-b-c	18.5	$6.9\cdot10^5$	
	a-b-c	11.0	109	
	Electrolytic	1.4	<0.15	$<1.8\cdot10^{-3}$

(*) Decontamination treatments:

a- chemical pickling : 4.7 g/l HNO_3+13g/l HCl, 60°C;

b- 2g/l $KMnO_4$+2 g/l NaOH, 1h, >95°C;

c- 2g/l oxalic acid, 1h, 90°C.

To conclude, the decontamination tests have shown that to reach the total decontamination, i.e. to allow the unrestricted release of materials, the use of specific multistep process is necessary.

2.4 Secondary wastes generation

The radioactive waste which is to be expected during application of the suggested decontamination multi-step procedure in a nuclear power plant can be very different. In fact, in the area of primary water, the first treatment cycle has to be repeated three times and the second treatment cycle twice. On the contrary, one treatment cycle each is sufficient for the steam and condensate area respectively.

As example fixing the following conditions: 2,000 m^2 material surface, 100 m^3 system volume and 30 μm necessary material removal, the following radioactive waste quantities can be expected: primary water area about 13 tons of solid materials and 1,600 m^3 of waste waters; steam and condensate area about 5.5 tons of solid materials and 700 m^3 of waste waters.

3. RESEARCHES CARRIED OUT BY ENEL

ENEL (Italian Electricity Generating Board) is the operators of the Garigliano plant (BWR dual of 150 MWe). Since March 1982, the plant is definitely out-of-service after about 15 years operation (first critical shutdown in June 1963 and last one in August 1978).

A research decontamination programme for decommissioning purposes was therefore started at the end of 1980. Research activities are developed by ENEL-CRTN (Nuclear and Thermic Research Centre) in close connection with ENEL's directives and with the technical co-operation of CISE of Milan.

The aim of the research activities was the identification, development and assessment of vigorous decontamination techniques. The work involved the design and construction of a special test loop (DECO loop) and experiments using contaminated samples from the Garigliano reactor.

3.1 Test materials

The contaminated test materials were taken from the Garigliano plant. The test pieces belonged to the primary cooling loop and were operated at high temperature (280°C) in oxidant conditions (oxygen level about 150 ppb).

In particular the following pieces have been taken and employed in the experiments: - parts of 24" pipe welded to the safe-end; some of them had already been decontamination in the plant by a commercial soft decontaminant and brushing; - parts of 2-3" tubes (by-pass lines, drainage lines, etc.); - separation plate of the channel head of the Secondary Steam Generator A(SSG/A). All these pieces are in AISI 304 type stainless steel. The dose rate, measured in March 1980, was found to range $5 \cdot 10^{-4}$ Sv/h (50 mR/h), for the predecontaminated pieces, to $4 \cdot 10^{-2}$ Sv/h (4 R/h) for the separation plate of SSG/A.

The contamination on the separation plate and on 24" pipe is greater than that on the 2" tube. The most significant radionuclide is 60-Co; the 137-Cs was present too. On the separation plate sample the presence of 22-Na, 54-Mn, 57-Co, 106-Rh and 144-Ce was also evaluated. Alfa contamination was monitored too; it could be connected with some fuel failures seen in the Garigliano plant. The presence of 239+240-Pu, 238-Pu, 241-Am and 244-Cm was verified.

Metallographic investigations of contaminated specimens, carried out in the hot laboratories of Ispra JRC-LMA (8), show that morphology and thickness of oxide were markedly different in the examined component parts. Nevertheless, the oxide composition was found to be quite uniform and it corresponds very closely to that of a Nickel-Ferrite spinel.

3.2 Test reagents

The choice of a decontamination reagent is generally affected by the following considerations: the aggressiveness towards the contaminated oxide to be removed; the compatibility with all the base materials which constitute the systems needing decontamination and the need to treat the secondary wastes.

For the experiments indicated in this report, concerning decommissioning, the aggressiveness of the reagent was mainly considered. Consequently strong mineral acid solutions were selected. Such solutions are similar to those used in stainless steel pickling and were first calculated to have an ionic strength of 5 m. Later, less concentrated solutions in order to evaluate the minimum concentrations which give high decontamination factors in a reasonable time have been used.

3.3 Experimental facilities

The experimental device for static tests is an electrochemical cell put in a Teflon beaker, with a volume of about 1 l of test solution.

The dynamic runs have been carried out in the DECO experimental loop located in the ESSOR complex at Ispra JRC (Fig. 2). It is composed of a primary decontamination loop including a centrifugal recirculating pump, a heat exchanger with a bundle in inert material, an expansion tank and a test section containing the decontamination material. Moreover, a feedwater pipeline, a blowdown waste pipeline with liquid radwaste pre-treatment section, and an oil loop (secondary) to control the test temperature are included. The primary loop is assembled in a special safe containment glove-box. DECO loop is designed in order to ensure that the temperature of 100°C is maintained in the test section. The other maximum operating parameters are : pressure : 4 bar; flowrate : 24 m^3/h, content 15 l. The containment materials in contact with test reagents at temperatures above 35°C are in PTFE (Teflon).

In order to measure the corrosion potential and the corrosion rate during the runs, a special electrochemical test section, designed by CISE, was adopted.

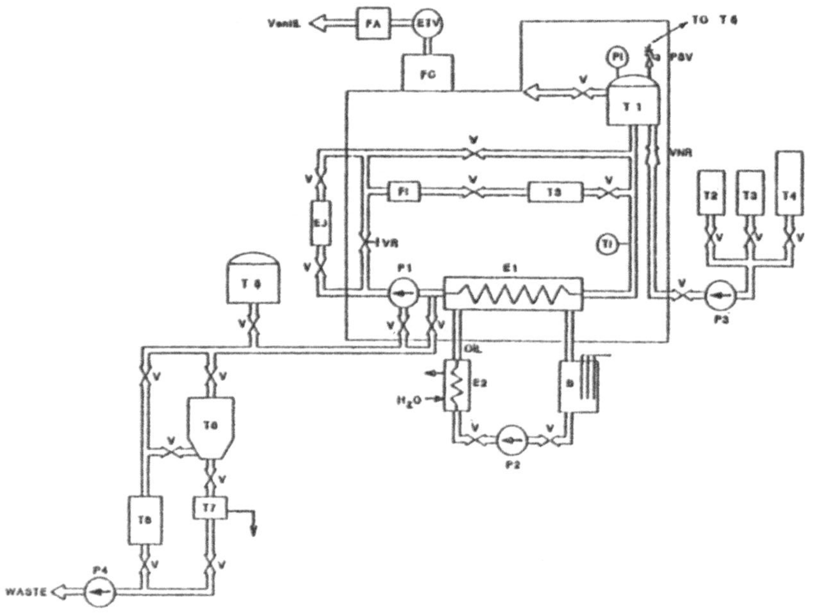

Figure 2 - Flow-sheet of the DECO loop

P1 Primary pump
P2 Feed exchanger pump
P3 Feed pump
P4 Waste pump
T1 Expansion tank
T2 Reagent tank
T3 Feed reagent tank
T4 Feed water tank
T5 Safety tank
T6 Neut. and floc. tank
T7 Gravity waste tank
T8 Gravity liquid waste tank
E1 Primary heat exchanger
E2 Secondary exchanger
B Boiler
TS Test section
EJ Jon exchanger
V Valve
VR Regulation valve
VNR Check valve
PSV Relief valve
ETV Centrifugal valve
FC Chemical filter
FA Absolute filter
PI Pressure indicator
FI Flow indicator
TI Temperature indicator

3.4 Test results and discussion

The results of carried out decontamination tests are reported in Table III, and for $HF+HNO_3$ and HCl solutions, are summarized in Figure 3.

TABLE III - Results of some hard chemical decontamination tests carried out by ENEL on Garigliano samples.

Test	Sample (*)	Reagent (%vol)	T (°C)	time (h)	Flow(°) (m/s)	DF(60-Co)	%dissolved activity	corrosion (mg/dm^2)
1	a	$1.5HF+15HNO_3$	79	0.67	st.	785	—	—
2	a	$1.5HF+15HNO_3$	28	4.17	st.	170	—	—
3	a	$4HCl+13HNO_3$	79	0.67	st.	76	—	—
4	a	$4HCl+13HNO_3$	31	2.00	st.	145	—	—
5	a	13.4HCl	75	0.50	st.	263	—	—
6	a	13.4HCl	33	2.00	st.	114	—	—
13	b	$1.5HF+15HNO_3$	88	0.50	st.	2420	31	920
16	b	13.4HCl	97	1.17	st.	3860	93	1117
18	b	$4HCl+13HNO_3$	98	0.28	st.	1.2	91	19
24	b	$4HCl+13HNO_3$	83	0.42	st.	1.1	97	—
25	b	$0.5HF+5HNO_3$	84	1.67	st.	2280	20	1120
26	b	4.5HCl	91	0.28	st.	537	80	137
28	b	$1.5HF+5HNO_3$	80	0.28	st.	2250	18	785
29	b	0.75HCl	95	0.53	st.	583	79	78
31	c	$1.5HF+5HNO_3$	80	0.38	st.	1530	19	1700
32	c	0.75HCl	90	0.73	st.	251	89	100
34	c	$1.5HF+5HNO_3$+Rod(°°)	84	0.65	st.	929	48	2075
35	c	1+1HCl+Rod(°°)	85	17.65	st.	1200	82	440
37	c	$4HCl+13HNO_3$	85	0.35	st.	1.7	—	—
39	b	$0.35HF+5HNO_3$	41	19.90	st.	4320	27	3182
40	b	0.85HCl	42	71.67	st.	2430	56	615
001	b	$1.5HF+5HNO_3$	80	0.53	1	7500	—	—
005	b	$0.35HF+5HNO_3$	40+80	9+5	1-4	3670	—	—
006	b	4.1HCl	40	4.83	4	2610	—	720
008	b	0.85HCl	40	11.75	4	2.5	—	344
009	b	$4HCl+2.5HNO_3$	40	2.10	4	5750	—	2069
010	b	$0.75HF+2.5HNO_3$	40+80	5.5+2	4	2360	—	—

(*) a- 24 in. primary water pipe near safe-end (after predecontamination);
b- 2 in. primary water tubes;
c- separation plate of the waterbox of SSG/A.

(°) st = static test.

(°°) Rod = Rodine 31/A; special inhibitor.

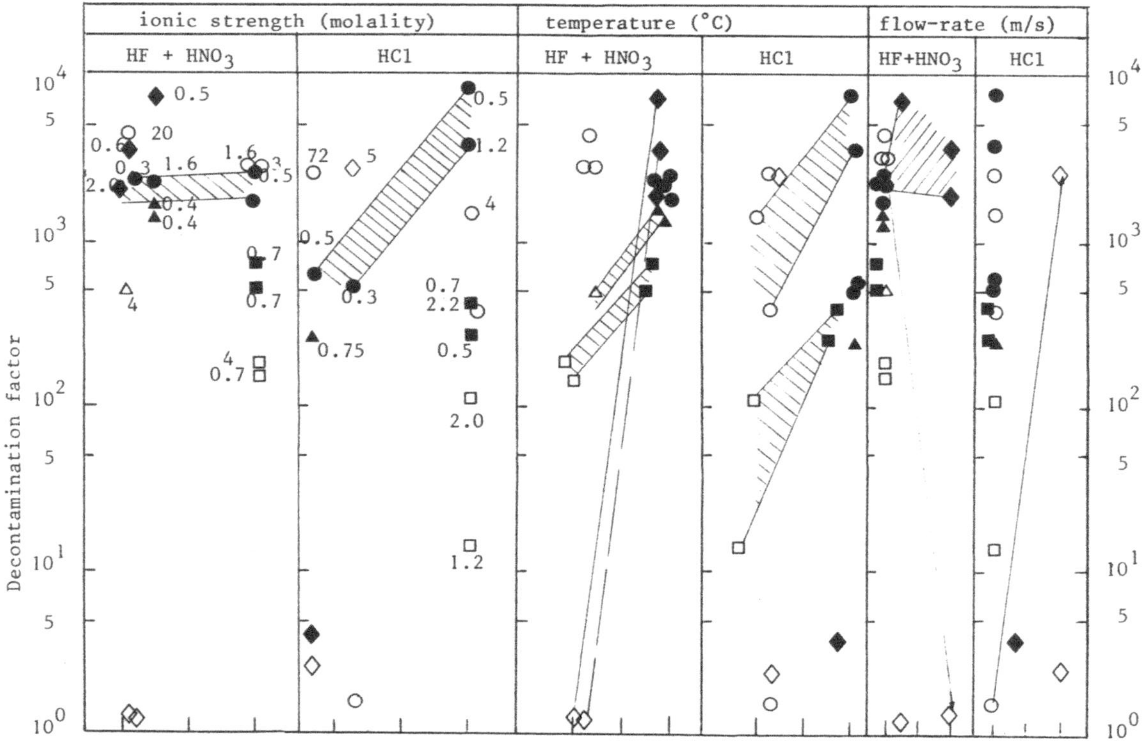

Figure 3 - DFs as a function of ionic strength, temperature and flowrate (symbols - circle : 24" pipe; square : 2" tube; triangle : SGG Plate; rhomb : DECO loop runs).

Decontamination factors. Considering the 60-Co radioactivity, the DFs are clearly in two different fields :

- DF > 200 when the oxide layer is totally removed aand the final surface of the specimen is shiny. DFs greater than 1000 have been generally obtained and residual radioactivity is in accordance with the unrestricted release limits;
- DF > 2 when the oxide is not removed and the final surface of the specimen is similar to the initial one. The low DFs have generally been obtained using unproper solutions (4 % vol HCl + 13 % vol NHO_3), solutions at very low concentration, or organic reagents (citric, oxalic).

The DFs concerning the gross-alpha radioactivity are in quite good agreement with those of the 60-Co radioactivity, even if one-to-one correspondence is not shown. Nevertheless, in the cases of low DFs, the gross-alpha DFs are always greater than the 60-Co DFs and this means that the alpha contamination is mainly present in the outer layers of the contaminated oxide.

To conclude the considerations about the DFs, the following solutions are considered the best and can effect the total detachment of the contaminated oxide and very high DFs:

- hydrochloric acid: 4-5% vol at low temperature (30°C), in about 5-6 hrs; 0.7-1% vol at high temperature (80°C), in about 4 hrs;
- hydrofluoric + nitric acid: 1.0-1.5% vol HF+2.5-5% vol HNO_3 at low temperature (30°C), in about 4-6 hrs; 0.3-0.5% vol HF+2.5-5% vol HNO_3 at high temperature (80°C), in about 5 hrs.

Concerning the flowrate it appears to need only a good recirculation of the solution and a high flowrate is not necessary.

Radioactivity trends and balances. The on-line radioactivity trends have shown different oxide dissolution behaviour depending on the test solutions. In fact, in the HF+HNO_3 solutions the detachment of the oxidelayer, together with a quick decrease of the radioactivity, was clearly observed, while in the HCl solutions a more homogeneous decontamination was observed.

The times corresponding to the oxide detachment in the HF+HNO_3 solutions and the decontamination slope in the HCl solutions, are influenced by the ionic strength of the solution (Fig. 4) and by the test temperature (Fig. 5).

The data of the radioactivity balances shown that during decontamination with HCl solutions, the oxide layer is mainly dissolved, therefore the radioactivity present as ionic phase is greater than the radioactivity fixed onto the filter: 50÷55% of the radioactivity is dissolved at low

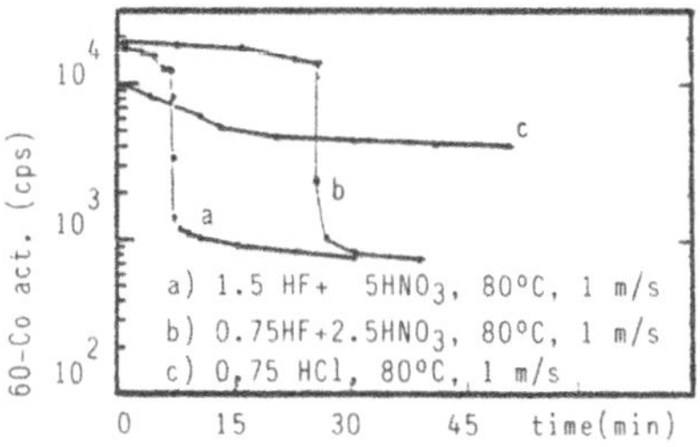

Figure 4 - Evolution of radioactivity with time during loop runs.

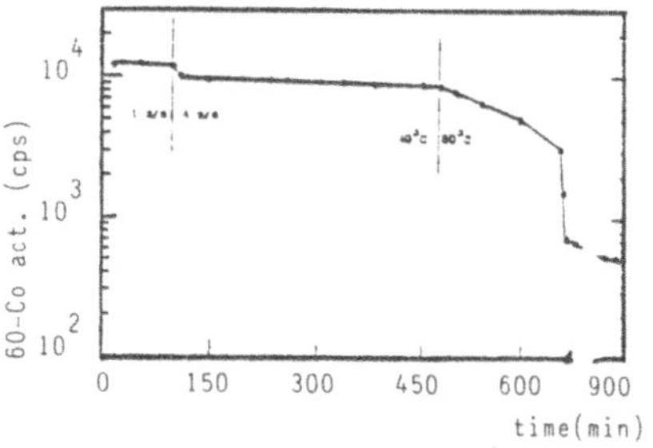

Figure 5 - Evolution of radioactivity with time during loop run.

temperature and 75÷80% at high temperature). On the contrary during decontamination with $HF+HNO_3$ solutions, the oxide is mainly taken away as particulate, (radioactivity fixed onto the filter is about 75÷80%).

Effect of ultrasound. During some static tests, ultrasonic treatments, 2 min. long, have been applied. The results showed the effectiveness of the mechanical action of ultrasound subsequent to the chemical action of reagents.

In any case the mechanical action of ultrasound appears to decrease the decontamination time, although the final DFs do not seem to be significantly affected.

Corrosion potentials. The trends experimentally recorded both during static tests and AISI-304 stainless steel corrosion experiences are reported in Figure 6. The behaviour suggests that during the first process step the potential assumes a value ranging between the equilibrium potential for the oxide dissolution and that of the metal dissolution. Practically, it is a mixed potential referring to the reductive reaction:

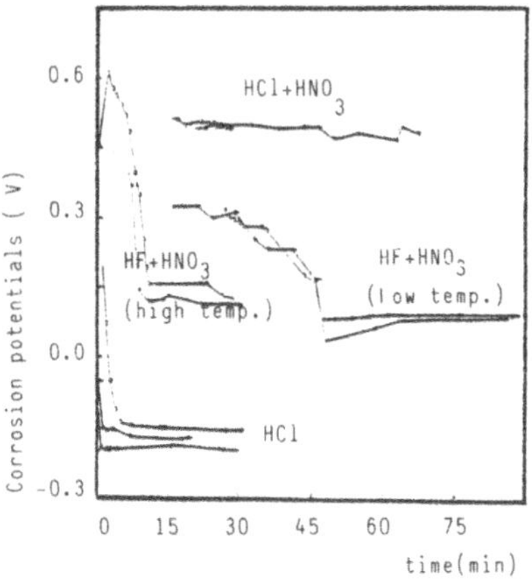

Figure 6 - Evolution of corrosion potentials with time (9).

$$Fe_2O_3+2\bar{e}+6H^+ \rightleftarrows 2Fe^{2+}+3H_2O$$

When the stainless steel corrosion potential is reached, the specimen undergoes active dissolution.

Potentiodynamic polarization curves. These curves had taken a significative part in the experiences; the aim was the identification of the base aspects connected with the chemical decontamination processes. Some controversial questions have been understood by means of them. In particular the non-efficiency of the 4% HCl+13% HNO_3 solution has been explained with the intersection of the cathodic and anodic curves in the passivity zone, probably due to the high ratio between HNO_3 and HCl.

The effects of the flowrate and temperature have been reviewed with the hydrodynamic laws and it was concluded that they can influence the intersection between cathodic and anodic curves causing different behaviours(9).

Dissolution/corrosion rates. The trends monitored on-line by the linear polarization method appear to show that the attack mechanism is at first an oxide reduction followed by an active corrosion of the base metal.

In any case the polarization resistance can be a good sign of the radioactivity removal. This is clearly shown by the example reported in Figure 7, where the 60-Co radioactivity behaviour is reported together with the linear polarization resistance performed in a run on the DECO loop. During the first step, when the 1/Rp is low, because of the presence of a compact oxide layer, the activity progressively decrease; however, during the second step the corrosion rate becomes equal to that of the stainless steel and simultaneously the activity falls to background values.

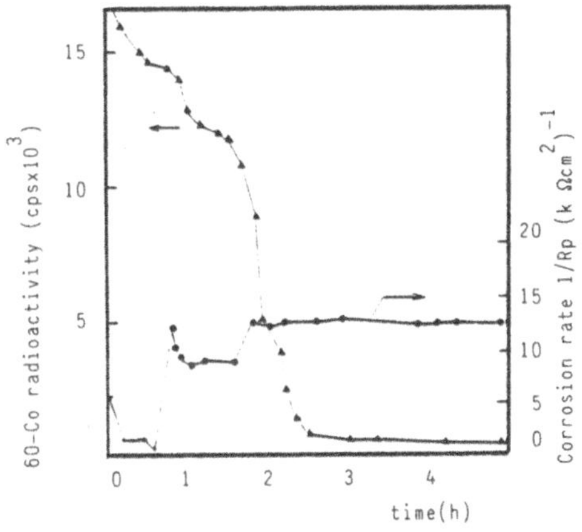

Figure 7 - Evolution of radioactivity and 1/Rp with time.

3.5 Spent decontaminant treatment

The results of preliminary tests performed by metallic precipitation with hydrate agents gave the following indications: when decontamination reagents are mineral acids, a good decontamination of solutions, after particulate removal, can be obtained with ionic precipitation with a proper hydrate agent (sodium hydrate was found to be more effective than ammonia); in this case 60-Co which represents almost the total activity, appears to be dragged down with Fe; DFs greater than 100 have been usually obtained. When decontamination reagents are organic acids, such as oxalic or citric, the hydratation and precipitation-flocculation with sodium hydroxide was not effective; DFs ranging from 2 to 5 were measured. In this case treatment on ion-exchange resins appears to be the only way; DFs of about 100 were obtained in this case.

Extrapolating the data, one finds a production of about 0.5-3 kg of dry wastes in the strong decontamination of a piece of area 10 m^2 and consequently the production of 1.6-10 kg of cement-immobilized waste form.

4. RESEARCHES CARRIED OUT BY KGB

The KGB (Kernkraftwerke Gundremmingen Betriebsgesellshaft GmbH) is the operator of the nuclear power plant Gundremmingen. It is a foundation of Kernkraftwerk RWE and Bayernwerke.

The unit A of plant, is a BWR (250 MWe) of the first generation which has been running for approximately 11 years from 1966 to the beginning of 1977. It had been shutdown for an incident which happened in January 1977 after a short circuit in the grid, involving subsequential damages to the plant. In 1980 the decision of decommissioning the plant was taken and the necessary planning work initiated. In particular, for the purpose of saving space, the plant components contained in the turbine building should be dismantled and to a large extend free-decontaminated (10).

A research activity was therefore started. The main objective was to identify, develop and assess the most suitable and economic decontamination procedures for decommissioning purposes with particular reference to BWRs.

4.1 Test materials

All contaminated test materials were taken from the KRB unit A reactor plant. Samples of horizontal ferritic steel pipes of 500 mm diameter belonging to the condensate duct and to the steam duct of primary circuit were employed for experimental tests and examined.

On the condensate pipe, an inner surface oxide layer of 100 mg/cm^2 was found; the contamination increased from the top (0.23 kBq/cm^2) to the bottom (0.73 kBq/cm^2) of the pipe; the contribution of individual radionuclides to the total contamination varied with the location within the following ranges: 60-Co: 16 to 55%; 134-Cs: 3 to 6%; 137-Cs: 41 to 77%. On the steam pipe, an oxide layer of about 1.6 mg/cm^2 and contamination of 17 kBq/cm^2 were measured, except for a 10 cm large streak at the bottom of the pipe, where a thicker oxide layer (22 mg/cm^2) and higher contamination (60 kBq/cm^2) were found; this streak is probably due to the presence, during operation of the system, of a rill of water condensed from the saturated steam; 60-Co accounts for 99% of the contamination of the steam pipe.

4.2 Chemical decontamination tests

In preliminary screening experiments, 20 organic and inorganic acids have been tested on 4 cm x 4 cm samples from the condensate duct. Tartaric acid and citric acid were found to be comparable to the best inorganic acids in decontamination efficiency and have, because of their unproblematic application, been selected for further studies.

The tests have shown that no further decontamination effect is achieved with initial solution after 1h of treatment. A concentration of at least 5% acids is needed to obtain significant decontamination rates. With an oxidation pretreatment (APAC type) and two consecutive treatments (3h, 92°C) with 5% tartaric acid, condensate samples have been decontaminated from initial levels up to 10^3 Bq/cm^2 to residual levels of 2 Bq/cm^2 or less.

A typical course of activity reduction for soft chemical decontamination methods is shown in Figure 8. The activity reduction is stagnating after a short time. The decontaminant-solution has to be renewed; therefore, the usual two-step chemical decontamination procedures and attack by aggressive acids would require, at least on primary loop, pipes below the limit for unrestricted release. Nevertheless this procedure generates a large amount of secondary waste.

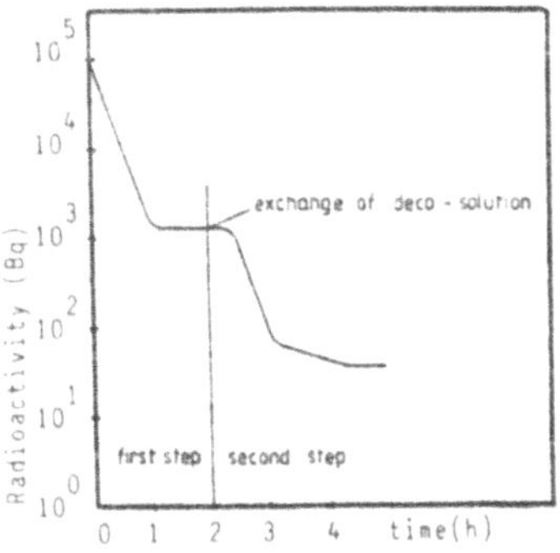

Figure 8 - Radioactivity reduction during chemical decontamination.

Among the inorganic acids tested the best results have been obtained with HCl and H_2SO_4 acid with some mixture.

4.3 Electrochemical decontamination

In the last years electrochemical decontamination techniques have been investigated and described (11,12). The experiences of conventional electropolishing can be adapted. A standard decontamination facility is shown in Figure 9. The piece is the anode, the tank, or specific Cu piece, is the cathode. Phosphoric acid (70%) was chosen as electrolyte.

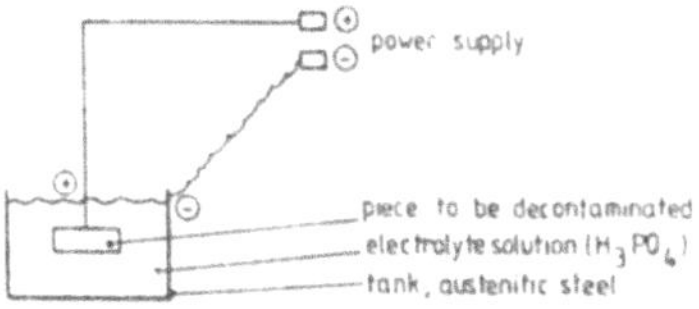

Figure 9 - Electrodecontamination equipment.

Precursing laboratory tests were carried out to optimize current densities, treatment time and secondary waste handling. At voltages between 5-20 V and current densities of 0.1-0.5 A/cm^2, an effective decontamination could be reached.

This reference method did not in all cases achieve complete oxide removal. Pre-treatment with commercial detergents proved to be very effective on ferritic steel samples from the condensate duct, removing 80-90% of 60-Co and 30-40% of Cs. A subsequent electrolytic treatment with 23.4 A/dm^2 removed almost all remaining contamination.

Stirring the solution vigorously did not significantly improve the decontamination. Reducing the phosphoric acid concentration from 70% to 40% resulted in preferred attack of surfaces where oxide layers were thin, and in higher residual contamination. A mixture of concentrated sulphuric acid and 70% phosphoric acid brought no improvement, even when vigorously stirred.

Tests with and without detergent pre-treatment were repeated on samples from the primary steam duct. Electropolishing is very efficient on this material, which has a higher chromium content. On these samples, the total decontamination factor was slightly increased by the pre-treatment.

Analyses of the electrolyte showed a steady increase of the metal content. No reduction of efficiency was observed with iron contents up to 23 g/l.

Figure 10 shows the decontamination effect on specimens of the condensate line at different current intensities. It is demonstrated that the decontamination velocity and current intensity take a parallel course.

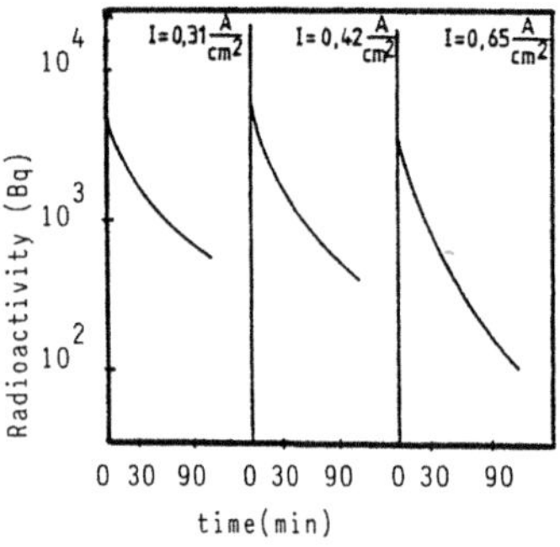

Figure 10 - Decontamination at different intensities of current.

137-Cs proved to be harder removable than 60-Co as shown in Table IV. This observation is in contrast to the known solubility of Cs and can be explained by the activity build up behaviour during plant operation.

Due to fuel failures during the start-up phase high 137-Cs concentration were included into the first oxide layer. In the following years of operation 60-Co was deposited on this first layer. The decontamination shows that after 90 min of electrochemical treatment the surface is about totally clean except some slight 137-Cs containing oxide spots, that can be removed mechanically. This phenomenon verifies that the Cs had not penetrated into the base material.

TABLE IV - DFs and current densities in electrochemical tests carried out by KGB on Gundremmingen samples.

Current densities	Decontamination factors		
(A/cm^2)	Total	137-Cs	60-Co
0.32	8	4	74
0.42	11	6	54
0.65	22	13	77

The iron load of the electrolyte bath seems to have no impact on effectiveness of decontamination. When saturation of iron concentration was reached, the additional iron input due to the electrochemical operation precipitated as iron phosphate sludge. As the radioactivity coprecipitated with the sludge, the decontamination of the solution did not exceed some 10^3 Bq/g. The sludge yelded 10^4-10^5 Bq/g.

For waste disposal the solution is neutralized and at pH 8 the iron phosphate and the absorbing radioactivity precipitate and are filtered off. Additional flocculation precipitation can be applied if necessary.

4.4 Large scale electropolishing

On a technical scale many tons (≃ 100) of material have been treated up to now following the above described procedure. The steel treated was of the type St. 37 and Cr Mo 44. Two tanks of 3000 l volume each and a rectifier plant of 6000 A are used.

The expected decontamination factors could be reached after 2 hours of treatment with current densities of 0.1-0.4 A/cm^2 without dumping the solution. An abrasion of the surface of 100-500 μm took place.

Regarding secondary waste evaluation for 100 tons of metal treated, 2 to 4 tons of contaminated iron-phosphate will be produced.

In particular two components of the KRB-A reactor have been treated to perform the required demonstration tests. The feed water pump of 15.6 tons weight, contaminated to 11 Bq/cm^2, treated for unrestricted release (less than 0.37 Bq/cm^2) in 200 hours. The primary steam valves of the turbine bypass system with a weight of 8.2 tons needed 108 hours to lower the measured $1.4 \cdot 10^3$ Bq/cm^2 to unrestricted release values.

5. RESEARCH CARRIED OUT BY CEA

The CEA-CEN Cadarache (Commissariat à l'énergie atomique - Centre d'étude nucléaire) has been working for a long time in the field of decontamination, also for decommissioning purposes, in the frame of the French nuclear programmes.

The aim of this research was to develop highly effective methods for steel decontamination for decommissioning purposes.

Chemical decontamination methods have been studied with the aim to provide very active scouring baths enabling the contaminated surface to be laid bare, without fears of corrosion damage. In order to limit the concentration of employed chemicals, the chemical action has been accelerated or amplified by electrolytic action.

The choice of a decontamination process for decommissioning purposes of LWRs depends mainly from the nature and superficial features of the base metal, from the amount and characteristics of the radioactivity, from the physico-chemical features of the corrosion product deposite and, finally, from the amount and kind of the secondary wastes produced during decontamination.

5.1 Test materials

The decontamination experiments have been carried out on representative specimens of the primary loop of BWRs and PWRs.

In particular for the tests on BWRs, specimens belonging to a safety loop of the feedwater system of German ISAR reactor have been used. The base material was ferritic steel. This alloyed steel (DIN 17 MnMo V 64) showed, on the exposed surfaces, hematite layers by steel corrosion in oxidizing environment (five years operation time) (13). The initial contamination ranges from 0.1 to 1 $\mu Ci/cm^2$, the main emitters found are 60-Co at about 60% and 54-Mn at about 40%. The dose rate ranges from 20 to 100 mrad/h.

For the experiments on PWRs, specimens coming from the auxiliary piping of the primary loop of CAP (Chaudière Avancée Prototype) at Cadarache, have been used. The base material was stainless steel. The exposed surface has been found to be covered by an oxide layer mainly composed of chromium and nickel-ferrite. The initial contamination is about 4 $\mu Ci/cm^2$ mainly constituted of 60-Co (more than 99%).

5.2 Test reagents

Decontamination tests have been performed using various bath: acid, alkaline, oxidizing, oxidizing-alkaline bath, etc.

Preliminary test on non-active specimens have been performed in order to evaluate: the aggressivity of the baths by measuring the density of the corrosion current, the chemical erosion, the electrochemical actions, etc.

The radiochemistry purification of the effluents, taking into account the toxicity of the effluents and the amount of solid waste produced, has been a fundamental stage in the choice of the best decontamination reagents.

5.3 Decontamination of BWR

Chemical decontamination. The results of the decontamination tests carried out in beakers, with a simple stirrer or with ultrasounds, are reported in Table V.

TABLE V - Results of chemical decontamination tests carried out by CEA on BWR ferritic steel.

Reagent/Conc. (moli/l)	T °C)	Erosion rate (μm/h)	Ultra-sound	DFs after 1-4 hours 1	2	3	4
2 H_2SO_4	25	0.26	NO	9	45	47	50
2 H_2SO_4	35	2.9	YES	1.9	19.8	>860	-
2 HF	25	1	NO	25	60	80	90
2 HF	50	1.85	YES	19.8	>935	-	-
2 HCl+0.5HF	25	0.75	NO	2	11	120	>120
2 HCl+0.5HF	30	5.8	YES	2	162	>748	-
2 HCl+0.1H_2O_2	25	13	NO	1.4	2.1	25	50
2 HCl+0.1H_2O_2	40	38	YES	3.1	178	>785	-
2 HCOOH+0.1Formol	25	0.17	NO	2.1	7.8	19.5	30
2 HCOOH+0.1Formol	30	0.8	YES	2.5	3.2	>1122	-

The greatest DFs have been obtained with strong acids as HF or HCl, with or without an oxidizing medium as H_2O_2. Lower DFs have been obtained with H_2SO_4 and with formic acid plus as reductant agent as "formol"(14).

The application of ultrasounds caused a significant increase in the final DFs and a residual activity lower than $1 \cdot 10^{-4}$ $\mu Ci/cm^2$ was generally obtained on simple specimens.

Electrochemical decontamination. The results of the tests are reported in Table VI. The specimen to be decontaminated can be placed either as cathode or as anode.

The lowest residual radioactivity has been obtained with a two-step process: an electrochemical reduction in order to fragilize and detach the oxide layer, followed by an oxidation that removes the residual contamination by the metal erosion-dissolution.

Spent decontamination treatment. The radiochemical treatment of the spent sulphuric decontaminat is possible by iron precipitation with sodium hydroxide, with pH ranging from 10 to 11 as shown in Figure 11.

TABLE VI - Results of electrochemical decontamination tests carried out by CEA on BWR ferritic steel (current density 100 mA/cm^2).

Reagent/Conc. (moli/l)	Specimen (polarity)	Erosion rate (μm/h)	DFs after 10-40 minutes 10	20	30	40
2 H_2SO_4	- cathode	0.4	83	115		
	+ anode	60			203	203
0.16 $H_2C_2O_4$	- cathode	<0.1	1.6	1.7		
	+ anode	21			3.9	4.6
1 HCOOH	- cathode	0.1	43	65		
	+ anode	85			92	229

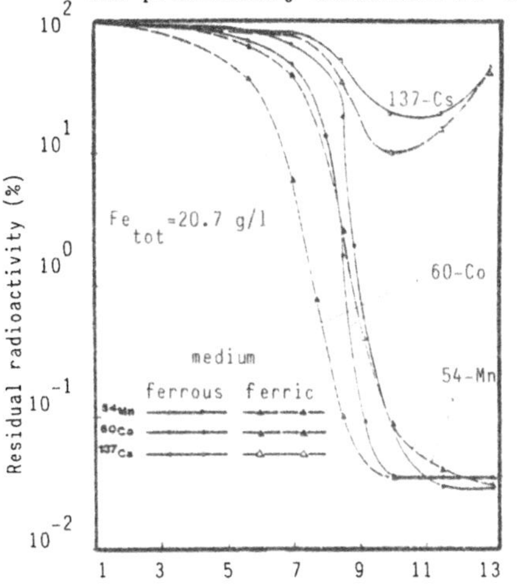

Figure 11 - Effect of the pH in radionuclides coprecipitation.

The preliminary oxidation of the Fe^{++} by H_2O_2 does not improve significantly the results. On the contrary, as shown in Table VII, the preliminary oxidation of the formic spent acid before the neutralization, allows to decrease the amount of the secondary wastes produced.

5.4 Decontamination of PWR

Chemical decontamination. The oxide formed in reductive conditions, as in PWR, can be removed only with a two-step process (15).

The first treatment works for the fragilization of the oxide (16). It is particularly effective with an alkaline permangate medium at high temperature. The use of an oxygen-ozone medium (17), being so the situation that produces the minor secondary wastes, has a low efficiency and could require a very aggressive second step.

The second step is a chemical erosion of the base metal in acid medium. The best results have been obtained with the fluonitric mixture with ultrasounds as shown in Table VIII.

Electrochemical decontamination. Among the checked reagents, only the oxalic acid allows a total decontamination by electrochemical way as shown in Table IX. The piece to be decontaminated is always placed as anode.

Spent decontamination treatment. The treatment of the spent decondaminant (18) is possible by neutralization with calcium hydroxide as shown in Figure 12. In fact it allows to lower both the F^- concentration and the residual radioactivity under acceptable limits (10 mg/l and 15 $\mu Ci/m^3$ for γtot respectively) as shown in Table VII.

The treatment of the spent oxalic electrolitic decontaminant is possible by neutralization with calcium hydroxide too. A preliminary electro-

TABLE VII - Results of tests concerning the spent decontaminant treatment carried out by CEA

Spent decontaminant (moli/l)	Treatment	Neutr.	pH	Dried products (g/l)	Radioactivity($\mu Ci/m^3$) 54-Mn	60-Co	137-Cs	F^- (mg/l)
1 HCOOH	As-received		2.4		35.1	67.5	8.1	
	Neutralization	NaOH	12	10.5	<2.7	< 5.4	< 8.1	
	Elect. ox.+neutr.	NaOH	12	5.1	< 2.7	< 5.4	< 8.1	
0.16 $C_2O_4H_2$	As-received		1.3		297	10945	40.5	
	Neutralization	$Ca(OH)_2$	12.7	28.5	< 5.4	40.5	8.1	
	Elect. ox.+neutr.	$Ca(OH)_2$	11.6	3	< 5.4	21.6	< 8.1	
2 HNO_3+0.5 HF	As-received				154	12700	48.6	1000
	Neutralization	NaOH + $Ca(OH)_2$	2.1 / 9	62	< 4	< 4	< 5.4	6.8
	Neutralization	NaOH + 3.5 g/lof $CaCl_2$ + $Ca(OH)_2$	2.1 / 9	57	< 4	< 4	< 5.4	7.1

TABLE VIII - Results of chemical decontamination tests carried out by CEA on PWR samples

Pre-treat. (*)	Reagent(moli/l) HF + HNO_3	T (°C)	Erosion rate (μm/h)	Ultra-sound	DFs after 1-8 hours 1	2	4	8
a	0.05 + 2	25	—	NO	—	530		
a	0.05 + 2	85	—	NO	900			
b	0.05 + 2	25	0.11	NO	—	140	240	400
b	0.05 + 2	85	0.99	NO	230			
a	0.5 + 2	25	—	NO	—	1500		
a	0.5 + 2	85	—	NO	10000			
b	0.5 + 2	25	0.3	NO	—	630	860	
b	0.5 + 2	85	6.07	NO	2000			
a	0.05 + 2	29-49	—	YES	(**)			
b	0.5 + 2	25	0.10	YES	660			

(*)Pre-treatment a : 0.5% $KMnO_4$ + 2.5 ml/l NaOH ; 1 h ; 85°C ;
b : oxygen ozone 2.5% (100 l/h); 7 h ; 25°C .

(**) DFs = 470 after 0.03 h; 780 after 0.08 h; 1200 after 0.16 h and 4500 after 0.5 h.

TABLE IX - Results of electrochemical decontamination tests carried out by CEA on PWR samples (current density 100 mA/cm^2).

Reagent/Conc.	Specimen (polarity)	Erosion rate ($\mu m/h$)	DFs after 2-60 minutes				
			2	5	10	30	60
2 HCOOH	+ anode	1.6			1	50	150
0.16 $H_2C_2O_4$	+ anode	11.6	12	36	890		

litic oxidation allows to lower the calcium hydroxide need and consequently the amount of the primary wastes generated, without change in the total effectiveness, as shown in Table VII.

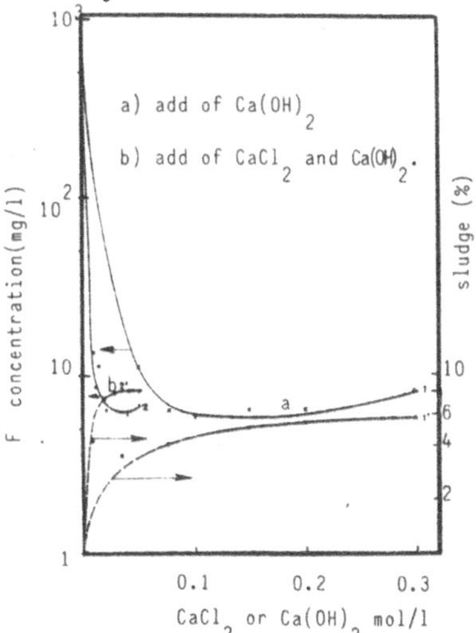

Figure 12 - Elimination of F^- ions in spent decontaminants.

6. CONCLUSIONS

The R&D activities, carried out on the chemical and electrochemical decontamination of operated materials coming from nuclear power plants, show that the total decontamination of the contaminated surfaces appears to be possible. Moreover, the researches have led a growing sensitivity towards the decontamination problems and an improvement of the basic know-how which is necessary to start extended decontamination actions in the framework of the decommissioning programmes.

In some cases the reached DFs are sufficient for obtaining residual activity levels in compliance with the unrestricted release limits, where these limits are existing.

Taking into account the differences existing in the investigated nuclear power plants, and also in a plant itself, a single chemical and/or electrochemical decontamination process cannot be generalized. In particular the chemical and radiometric features of the contamination which covers the surfaces of the cooling and auxiliary loops affect strongly the choice of the decontamination process.

The choice of the best decontamination procedure, supported by experimental investigations, cannot disregard the economical considerations and the amount and characteristics of secondary wastes. It is important to note that the problem of the spent decontaminant treatment appears to need further investigation.

The electrochemical decontamination has already been extensively employed with good results in the decontamination of components and lines of the turbine building of the KRB-A Gundremmingen reactor; nevertheless it does not appear to be suitable for the total decontamination of some large or very small components i.e. heat-exchangers, condenser, etc. In these cases the strong chemical decontamination seems to be the best way. In fact,the experimental tests with inorganic acids have shown very good DFs. Nevertheless these processes have been extensively checked only in laboratories and need a confirmation in real-scale applications. The use of a soft pretreatment (i.e. with LOMI), in order to remove the contamination oxide layers, followed by a treatment with strong mineral solution, in order to remove some metal layers, has proved to be another interesting chemical procedure.

REFERENCES

1. **NELSON, J.L. and DIVINE, J.R.** (1981). Decontamination Processes for Restorative Operations and as a Precursor to Decommissioning: A Literature Review. NUREG/CR-1915; PNL 3706.
2. **OSTERHOUT, M.M. ed.** (1980). Decontamination and Decommissioning of Nuclear Facilities. Plenum Press, New York.
3. **AYRES, J.A.** (1970). Decontamination of Nuclear Reactors and Equipment. The Ronald Press Company. New York.
4. **LANZA, F. et al.** (1979). Contamination Mechanisms and Decontamination Techniques in LWRs. EUR 6422 EN.
5. **MANION, W.J. and LAGUARDIA, T.S.** (1981). Decommissioning Handbook. DOE/EV/10128/1.
6. **BRADBURY, D. et al.** (1982). Decontamination of Systems of BWRs and PWRs Based on LOMI Reagents. Int. Conf. on Decontamination of Nuclear Facilities, ANS-CNA, Niagara Falls, September 19-22.
7. **AGOSTINELLI, A et al.** (1982). Vigorous Decontamination Tests of Steel Samples. Int. Decommissioning Sym. Seattle, W, USA, October 10-14.
8. **BREGANI, F. et al.** (1983). Metallographic Characterization of Contaminated Oxide Films of the BWR Garigliano Nuclear Poser Plant: Relation to Decontamination, JRC-Tech. Note N° 1.06.00.83.101.
9. **PASCALI, R. et al.** (1983). Hard Chemical Decontamination of Parts of the Primary Loop of Serviced Nuclear Plants. 3rd Int. Conf. on Water Chemistry of Nuclear Reactor Systems. Bournemouth, UK, October 17-21.
10. **WATZEL, G.V.P. and STANG, G.** (1982). Decommissioning of Unit A of the Nuclear Power Plant Gundremmingen and its Conversion to a Condition of Protective Storage. Ibid ref. 7.
11. **EICKELPASCH, N. and LASCH, M.** (1983). Electrochemical Decontamination Experience at Gundremmingen Power Plant. Ibid ref. 9.
12. **ALLEN, R.P. et al.** (1978). Electropolishing as a Decntamination Process: Progress and Application; PNL-SA-6858.
13. **JOHSON, A.B. et al.** (1980). Nature of Deposits on BWR and PWR Primary System Surfaces. Relation to Decontamination. 2nd Int. Conf. on Water Chemistry of Nuclear Reactor Systems. Bournemouth, UK, October 13-17.
14. **SCHENKER, E et al.** (1982). A New one step Decon. Proc. Ibid ref. 6.
15. **DARRAS, R.** (1980). Chimie et radioactivation des circuits primaires des centrales électronucléaires à eau sous pression. DCAES/SCECF-80/37.
16. **PANZER, G.P. and ZEGGER, J.L.** (1960). US Pat. 3.013.909.
17. **AECL** (1980). Procédé de décontamination des réacteurs. N° 80080.10.
18. **SHIMOMURA, S.** (1976). Brévet Japonais N° 51.10658.

DECONTAMINATION OF CONCRETE, WITH PARTICULAR REFERENCE TO FLAME SCARFING

W. EBELING, B. BOEDEKER*, K. ROSE* and K.H. SCHALLER**
Salzitter AG
* Gg. Noell GmbH
** Commission of the European Communities

SUMMARY

When decommissioned nuclear power stations are demolished, the contaminated surface layers must be removed, if it is not intended to dismantle and store the entire structure. The economies of this operation are important since the contaminated areas can be very large.
The aim of this report is to give an outline of existing methods of concrete decontamination. Both stripping and non-stripping methods are described, though the latter are relatively unimportant for the purposes of demolition.
Particular reference is made to flame scarfing, which has been tested by Salzgitter AG/Noell as part of the Community research programme "Decommissioning of nuclear power plants". The effectiveness of flame scarfing as a method of decomtaminating concrete has been established by tests at the decommissioned nuclear power plant Gundremmingen KRB-A.

1. INTRODUCTION

When demolishing decommissioned nuclear power plants particular attention should be paid to contaminated areas, which in the case of concrete means the biological shield, which has the highest radioactivity level. However, the concrete surfaces of the facility, which are contaminated to varying depths, entail important economic considerations since they can cover tens of thousands of square metres.

Concrete decontamination serves two purposes :

- to reduce the exposure of personnel;
- to reduce the activity of the surface contamination to a level which permits unrestricted use of previous controlled area, i.e.to permit material from a demolished plant to be treated as normal building waste.

This survey concentrates on methods involving stripping, since measurements have shown that other methods cannot adequately remove contamination which penetrates porous material.

Surface stripping methods have been applied ever since concrete has been used, and one of the most effective tools for this is still the chisel.

2. DETERMINING THE NATURE OF THE CONTAMINATION

During tests at the decommissioned nuclear power plant Gundremmingen KRB-A core samples were drilled and analysed in the radiation protection laboratory by gamma spectrometry to determine the nature and depth of the contamination. Cs-137 and Co-60, and traces of Cs-134, were found to be present. The isotope Cs-137 accounted for about 90 % of the radioactivity.

As shown in Fig. 1, contamination with a high proportion of Cs-137 penetrates to a greater depth than contamination in which Co-60 is predominant.

specific activity (Bq/g)

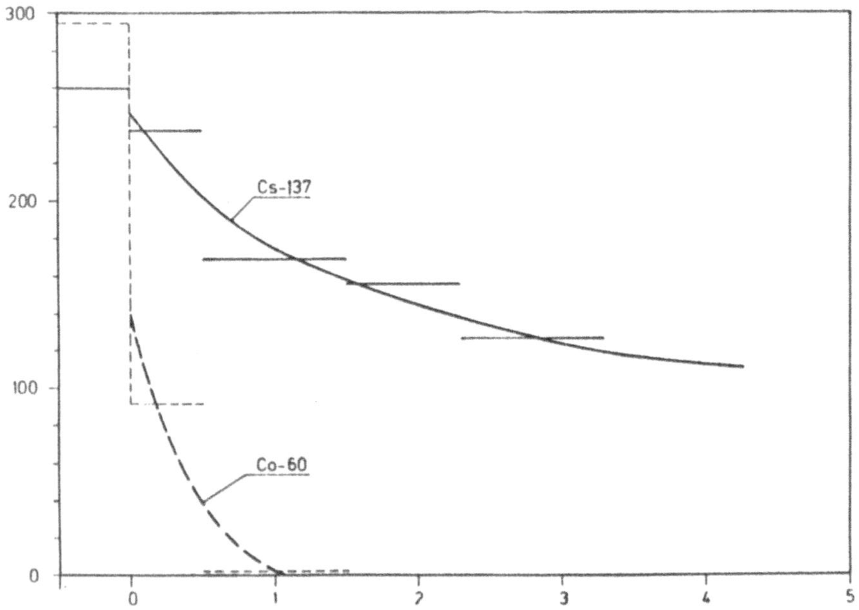

Decontamination seal Depth of penetration (mm)

Fig. 1 : Penetration in concrete of Cs-137 and Co-60 according to (16)

Depth of penetration is determined by the nature of the concrete, its porosity, the duration of exposure and the composition of the contamination. There is no doubt that one reason for the greater penetration of Cs-137 is that caesium is usually present as a dissolved salt in condensation water and fluid cleaning agents, while cobalt is normally present in undissolved form. The caesium ions present, being more alkaline than potassium and sodium ions, are assumed to exchange with the potassium and sodium salts in the concrete.

Penetration depth is not affected by the intensity of radioactivity.

3. EXISTING TECHNIQUES

3.1. Removal of surface contamination

3.1.1. Scrubbers with liquids

The effectiveness of mechanical scrubbers in conjunction with acid and chemical cleaning agents has already been examined (8).

An industrial scrubber (0.5 m) with medium-hard, wear resistant nylon pads operating on 110 V was tested. A 10 % solution of phosphoric acid was applied to one floor surface and a liquid chemical cleaning agent to another; after scrubbing, and washing with small amounts of water the liquids were drawn off and poured into drums. The acid collecting vessel contained a neutralizer and the container for the chemical cleaning solution a foam inhibitor.

Decontamination factors of around 400 were obtained. The amount of liquid waste produced by this process is approx. 8 l/m^2. Solid waste (scrubber pads) is negligible.

Three workers can decontaminate about 1.5 m^2 per hour.

3.1.2. Decontamination with foam

The foam process was introduced on a routine basis by ESG for the decontamination of "hot cells", glove boxes and obsolete nuclear facilities during decommissioning (6).

The process was shown to be highly effective for non-porous surfaces and works best when the foam is sprayed on hot (at around 50°C).

A compressor with a maximum pressure of at least 70 bars is recommended for continuous spraying. Before the surfaces are rinsed, the foam should be removed with a dry/wet dust extractor. An anti-foam agent should be added to the rinsing water.

Keeping the foam on vertical surfaces for a sufficiently long period can be problematic. The main advantages of the process are the small amount of liquid waste, ease of application and the fact that it can be used with a large number of chemicals.

The process can be applied at a distance of up to 6 m from the surface to be cleaned. Moreover, it does not damage the surface and is not dangerous or explosive.

3.1.3. Decontamination with freon

Freon-113 has been used as decontamination agent in a new process (6). Its advantages are highly effective cleansing, speed of application, compatibility with a wide range of materials, the small amount of radioactive waste produced and reduced treatment of materials and waste.

The freon is sprayed on to the contaminated surface under high pressure, thus loosening the contamination. There is a vacuum cover on the surface being treated so that the contamination can be extracted with the freon and subsequently separated.

3.1.4. Removable foil

This is sprayed on with a pump (8), which gives the spraygun a pressure of approx. 155 bars. The foil is between 0.8 and 1.3 mm thick and takes 24 h to dry. After drying it is cut into strips with a knife, removed and can then be packaged immediately.

Decontamination factors of around 100 are obtained. An area of 200 m^2 produces enough waste to fill a 200 l drum. 10 m^2 can be treated per hour by a three-man team.

3.2. Stripping methods

3.2.1. Chipping hammers

The hardened steel chisels of these devices are set in motion by compressed air or a hydraulic fluid (1). There are two types available. Heavy models, weighing between 15 and 45 kg, are used mainly for ripping up floors and attain speeds of about 1 500 strokes per minute. The lighter models, weighing between 7 and 15 kg, operate at around 2 000 strokes per minute and are used primarily for walls and ceilings. Being very cumbersome, these tools are recommended for use only in decontaminating small areas.

The hydraulic ram, which was used successfully in the SRE decomissioning project (2), is also fitted with chisels.

3.2.2. Drilling and hydraulic blasting

This technique was developed to remove contaminated concrete surfaces up to a depth 50 mm without destroying the inner structure of the concrete. Holes of 25 - 40 mm diameter are drilled in the concrete, into which expandable tubes of the same diameter are inserted. A hydraulically driven conical bit splits the end of the tube causing the fingers of the tube to spread outwards and blast the concrete (Fig. 2). The positioning of the holes is such that the blasted areas overlap. The removal rate is approximately 7 m^2/h.

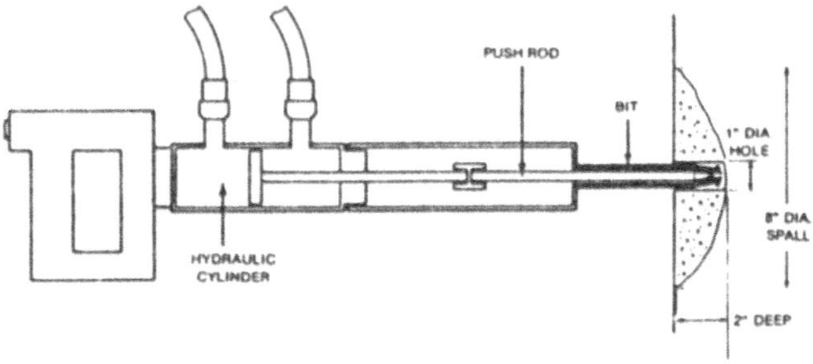

Fig. 2 : Drilling and hydraulic blasting

3.2.3. Jet guns

These include the glycerine gun and the water cannon (1 and 5).

The glycerine gun (Fig. 3) has a short, smooth barrel, at the end of which is a nozzle with a funnel-shaped shield. The gun fires glycerine capsules in conventional cartridges with gunpowder. Craters 75 - 100 mm in diameter and 20 mm deep are produced. The dust is bound by the glycerine and removed by suction fron the shield together with the broken concrete. About 0.7 m^2 concrete can be removed per hour, for which approximately 270 shots are needed.

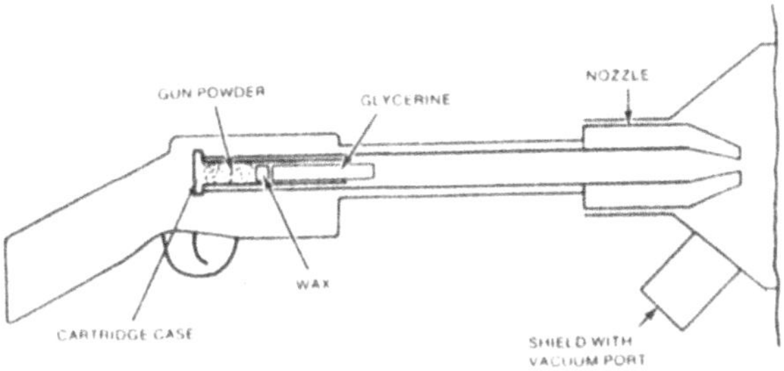

Fig. 3 : 458 magnum glycerine gun

In the water cannon (Fig. 4) a piston is driven by compressed gas, forcing a small amount of water through a nozzle. The gas is compressed hydraulically. As in the case of the glycerine gun, the nozzle is encased in a funnel-shaped shield, in which the rubble is collected and removed by suction. After each shot more water is injected into the chamber next to the piston. The water cannon works more slowly than the glycerine gun, approximately 0.4 m^2 being stripped per hour.

Because of their slowness, jet guns are suitable only for use on small areas (5).

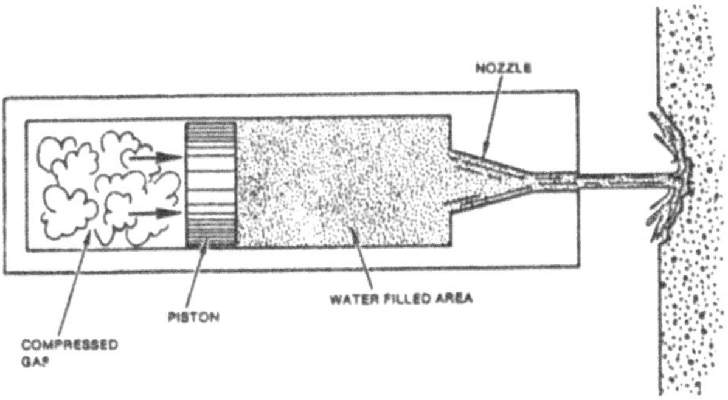

Fig. 4 : Water gun

3.2.4. Grinding

Grinding can be used when only thin layers are to be removed. It is preferable to use coarse-grained materials, in particular the water-cooled diamond grinders (Fig. 5).

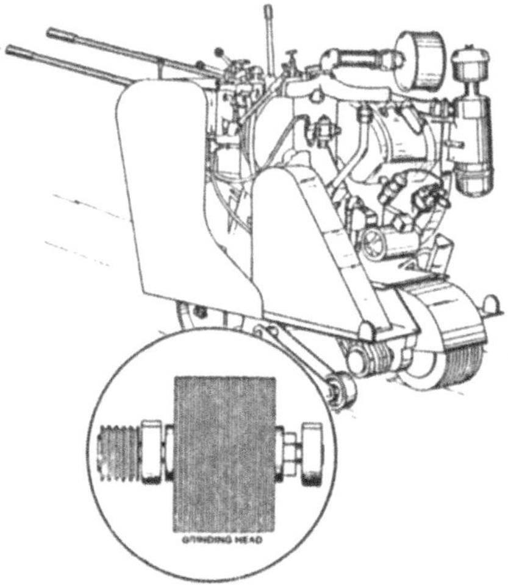

Fig. 5 : Floor grinder

The cooling water is injected into the middle of the grinding head, thus preventing excessive dust formation. Grinders should be equipped with local suction devices and dust should be removed by HEPA filters. Layers about 10 mm deep are removed using such equipment.

Floor grinders and hand grinders have been successfully used for concrete decontamination at the San Onofre 1 nuclear power station.

3.2.5. Spike hammers

These are suitable for removing thin layers of contaminated concrete. They incorporate several pneumatically driven pistons whose heads may be fitted with 5 or 9 tungsten carbide spikes. Different models are used for floors and wals. The travelling model for removing floor surfaces can have 5 to 11 staggered pistons and a local dust extraction unit. The type used for walls has three pistons arranged side by side. Stripping speeds are about 4.5 m^2/h per piston for the floor model and between 7 and 11 m^2/ for the wall model.

A floor model with 7 pistons was used in the SRE decommissioning project (1 and 9). Similar equipement was used in the decontamination of the DP West Plutonium plant at Los Alamos (3) and at the Frankford Arsenal (4).

3.2.6. High-pressure water lances

These have already been used successfully to strip concrete surfaces on bridges.

This led to the first tests aimed at using high-pressure water lances to decontaminate concrete surfaces (5). It is important that the contaminated material should be collected with the water and processed to minimize the volume of waste. The tests showed that at depth of approx. 3 mm stripping speeds of approx. 12 m^2/h could be achieved. The system operates at a pressure of around 240 MPa. The size of the concrete particles removed exceeds 80 μ m, and so they can be separated off without difficulty using hydrocyclones and conventional filter systems.

3.3. The flame scarfing process

Flame scarfing has long been used to treat surfaces in order to produce a clean, dry base for paint and other surface coatings. It has been used on both steel and concrete surfaces (10,11 and 12). Flame scarfing is not only used for stripping, but is also effective in producing fluting in concrete surfaces, for example in order to increase the friction coefficient of a road surface and so reduce the danger of skidding (10 and 11).

Flame scarfing is carried out using oxy-acetyline burners, which are available as hand and machine burners. The nozzles are arranged side by side at an angle to the direction of feed. Hand burners have burner widths of 50 to 250 mm and are used mainly for vertical surfaces. Machine burners, with widths of 500 to 1 000 mm, are used for large surfaces such as roadways and large floor areas (10 and 11).

Burners are moved over the concrete surface at a constant rate of feed, the temperature of the flame being around 3 200°C. The flame shows an excess of oxygen because in this process the kinetic energy of the oxygen jet is converted to thermal energy (10).

Since concrete, unlike metals, is a relatively poor heat conductor, most of the energy from the flame is held back on the top layer (1 - 2 mm). The energy produced gives the crystalline surface a kind of "heat shock". The large crystalline minerals beneath the surface react to the sudden intrusion of energy by changing their inter-crystalline structure. The crystal axes are displaced almost immediately. This releases intercrystalline forces which destroy the crystal structure. If a group of crystals releases inter-crystalline forces in the concrete surface, the mechanical effects influence particles which have been subject to little or no excitation. The forces cause pieces which can measure up to several square centimetres to erupt from the surface. This process might be termed "spattering".

In contrast to the process which "coarse" crystals undergo, no mechanical effects are observed in highly crystalline and amorphous minerals; instead, a melting process occurs. The flame produces glass-like molten beads which lie more or less loosely on the surface.

Both spattering and melting use up almost all the thermal energy of the flame, so that no heat penetrates the material. The energy remains in the spattered and molten material, which no longer forms part of the concrete (10 and 11).

After flame scarfing the loose particles and remnants of combustion must be removed from the treated surface. Circular brushes or cleaning machines, which can be fitted with steel wire brush rollers or beater rollers, are recommended for this task (11).

3.3.1. Tests on non-radioactive concrete samples

Preliminary tests were carried out on non-radioactive concrete samples in order to optimize the working parameters. The purpose was to obtain information on the positioning of the burners, the rate of feed and gas pressures.

The following optimum gas pressures were obtained for a 250 mm wide fan jet burner :

$$C_2H_2 : \quad 0.5 \text{ bar} < p < 1 \text{ bar}$$

$$O_2 : \quad 6 \text{ bar} < p < 8 \text{ bar}.$$

With regard to the positioning of the burners, the distance between the burners and the concrete surface and their angle of inclination are of importance. They were most effective at a distance of about 5 mm and at an angle of inclination of around 30°.

The ideal rate of feed for uncoated concrete surfaces is approx. 1 m/min. A 50 % reduction in the rate of feed in necessary for coated surfaces to ensure complete combustion of the decontamination coating.

Further tests were carried out on non-radioactive concrete surfaces to obtain information on the effectiveness of the process and on dust and aerosol formation with a view to applying the process in the controlled area of a nuclear power station. Various types of filter for aerosol removal were tested.

The lay-out of the test plant is shown in Fig. 6. The test cabin (2) is divided into two sections - the burner room and the control room. The burner room can be surveyed from the control room through a large glass panel. The radial ventilator (7) and the burner feed control are actuated from the control room. Next to the burner room is a frame-mounted filter which prevents foreign bodies entering the test cabin with the extracted air. Connected to the test cabin is an extraction line (3) with three successive filter units, in which the dust and aerosols resulting from flame scarfing are collected. The first stage consists of a bulk drum filter (4) for preliminary filtration. The intermediate stage is a bank of box filters (5) incorporating, for secondary filtration, an aerosol filter class EU 7, DIN 24 185 (14) and, for fine filtration, an aerosol filter class S, DIN 24 184 (13). For very fine filtration there is a drum filter incorporating an aerosol filter class S, DIN 24 184 (13).

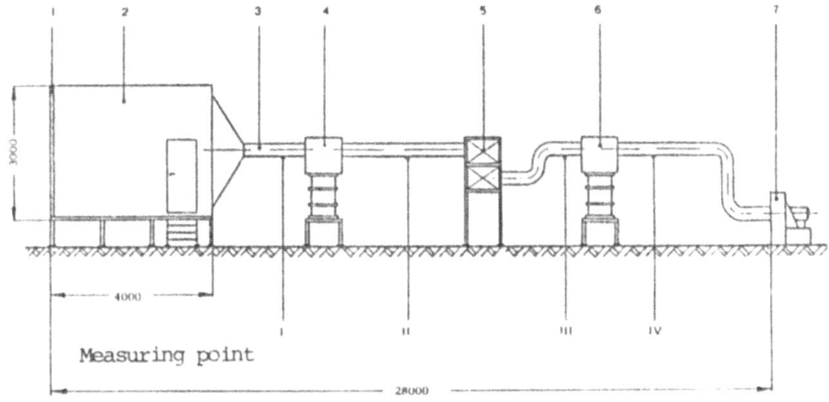

1. Frame-mounted filter (suction filter)
2. Test cabin
3. Extraction line
4. Bulk drum filter
5. Bos filter
6. Aerosol drum filter
7. Radial extractor fan

Fig. 6 : Flame scarfing test station showing arrangement of filters and measuring points

3.3.1.1. Stripping tests

The tests alreday carried out have shown that the flame scarfing of coarse-grained concrete (grain size 0-32 mm) produces stripping depths of 2.5 - 3 mm, while in the case of fine-grained concrete (grain size 0 - 8 mm) depths of between 1.4 and 1.5 mm are reached in one flame scarfing operation. This result was as expected, since coarse particles obviously allow a greater depth of penetration than fine particles.

3.3.1.2. Filtration of dust and aerosols resulting from flame scarfing

In order to evaluate flame scarfing as a decontamination process it is very important to know the grain size of the dust and aerosols emitted and the effectiveness of the dedusting equipement used. For this reason dust content and grain size were determined both in the raw gas and downstream of the individual filter stages.

As recommended by the filter manufacturer, the tests were carried out at an extraction flow rate of 3 000 m^3/h. However, similar filtration efficiency was obtained in tests with flow rates of 1 500 m^3/h.

During flame scarfing average dust concentrations in the raw gas are approx. 100 mg/m^3, while the average raw gas dust concentration during subsequent brushing is around 1 060 mg/m^3. By means of local dust extraction (using extractors with a slit-shaped nozzle) it was possible to reduce the dust content of the raw gas during brushing to about 350 mg/m^3.

The dust content of the raw gas could be reduced still further by means of an annular extractor. Exposure of operating personnel and the spread of contamination could thus be minimized.

A large proportion of the dust and aerosols is collected in the preliminary drum filter (approx. 80 % in the case of flame scarfing and approx. 92 % in the case of mechanical cleansing). The efficiency of the bank of box filters is between 99.91 and 99.99 %. This performance was improved by the addition downstream of the aerosol drum filter. Using all filters, efficiency was computed to be between 99.991 and 99.999 %.

Fig. 7 shows the particle distribution in the raw gas for flame scarfing, brushing and brushing with local dust extraction. The differences between the curves are very slight. Maximum values are for particle diameters of 0.6 to 0.7 μm.

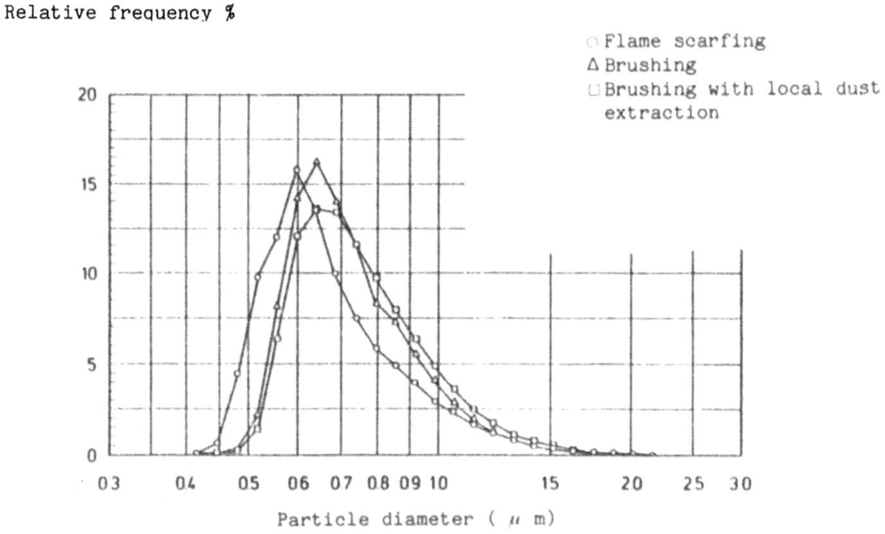

Fig. 7 : Particle distribution in raw gas (measuring point 1)

3.3.1.3. Dust storage capacity of preliminary filters

An important economic feature is the useful life of filters. If solids are deposited in a filter, its lacunarity, i.e. the ratio of void to the overall filter volume, decreases. There is thus an increase in the resistance of the filter and in the loss of pressure at the filter.

The dust storage capacity of a filter is therefore measured by the increased loss of pressure as a function of the amount of dust trapped. All filters have a maximum permissible value for loss of pressure. The sooner this limit is reached, the shorter the filter's useful life.

The dust storage capacities of the two preliminary filters are compared in Fig. 8, which shows the loss of pressure Δ p as a function of the amount of dust trapped.

Loss of pressure at the bulk drum filter acculerated by about 30 % with approx. 2 700 g dust, while in the case of the box filter an increase of around 330 % was recorded for approx. 1 000 g of dust.

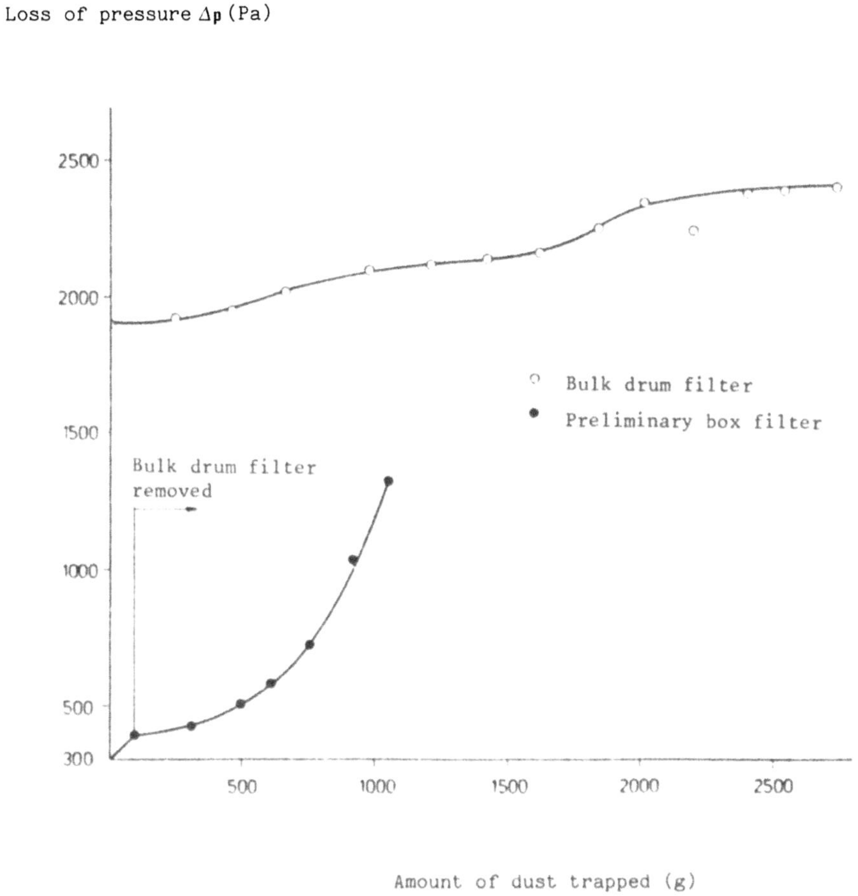

Fig. 8 : Loss of pressure Δp at the bulk drum filter and at the preliminary box filter as a function of the amount of dust trapped

It was not possible to use the filters to full capacity during the tests. However, the sharply accelerating drop in pressure clearly indicated that the dust storage capacity of the box filter was almos exhausted, while the storage capabilities of the bulk drum filter cannot as yet be forecast.

The tests show that the drum filter, because of its condiderably greater dust storage capacity, is to be preferred to the box filter for the cleaning of raw gas in the flame scarfing of concrete.

3.3.2. Tests on contaminated concrete surfaces

Flame scarfing tests on real contaminated concrete surfaces were carried out at the solid waste store of the decommissioned nuclear power plant Gundremmingen KRB-A. From 1966 to 1977 this store was used for the reduction, packaging and storage of solid radioactive waste.

When the tests were carried out the entire store was free from mobile activity and contamination which could be removed by rubbing.

Since there was no fixed exhaust air system available for the tests, a temporary air circulation system had to be set up using a mobile filter unit.

3.3.2.1. Decontamination effect of flame scarfing

A limit of 10^{-5} $\mu Ci/cm^2$ is required to make a controlled area or supervised operational area fit for unrestricted further use (15). To achieve this, 4 - 5 mm concrete had to be removed.

By applying the parameters optimized in cold tests, a layer 1 - 1.5 mm deep was stripped per operation at a feed rate of approx. 1 m/min. After each flame scarfing test the debris was mechanically removed with simultaneous local dust extraction.

The tests showed that four runs were sufficient to decontaminate the concrete surface to below the above mentioned limit. This applies to the entire area treated in the store. It was also found that the depth of penetration of the contamination was unrelated to the radioactivity level.

3.3.2.2. Aerosol activity in flame scarfing

In some tests involving a mobile dust sample collector, dust trapped by measuring filters was radiologically analysed. Aerosol activity in the ambient air was measured on the three floors of the store before and after flame scarfing, as well as after brushing. During these tests no local extraction was carried out during brushing in order to determine whether activity in the debris rose during flame scarfing.

Operation / Place	Initial measurement	Flame scarfing	Brushing
Floor -2	$1,3 \times 10^{-11}$	$4,2 \times 10^{-12}$	9×10^{-12}
Floor -1	$1,1 \times 10^{-11}$	$2,5 \times 10^{-11}$	$3,4 \times 10^{-10}$
Approach area	$1,0 \times 10^{-11}$	$5,1 \times 10^{-11}$ $1,2 \times 10^{-10}$*)	no measurement

Tab. I : Aerosol activity in Ci/m^3 in flame scarfing and brushing tests
*) Measurement repeated after 10 minutes

Tab. I indicates that owing to the heat produced during flame scarfing aerosol activity increased from the bottom to the top of the store. The marked increase in aerosol activity during brushing suggests that activity increased in the debris.

3.3.2.3. Recontamination of surfaces

Recontamination was tested by marking points on the wall and floor near the surfaces to be treated by flame scarfing, at which smear samples were taken before and after the flame scarfing tests. The results showed surface recontamination to be insignificant even after four flame scarfing tests.

Although the flame scarfing and brushing tests resulted in negligible surface recontamination, it is recommended that treated surfaces should be sealed with a suitable material after measurements have been taken in order to prevent possible recontamination. If it is necessary to decontaminate ceilings and walls, upper areas should always be treated first.

3.3.2.4 Assessment of decontamination costs

The costs resulting specifically from the thermal decontamination of concrete sufaces were assessed on the basis of the tests at the KRB-A solid waste store. Initial outlay was not taken into consideration.

Labour accounted for most of the cost of decontamination by flame scarfing, materials and energy accounting for a much smaller proportion. A skilled worker to operate the burner and a radiation protection technician, who could also assume responsibility for fire protection, were needed for the work.

In order to calculate labour costs it is necessary to know the times required for the various operations. The time needed to decontaminate a concrete surface of 1 m^2 using a mechanical burner 1 m wide may be broken down as follows.

	Skilled worker	Radiation protection technician
4 x flame scarfing operations	5 min	5 min
4 x brushing and dust extraction operations	20 min	20 min
Measurement of area	-	25 min
Sealing off of area	-	5 min
Changing clothing, personal radiation protection precautions, set-up time, rest period	60 min	30 min
	85 min	85 min

The following costs result :

a) Labour costs :

1 skilled worker (DM 60 per hour)	DM 85
1 radiation protection technician (DM 80 per hour)	DM 113
	DM 198

b) Materials :

gloves, shoe coverings, breathing masks, foil, adhesive tape	DM 20

c) Gas :

2.2 $m^3 O_2$	DM 4
2 $m^3 C_2H_2$	DM 2
Total costs per m^2	DM 224

4. COMPARISON OF EXISTING TECHNIQUES

Chipping hammers may be used to decontaminate small areas, especially where contamination has penetrated very deeply. For reasons of cost, fluid guns should only be used on small areas which are difficult to reach. Drilling and hydraulic blasting is by far the most suitable method for decontaminating walls. Travelling grinding machines, spike hammers and flame scarfing burners may be used for stripping large floor areas. Portable models also exist for walls and small areas.

Process	Personnel required	Stripping speed m^2/h	Stripping speed m^3/h	Specif. primary waste vol. m^3/m^2	Specif. costs DM/m^2	Specif. costs DM/m^3
Chipping hammer (40 kg, 1300 stokes/min)	3	-	2	-	-	129
Drilling and hydr.blasting	2	7	-	0,022	100	-
Glycerine gun	2	0,7	-	0,0125	830	-
Water gun	2	0,4	-	?	?	-
Grinding	2	20-30	-	0,01	147	-
Spike hammer (7 pistons)	2	30	-	0,01	5	-
High pressure water lance	?	12 6	-	0,003 0,006	?	-
Flame scarfing*)	2	0,7	-	0,005	224	-

Tab. II : Comparison of various concrete decontamination techniques
*) 4 x flame scarfing and 4 brushing operations, measuring, sealing off

It is not easy to compare the cost of the various processes, since the available figures come from different sources and are to some extent produced according to different criteria.

For example, final storage costs are not considered. The cost of final storage of a 20 l drum is at present around DM 3 000. This would result in the following estimated final storage costs per m^3 of decontaminated surface (assuming a container filling factor of 0.5):

drilling and hydraulic blasting	DM 660;
spike hammering	DM 300;
flame scarfing	DM 150.

Similarly, the comparison does not take account of the unit cost of filters, which depends on the useful life of filters and can be considerable.

The cost of measuring and sealing off the area to be treated was considered only in the case of flame scarfing, which means that other unit costs would arise.

Analysis of the available data (Tab. II) suggests that the spike hammer is the most economic stripping tool. If the minimization of waste is the prime consideration, however, flame scarfing should be given particular consideration.

5. CONCLUSIONS

All the methods described are suitable for decontamination concrete surfaces. For certain tasks, e.g. the decontamination of walls or large floor areas, the number of possible methods is limited.

Particular consideration must in future be given to the minimization of waste. Flame scarfing, which makes it possible to remove thin layers, is the most suitable method for this purpose. The present state of knowledge does not permit an evaluation of the high pressure water lance method. If contamination has penetrated more deeply, the spike hammer is the most appropriate tool.

Improving filter technology is an important task for research. The dust and aerosols produced can be controlled using existing filters, but it is necessary for economic reasons to increase the useful life of preliminary filters.

BIBLIOGRAPHY

1. Decommissioning Handbook. Prepared for U.S. Department of Energy by Nuclear Energy Services, Inc.
2. KITTINGER, W.D. et al. (1982). Lessons learned in decommissioning the Sodium Reactor Experiment. International Decommissioning Symposium, Seattle.

3. GARDE, R. et al. (1982). Los Alamos DP West Plutonium Facility decontamination project. International Decommissioning Symposium, Seattle.

4. TORRISI, S.P. (1982). Decontamination of the Frankford Arsenal. International Decommissioning Symposium, Seattle.

5. ALLEN, R.P. (1982). Development of improved technology for decommissioning operations. International Decommissioning Symposium, Seattle.

6. FOWLER, D.E. (1982). Decontamination application with Freon. Decontamination of Nuclear Facilities, International Joint Meeting ANS-CNA, Niagara Falls.

7. HARRIS, J.M. et al. (1982). A foam process for application of decontamination agents. Decontamination of Nuclear Facilities, International Joint Meeting ANS-CNA, Niagara Falls.

8. BENGEL, P.R. and MASON, R.C. (1982). Results of the Gross Decontamination Experiment and implication for future decontamination activity. Decontamination of Nuclear Facilities, International Joint Meeting ANS-CNA, Niagara Falls.

9. BRENGLE, R.G. (1979). The decontamination of concrete. Decontamination and Decommissioning of Nuclear Facilities, Proceedings of the ANS-Meeting, Sun Valley.

10. HILL, F.W. Thermische Oberflächenbehandlung zum Erzielen optimaler Konservierungsbedingungen auf Beton. Schweissen und Schneiden 25 (1973) 7, 254-257.

11. BECKER, W. Stand der Anwendung des Flammstrahles von Stahl und Beton. Schweissen und Schneiden 21 (1969) 3, 103 - 110.

12. HILL, F.W. Flammstrahlen von Beton und Naturstein. Sonderdruck 14 der Firma Linde AG, Höllriegelskreuth.

13. DIN-Norm 24 184, Typprüfung von Schwebstoffiltern.

14. DIN-Norm 24 185, Prüfung von Luftfiltern.

15. Verordnung über den Schutz vor Schäden durch ionisierende Strahlen (Strahlenschutzverordnung), Bundesgesetzblatt Z 1997 A, Bonn 1976.

16. ROSE, K. (WS 1981/82). Ermittlung von Kontaminationen zwecks Planung des Abbaus von Kernkraftwerken, Diplomarbeit Fachhochschule Lübeck.

SPECIFIC DECONTAMINATION METHODS : WATER LANCE, EROSION BY CAVITATION, APPLICATION OF GEL-BASED DECONTAMINANTS

D. BOULITROP, CEA-Saclay, J.P. GAUCHON, CEA-Cadarache and Y. LECOFFRE, Alsthom Atlantique Neyrtec

Summary

Decontamination methods involving spraying are suitable for use in the decommissioning of nuclear installations because of their compatibility with remote control methods, which make it possible to treat materials on the spot with no irradiation of personnel. They do not give rise to contamination from aerosols but produce a large amount of radioactive waste, and their effectiveness is sometimes limited.

The erosive and decontaminating effects of high-pressure water lances have been improved at the CEN-Cadarache by increasing pressure (up to 600 bars) and adding chemical reactants prior to spraying. Alsthom Atlantique Neyrtec at Grenoble has developed a system of erosion by cavitation which can be adapted for decontamination purposes. At the CEN-Saclay a comparative study on the incorporation of decontaminants in gel-based media has made it possible to select compounds which are effective in decontaminating the main materials used in the nuclear industry.

1. DECONTAMINATION BY THE SPRAYING OF WATER UNDER HIGH PRESSURE

The spraying of water under pressure, while effective in eroding materials, produces a large amount of waste. Its effectiveness can be improved by increasing pressure or by adding chemical reactants.

Tests on the erosion of mild and stainless steels and on the decontamination of ferritic steel from the primary circuit of a boiling water reactor were carried out at the Cadarache Centre d'Etudes Nucléaires at pressures ranging from 150 to 600 bars with and without prior spraying with a gel-based solution of sulphuric acid and hydrazine.

1.1 Experimental unit for decontamination using water under high pressure (hyperbar)

The experimental unit which we used in the tests was designed on the basis of the decontamination technique using a high-pressure water lance at pressures of between 150 and 600 bars and takes account mainly of the following :

- effectiveness as determined by the principal parameters (pressure, flow-rate, nozzle-target distance, angle of impact, etc.);
- the volume of effluent produced and the ease of treatment;
- the amount of solid waste obtained;
- the most suitable applications of the technique.

The decontamination unit using 'hyperbar' water comprises the following :

- a reception area in which contaminated materials can be checked and processed before spraying (dismantling, cutting, preliminary spraying, etc.);
- a spraying cabin (figs. 9 - 10), which is a sealed chamber in which the operator, wearing ventilated clothing, carries out spraying manually;
- a spraying cell (figs. 9 - 11), also a sealed chamber, but in this case spraying is controlled from outside;
- a waste treatment unit (fig. 12) comprising two candle-shaped filters and one crushed resin filter.

1.2 Effect of a high pressure water lance on a metal surface

1.2.1 Characteristics of the nozzle used

All the experimental data were obtained using a spraying cell and a nozzle producing a flat jet at an angle of 30° to the horizontal and an aperture diameter of 1.8 mm (form 19, index 2.3, manufacturer Socofren). Flow-rates were 46 l/min at 600 bars and 24 l/min at 150 bars.

When the sample was sprayed longitudinally at a nozzle-target distance of 5 cm, the width of the eroded surface was 2.2 cm at a pressure of 600 bars.

1.2.2 Types of measurement

We used stainless steel samples (10 cm x 10 cm; NFA: Z2 CN 18-10 - AISI 304 L) to determine the effectiveness of this technique on metal surfaces widely used in nuclear power stations.

In the case of mild steel, the following criteria for comparison were adopted to assess the effect of the sprayed water on the samples :

- mean erosion (g/m^2), i.e. the ratio of target weight loss to the surface area of the section eroded;
- the depth of penetration Rp (μm), indicating the difference between the mean line obtained by planing the roughness profile (troughs interspersed with peaks) and the envelope line through the peaks.

In the case of stainless steel, only the progress in penetration depth was considered.

Penetration depth was measured by means of a WlBe perthometer in conjunction with a R100 T perthograph, which uses a mechanical gauge. Four measurements were taken in various directions within the zone of erosion and on its periphery; only the significant mean values were used.

1.2.3 Parameters considered

We varied the following parameters in the tests :

- pressure (300-600 bars) and the corresponding flow-rates;
- nozzle-target (2.5, 5 and 10 cm);

- angle of impact (80° and 90°);
- the specific volume of water expressed in m^3/m^2 of surface sprayed.

In all tests the target was fixed to a trolley moving to and fro at a constant speed (2 cm/s) in front of the lance, which was kept immobile at the angle and distance chosen.

Under the constantly maintained experimental conditions, ant at a pressure of 600 bars and a nozzle-target distance of 5 cm, a single run represents a specific volume of 1 m^3/m^2.

1.2.4 Erosion of mild steel

The high-pressure water lance causes erosion of mild steel and considerable loss of weight (fig. 2) at a distance of 5 cm, a pressure of 400 bars and with a specific volume of 50 m^3/m^2. The same is true for 500 bars and 20 m^3/m^2.

Weight loss is not significant for specific volumes of up to 5 m^3/m^2 at 600 bars.

It should be noted that erosion is irregular because the speed of the water particles projected is greatest around the axis of the jet. Erosion is found to be most pronounced in that area.

The surface condition of these samples can be determined more accurately by measuring penetration depth (fig. 3). Similar conclusions are reached, however.

1.2.5 Erosion of stainless steel

In contrast to the mild steel samples, the initial roughness of the stainless steel samples was very slight, since micropeening produces a mirror-like surface which makes erosion easily observable. Erosion becomes apparent (fig. 4) at a target distance of 5 cm under conditions similar to those of conventional decontamination, i.e. at 400 bars and 5 m^3/m^2 and at 500 bars and 2 m^3/m^2.

At pressures of between 300 and 400 bars, changes in the penetration depth only become appreciable at specific volumes equivalent to about 10 decontamination cycles (20 to 50 m^3/m^2).

Changes in the roughness of the sample at 600 bars (fig. 5) remain slight at specific volumes of less than 20 m^3/m^2 and at target distances greater than 6 cm.

At a target distance of 10 cm there is no change in the surface condition of the sample, even when the specific volume of the water jet increases to 50 m^3/m^2 (fig. 6). At a target distance of 5 cm, erosion of the sample with a specific volume of 50 m^3/m^2 results in an increase in roughness by a factor of around 5.

Similar phenomena are observed on a slightly more compressed scale (figs. 7 and 8) when the angle of impact is reduced to 80° ($0.70 \times Rp\ 90 > Rp\ 80 > 0.5 \times Rp\ 90$). Spraying at 90° produces surface battering.

1.3 **Decontamination by high pressure water lance of samples taken from the primary circuit of a BWR**

1.3.1 Characteristics of the samples

The samples from the water supply lines (auxiliary circuit) of the German ISAR boiling water reactor were of ferritic steel (DIN 17 MnMoV64) and were non-activated, having remained for about 5 years in the primary circuit.

Radiochemical activity was due mainly to roughly equal proportions of Mn-54 and Co-60 and was located in a deposit composed largely of hematite (1) due to the corrosion of the steel in an oxidizing environment (generaion of oxygen by radiolysis).

1.3.2 Parameters considered

The degree of erosion caused by the high-pressure water lance on a metal surface was examined. For an angle of impact of 80° at varying target distances (5, 10 and 25 cm), specific volume of sprayed water (1, 2, 5, 10 and 20 m^3/m^2) and pressure (150, 300, 450 and 600 bars).

The samples measuring 18 cm^2 (2 x 9 cm) were cut from a pipe 350 mm in diameter and fixed to the spraying plate of the spraying cell. The lance was kept stationary during the erosion tests and the sample was moved at a speed of 2 cm/s (at a target distance of 5 cm), the distance from the nozzle having been fixed beforehand.

Initial and final activities of the test samples were measured by means of a scintillation counter (SPP2) coupled to a scaler (scaling time 100 s).

Decontamination factors were calculated on the basis of the ratio of initial to final activities measured in cps.

The initial activity of the samples ranged from 2.2 kBq/m^2 and 7.4 kBq/m^2 (0.06 and 0.2 $\mu Ci/m^2$).

1.3.3 Decontamination of test samples covered with hematite by the high-pressure lance (cf. figs 13 and 14).

At a target distance of 5 cm decontamination factors increased fairly regularly by a factor varying from 2.5 to 3 when water pressure ranged from 150 to 600 bars.

At a distance of 10 cm decontamination factors increased by a factor of only 2 to 2.5 when pressure was increased from 150 to 600 bars, and at 25 cm the factor was close to 1 (1.1 to 1.3).

The specific volume of sprayed water had only a relatively slight influence on decontamination at all target distances. When the specific volume was increased from 1 to 20 m^3/m^2 the decontamination factors obtained increased by a factor varying from 1.4 to a maximum of 2.

Application at a pressure of 600 bars improved effectiveness only at target distances of between 5 and 10 cm. In this case manual spraying was virtually impossible (excessive recoil).

1.3.4 Combination of chemical action with the high-pressure water lance

Chemical action can be combined with the mechanical action of the high-pressure lance by dissolving the chemical agent in the water before spraying, by injecting the agent into the nozzle during spraying or by spraying the agent on to the sample beforehand.

We carried out preliminary spraying of the samples using 200 g/m^2 of a gel made from a solution containing 200 g/l sulphuric acid and 1 g/l hydrazine. These reactants were chosen with a view to weakening the layer of hematite (acid reduction).

After one hour of contact with the sample, the gel was removed in an initial series of tests by soaking for a sufficient period in water to avoid the mechanical effect. Decontamination factors varied from sample to sample from 2.5 to 4.

In the second set of tests we cleansed the samples with a high-pressure water jet. The following parameters were adopted : target distances - 5, 10 and 25 cm; pressures - 150 and 600 bars.

The combination of the gel and the high-pressure lance results in decontamination factors greater than those obtained by the lance alone by factors in the region of :

- 5 to 7 at a distance of 5 cm;
- 10 to 16 at a distance of 10 cm;
- 17 to 22 at a distance of 25 cm.

At the same pressure, the effectiveness of decontamination using the combination of gel and high-pressure lance does not vary appreciably;

- $80 < DF < 120$ with a pressure of 600 bars;
- $40 < DF < 80$ with a pressure of 150 bars.

Increasing pressure when washing the gel from the sample only slightly increases the effectiveness of decontamination (by a factor of 1.7 at a distance of 5 cm; 2.2 at 25 cm).

These results demonstrate the value of this process, which allows a substantial improvement in the effectiveness of decontamination by high-pressure water lance without any marked increase in the content of chemical pollutants in the waste.

1.4 Conclusion

The erosion and decontamination tests gave an indication of the suitable applications of the high-pressure water lance, taking into account the critical parameters of pressure and nozzle-target distance.

A pressure of 150 bars, provided by the normal commercialy available equipment, produces limited results on samples coated with hematite ($5 < DF < 10$).

A pressure of 600 bars does not appreciably increase the effectiveness of the process ($5 < DF < 34$).

On the other hand, preliminary spraying with gels containing chemical reactants results in much greater decontamination factors :

DF = 40 with a pressure of 150 bars;
DF = 80 with a pressure of 600 bars.

It should be noted that the increase in the decontamination factor using this type of sample is likely to be limited by the resistance to the jet of the oxide coating and not of the metal itself.

Because of the resistance of this oxide coating it is only possible to obtain decontamination factors which are roughly proportional to pressure in the case of target distances of 5 cm.

For all the above reasons, pressures of 600 bars appear to be advantageous in the following cases :

- decontamination treatment of narrow bore pipes (less than 100 mm in diameter) using a reaction nozzle;
- decontamination treatment of large bore pipes (diameter 600 to 700 mm) using a rotating head fitted with several nozzles.

2. CAVITATION EROSION

Mechanical abrasion of contaminated surfaces may also be carried out in a liquid medium, thus with no aerosol production, by means of cavitation. On the basis of the study conducted by Alsthom Atlantique Neyrtec at Grenoble cavitation equipment was selected and the suitability of this method of erosion evaluated.

2.1 Principle of the method and objectives of the study

Cavitation results from the local formation of vapour caused by partial vacuums in liquids flowing at high speeds. The vapour structures thus produced are caught in the movement of the liquid and are subject to abrupt recondensation ('collapse') when pressure rises.

The violence of the process can cause erosion to the adjoining walls, a phenomenon which, though normally undesirable, can be used for decontamination purposes. Surfaces are removed, thus reducing radioactivity, which is normally concentrated at the surface.

The objectives of the study conducted by A.A. Neyrtec was to demonstrate the feasibility of the process for such applications. The first stage was to produce a suitable hydrodynamic design. The method finally adopted was to create a flow with radial diffusion; this method gave the best removal rates and was easy to apply. The tests were carried out in a special test circuit at medium pressure (60 bars).

2.2 Cavitation erosion mechanisms

Collapse of the vapour cavities produced by the flow of liquid is a very rapid process, usually measured in microseconds when the general dimensions of the bubbles are of the order of mm. At the end of a collapse the wall speeds of these vapour structures are much greater (by two or three times) than the general speed of flow. Consequently, given the same impact, velocity cavitation is much more effective as a method of erosion than conventional non-cavitating water lances - even those with intermittent spray appliances. However, cavitation requires a much higher flow rate.

To achieve greater abrasive efficiency, the wall speeds at the end of collapse must be increased, and as many cavities as possible should be created so that as many as possible produce impact.

To improve the system it will be necessary to make a systematic study of pressures upstream and downstream, the general volume of flow ang geometrical factors.

Since there is no theoretical method of forecasting erosion in a flowing liquid, we applied the following conventional principles which have already been described in the literature on the subject :

- erosion is highly non-linear in terms of time. It consists of an "incubation" period with no loss of mass, a period in which erosion accelerates rapidly, followed by erosion at a more or less constant rate;
- the effect of the general speed of flow is very important. It is agreed that the rate of erosion produced by the flow is proportional to a power of the order of 5 to 9 of the velocity;
- little is known of the scale effects of geometry.

For the purpose of decontamination, a given layer of material must be removed at a minimum water flow rate.

The theoretical analyses carried out by Neyrtec concerning the hydrodynamics of the process suggest that overall dimensions should be reduced with a view to reducing specific power (2).

2.3 Experimental work

The aim of the first phase of the study was to determine the geometrical characteristics of the most abrasive cavitation and to measure the damage likely to be caused.

The general configuration of the cavitation vessel is shown in fig. 15. In this case a convergent-divergent flow was set up in a venturi tube, the upper wall of which formed the cavitation instrument. The lower wall was the material to be treated. Cavitation was produced in the neck of the venuti tube by means of an obstacle connected to the instrument.

To avoid excessivce cavitation and ensure the highest possible pressure in the collapse zone, pressure was again applied downstream of the cavity by means of an adjustable slide valve.

The first study made it possible to optimize the type of cavitation applied and the downstream pressure layer. Pitting cavitation was found to be the most abrasive, the curve for loss of mass reaching a maximum when the downstream counterpressure was varied. A downstream valve is therefore desirable.

Although upstrem pressure was relatively low (60 bars), substanatial erosion of stainless steel was noted after 1 hour, and after 8 hours of tests 1.2 mm^3 was removed from a duraliminium plate.

Once this first series of tests had demonstrated the feasibility of the process, a second study (EEC contract No DE-B-010-F) was conducted with a view to reducing the specific pressure to be applied and to obtain configurations which were geometrically easier to implement.

Two main types of configuration were adopted - one using rotational symmetry as illustrated in fig. 16, the other combining the effects of an axial jet and vortex flow (fig. 17).

The most abrasive configuration uses rotational geometry, with or without a central obstacle. Numerous tests were carried out on various materials under different flow conditions.

By applying the most extreme flow conditions, with an upstream pressure of 56 bars, a downstream pressure of 14 bars and a flow rate of 2.4 l/s, the following amounts of metal were removed :
60 mm^3 pure aluminium in 10 min and 0.3 mm^3 steel in 10 hours.

In terms of specific power, specific loss of mass may be deduced thus :

mg/(min x kW)		
duraluminium	6.8	10^{-3}
aluminium	1.0	
stainless steel	4.6	10^{-4}
carbon steel	3.9	10^{-3}
plexiglass	1.8	10^{-2}

2.4 Extrapolation to higher pressures

The loss of mass caused by cavitation of flow was considered proportional to a power of between 6 and 9 of the velocity, which corressponds to an upstream pressure of between 3 and 4.5. Specific erosion is thus proportional to a power of between 1.5 and 3 of the pressure, and an increase in pressure would improve the efficiency of the system.

Since a pressure of 60 bars results in a loss of mass for stainless steel of 2.8 mg/h, under identical conditions with an upstream pressure of 600 bars, erosion would be between 2.8 g/h and 88 g/h. Using the same nozzle, specific erosion would increase from 4.6 x 10^{-4} mg/(kW x min) at

60 bars to between 1.45 x 10^{-2} mg/(kW x min) and 0.45 mg/(kW x min) at 60 bars. On this basis, and by applying a pressure of 600 bars, erosion of a layer of 2 μm of stainless steel measuring $1m^2$ would be possible in 10 minutes to 5 hours.

2.5 Conclusion

The study be carried out by Neyrtec under contract for the EEC demonstrated the feasibility of a means of decontamination by cavitation erosion. Improvements in the second phase of the study resulted in an increase in specific erosion by a factor of 40.

This first study was conducted using a test circuit of limited performance. To continue, it would be necessary to use a high-pressure installation which would provide indications as to the removal rates which could be achieved under actual operating conditions, possibly using contaminated samples.

3. DECONTAMINANT GELS

Radioactive installations are often decontaminated by the spraying of aquous solutions. These produce a large amount of effluent, and contact between the items to be decontaminated and the chemical substances is of short duration. These drawbacks can be offset by applying the decontaminants in gel-based media.

At CEN-Saclay we have developed gel-based decontaminant compounds and assessed their suitability by carrying out comparative contamination experiments in the laboratory and by decontaminating samples of materials used in the nuclear industry.

3.1 Manufacture and physical properties of decontaminant gels

3.1.1 Glycerophtamic gels

Glycerophtalic gel is used as an acid support medium. We chose solphamic, oxalic and sulphuric acid.

Sulphamic acid, which is used mainly to decontaminate mild steel and iron (reduction of iron oxides), is added slowly at 120 to 130°C to the glycerophtalic gel, which is made up beforehand by adding phthalic anhydride to glycerine at that temperature. the compound has a viscosity of about 15 P (1500cp).

Oxalic acid, used for the pickling of stainless steels, is added hot to the glycerophthalic gel (viscosity of around 5P).

Sulphuric acid, used for the decontamination of alloys of copper, silver or nickel in association with nitric acid was in this case dissolved alone and cold in the proportion of 50 ml of concentrated acid (d = 1.83) per litre of gel. The compound thus produced has a viscosity of 10 P.

3.1.2 Glycerophosphoric gel

Phosphoric acid is added to substances used to decontaminate iron and mild and stainless steels.

Glycerophosphoric gel is formed by dissolving concentrated phosphoric acid (d = 1.70) in glycerine, after which it is heated to 100°C for about 1 hour. Gelling occurs on cooling (viscosity of around 0.7 P). One or two moles per litre of phosphoric acid (d = 1.70) or 100 ml/l of the phosphoric detergent 'Paracodine 120' (TM) were added to this gel.

3.1.3 Silica gel

Silica gel is used for decontamination, primarily as a support medium for nitric or sulphuric acid.

Gelling occurs at ambient temperature after concentrated nitric acid is added slowly to a solution of 0.4 mole/l of sodium silicate. The silica gel thus formed has a viscosity of 1.3 x 10^3P. The addition of 1.1 moles of concentrated nitric acid per litre of gel destroys the molecular structure and produces a colloidal suspension with a viscosity of around 0.5 P.

Inclusion of 50 or 100 ml of concentrated sulphuric acid results in a compound with a viscosity of between 1 and 10 P.

Silica gel containing sulphuric acid adheres well to stainless steel (1 500 mg/dm^2) and is almost totally removed by water (residual surface mass of 200 mg/dm^2).

3.1.4 Diopside gel

This gel is stable in an oxidizing and basic medium.

Diopside gel is formed by adding sodium silicate to an equimolecular mixture of calcium nitrate and magnesium nitrate in the presence of sodium. It has a viscosity of around 2 P.

Sodium hydroxide (40 g/l of gel) and potassium permanganate (10 g/l of gel) are added to the diopside gel thus formed. Viscosity decreases to about 0.5 P.

3.2 **Comparative experiments in the decontamination of samples contaminated in the laboratory**

3.2.1 Contamination in the laboratory

A known quantity of a radioactive solution of fission products or plutonium-239 is deposited on each representative sample of a material.

After drying by ventilation at ambient temperature, the activity A_0 is measured. The activity is easily removed from the sample plate by spraying with water. After ventilation drying the 'fixed' activity A_1 is measured under the same conditions as the activity deposited.

3.2.2 Decontamination by gel deposit

The decontaminant gel is deposited on the plate. After 24 hours of contact the gel is dissolved and removed by spraying with water or with solutions which destroy the gel. The residual activity A_2 is measured after drying. The decontaminant effect of the gel in question on the material being treated is represented by the decontamination factor, which is the quotient of the fixed activity A_1 divided by residual activity A_2.

3.2.3 Decontamination by soaking in an aqeuous solution

To compare the effectiveness of the decontaminants in a gel or used alone, identical contaminated plates are also soaked for 1 hour and then placed vertically in a decontaminating solution with no associated mechanical operations.

The decontamination is defined as for decontamination by gels.

3.2.4 Experimental conditions

Each test is carried out 3 times on 10 cm square plates. On the central area of each plate the following is deposited :

- either 1 ml of a solution of caesium-137, strontium-90 and yttrium-90 in a decinormal or centinormal nitric medium with an overall activity of 15 or 1.5 kBq (0.4 or 0.04 μCi);
- or 1 ml of a solution of plutonium-239 with an activity of 15 or 1.5 kBq (0.4 or 0.04 μCi) in a decinormal or centinormal nitric medium.

The radioactivity deposited (A_0), the fixed activity (A_1) and the residual activity (A_2) are measured under the same geometric conditions using a unit for measuring α or β radioactivity fitted with proportional counters (argon-methane).

For each series of 3 plates, the decontaminating power of the entire contaminant solution, the contaminated material, the gel, and the decontamination solution is represented by the geometric mean of the 3 decontamination factors.

3.3 **Results of the comparative decontamination tests**

During the study we added commercially available surface active agents to the gel compounds in an attempt to improve their performance. The carboxylic acids, in particular, are far less effective decontaminants of steels in glycerophthalic gel than in the aqueous solution.

3.3.1 Decontamination of stainless steel

All the test pieces were cut from the same stainless steel plate (Z2CN18-10 - AISI : 304L), thickness 0.4 mm.

The susceptibility (A_1/A_0) to contamination of this stainless steel is approximately 2% in the case of fission products and 29% for plutonium in decinormal nitric solutions.

Decontamination of fission products by immersion in an aqueous solution is most effective with oxalic, nitric and phosphoric acid, which yield decontamination factors of 108, 189 and 289 respectively.

The effectiveness of sulphuric acid is of the same order in an aqueous solution as when incorporated in glycerophthalic gel (decontamination factor (DF) = 102 in solution, DF = 165 in the gel).

The phosphating compounds, which are effective in decontaminating stainless steel by soaking (DF of around 200), remain effective when incorporated in glycerophosphoric gel (DF of 20 to 40).

Similarly for nitric acid, the decontamination factor of fission products is approximately 50 in silica gel and is increased to 400 by soaking in an aqueous solution.

Glycerophosphoric gel in combination with a commercially available phosphating detergent "Paracodine 120" (R) and with a surface active agent based on nonylphenol polyglycol ether "Arkopal N 100" (R) is as effective as silica gel with nitric acid and the same surface active agent (DF = 45).

The surface active agent in glycerophosphoric gel results in substantial decontamination of plutonium (DF = 200). Sulphuric acid in glyceropthalic gel (DF = 165) or in silica gel (DF = 114) is a better decontaminant of plutonium in a gel than as a solution for immersion.

Decontamination of plutonium by sodium in combination with potassium permanganate is good in an aqueous solution (DF = 50), and especially so in gel diopside (DF = 740).

In the case of the decontamination of fission products, the most effective viscous compound (DF = 125) is made up of 2% of isopropanol (in solution) with 20% of Marlox NP 109 (R) (produced by adding alkylphenol and alkylene oxide) and 50% Marlophen 86 S (R) (ethoxyl alkylphenols). This decontaminant has a viscosity equal to 600 cp (0.6 P) after 24 hours.

3.3.2 Decontamination of mild steel

In an aqueous solution we obtained decontamination factors in excess of 1300 using phosphoric acid and sulphuric acid for plutonium, and greater than 100 using these acids and sulphamic acid for fission products.

For fission products the most effective gel compounds contain nonylphenol polyglycol ether "Arkopal N 100" (R), 10 g/l of which is added to the glycerophosphoric gel (DF = 168) or the glycerophthalic gel containing 100 g/l of sulphamic acid (DF = 110).

In the case of plutonium, the nitric acid in silica gel, glycerophosphoric gel and the sulphamic acid combined with the surface active agent "Arkopal N 100" (R) in glycerophthalic gel produced decontamination factors in excess of 200.

As in the case of stainless steel, the reactivity of acids is greatly reduced by glycerophthalic gel, the inhibiting power of which can be offset by adding nonylphenol polyglycol ether.

3.3.3 Decontamination of an aluminium-magnesium alloy

In an aqueous solution decontamination factors for fission products are approximately 30 in the case of sulphuric acid, 40 in that of sulphamic acid, 60 in that of phosphoric acid and 90 in that of oxalic acid. In the case of plutonium the decontamination factors are approximately 60 using nitric or sulphuric acid and almost 6 000 using phosphoric acid.

Glycerophosphoric gel is effective both for fission products (DF = 70) and for plutonium. Where this aluminium alloy was contaminated with plutonium, we opted for glycerophosphoric gel alone (DF = 144) or combined with Arkopal N 100 (R) (DF = 6146) and the silica gel containing nitric acid (DF = 71) or sulphuric acid (DF = 810).

3.3.4 Decontamination of copper

Sulphuric acid, phosphoric acid alone or combined with the phosphating detergent Paracodine 120 (R) and nitric acid are good decontaminants of copper and yield decontamination factors generally in excess of 100 in a solution.

The silica gel combined with nitric or sulphuric acid, when used on copper, effectively decontaminates fission products (DF = 116 and 64) and plutonium (DF = 238 and 363).

The nonylphenol-polyglycol ether Arkopal N 100 (R) improves decontamination using glycerophosphoric gel (DF = 108 for fission products and 457 for plutonium).

3.3.5 Decontamination of plexiglass

Since it is difficult to fix radioactive contamination on this material, we used decinormal and more highly concentrated nitric contaminant solutions (0.4 Ci/ml - 1.5 x 10^4 Bq/ml).

Mean susceptibility to contamination is 2.8% for fission products and 15% for plutonium.

The decontamination factor obtained for fission products by soaking in a solution is approximately 220 in the case of oxalic acid and 120 for Parcodine 120 (R), a phosphating detergent.

In a gel medium, the highest decontamination factors are obtained with Paracodine 120 (R) incorporated in glycerophosphoric gel : DF = 87 for fission products and 66 for plutonium) and with the diopside containing sodium and permanganate : DF = 79 for fission products and 36 for plutonium.

3.4 **Area of application of gel decontamination processes**

Since the gel decontaminant compounds chosen have a viscosity of a few poiseuilles and are highly stable with time, they are compatible with the low-pressure spraying methods which are well suited for use in radioactive decontamination. The simplicity of the equipment used permits decontaminationin a hostile environment.

On the basis of 30% spraying efficiency, the cost in 1983 of treating a flat sheet of stainless steel one metre square would be as follows :

FF 150 per m^2 for glycerophthalic gel with acid;
FF 110 per m^2 for glycerophosphoric gel;
FF 8 per m^2 for silica gel with sulphuruc acid;
FF 16 per m^2 for diopide gel with sodium and permanganate.

The amount of effluent produced for stainless steel by spraying followed by the decomposition of the gel decontaminants is roughly as follows :

20 l/m^2 for glycerophthalic gel;
10 l/m^2 for glycerophosphoric gel, silica gel and diopside gel.

3.5 **Conclusion**

Glycerophthalic, glycerophosphoric, silica and diopside gel are all compatible with most decontaminants. The comparative decontamination study made it possible to select at least one compound yielding high decontamination factors (in excess of 50) in the laboratory for mild steel, stainless steel, aluminium, copper and plexiglass.

Owing to its suitability for use with remote control methods and to the low production of secondary effluents, the gel decontamination process permits in-situ decontamination of highly contaminated installations with no additional exposure of personnel.

The development of liquid erosion techniques by cavitation or by high-pressure spraying and the incorporation of decontaminants in gels contributes towards improving the effectiveness of the decontamination of decommissioned nuclear power stations.

REFERENCES

(1) Nature of deposits on BWR and PWR primary system surfaces. Relation to decontamination. A.B. JOHNSON, JrB. GRIGGS and F.M. KUSTAS. Pacific Northwest Laboratories and R.A. Show Electric Power Research Institute, Water Chemistry 11 BNS 1980, Paper 5.4

(2) Y. LECOFFRE, B. VALIBOUSE, J. MARCOZ, "Generator of cavitation Vortex", ASME Meeting Cavitation Erosion in Fluid Systems, BOULDOR (Colorado) - June 1981.

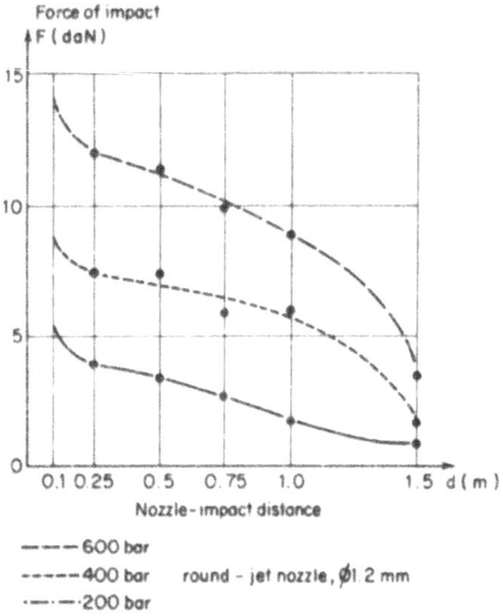

Fig. 1 - Force of impact as a function of nozzle target distance and pressure

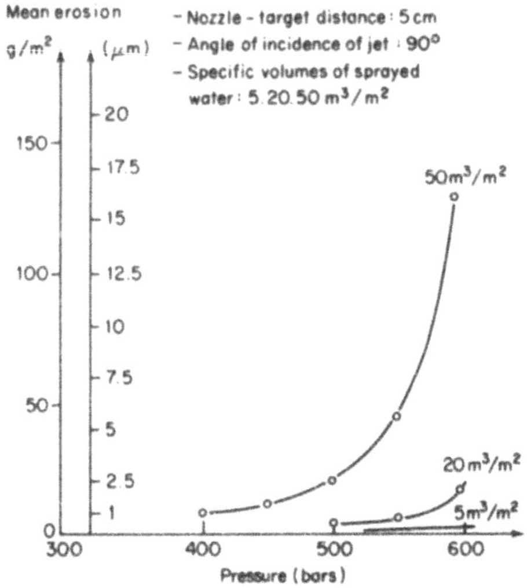

Fig. 2 - Mean erosion of a mild cold rolled steel (NFA 36401) as a function of sprayed water pressure

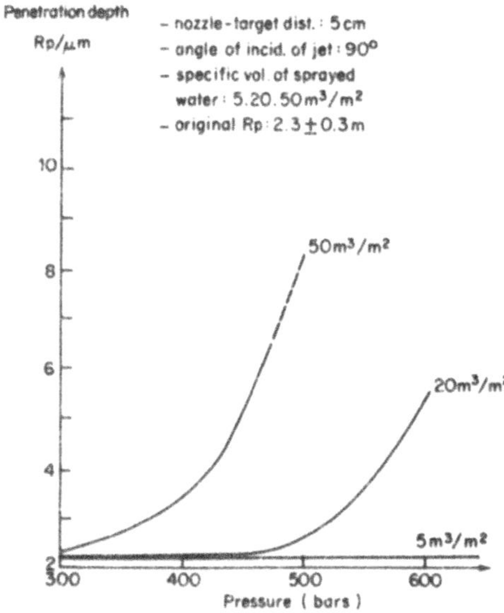

Fig. 3 - Penetration depth in mild cold-rolled steel (NFA 36401) as a function of sprayed water pressure

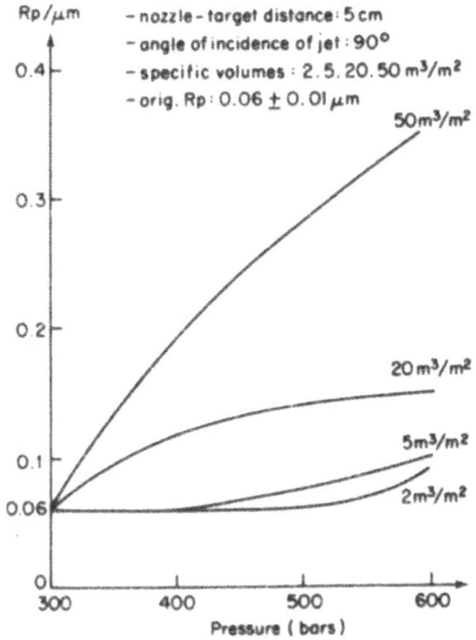

Fig. 4 - Penetration depth in a stainless steel (304 L) as a function of pressure of sprayed water

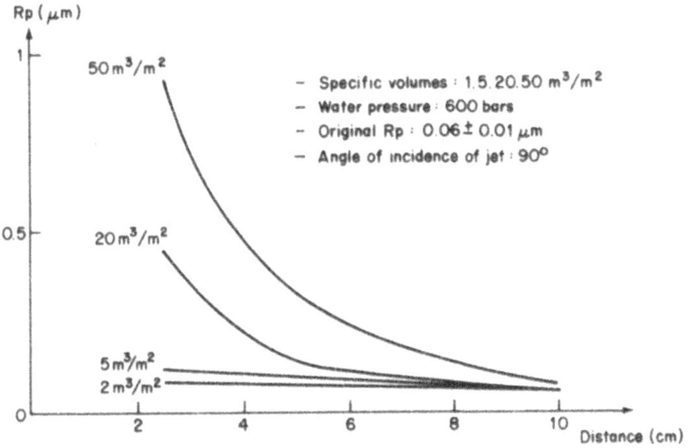

Fig. 5 - Penetration dept in stainless steel (304 L) as a function of nozzle-target distance

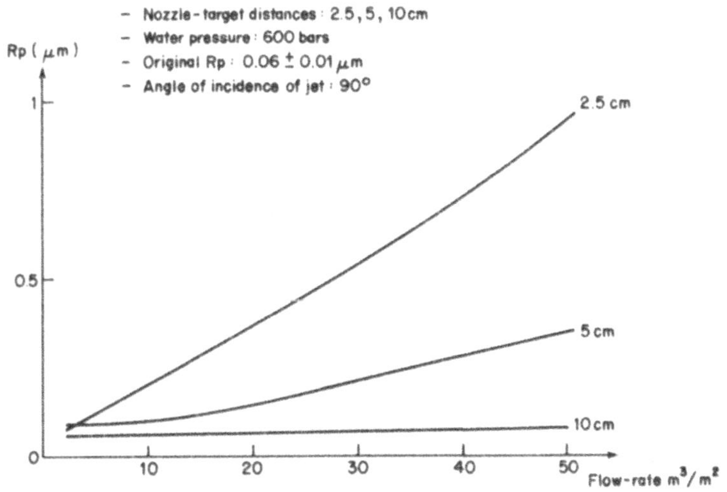

Fig. 6 - Penetration dept in stainless steel (304 L) as a function of the specific volume of sprayed water

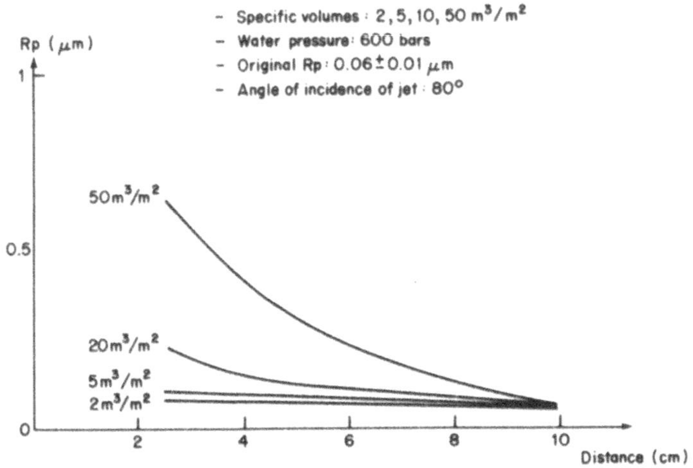

Fig. 7 - Penetration depth in stainless steel (304 L) as a function of nozzle-target distance

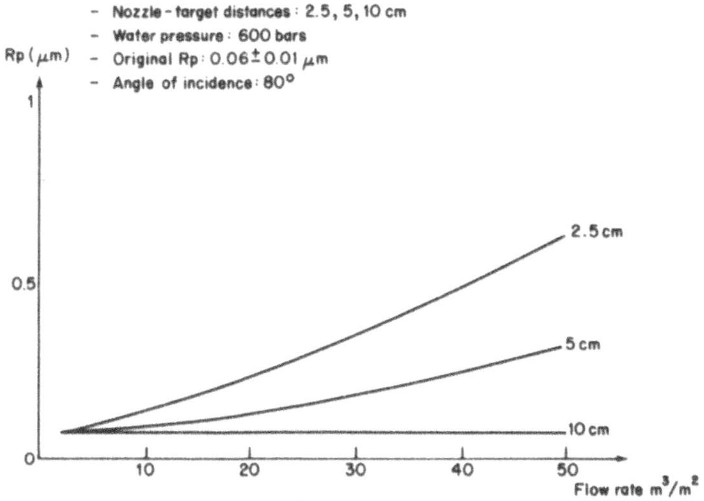

Fig. 8 - Penetration depth in stainless steel (304 L) as a function of the specific volume of sprayed water

DECONTAMINATION UNIT FOR PLANT MATERIALS USING WATER UNDER HIGH PRESSURE (HYPERBAR)

Fig. 9 - Spraying chamber

Fig. 10 - General view

Fig. 11 - Spraying cell

Fig. 12 - Waste purification

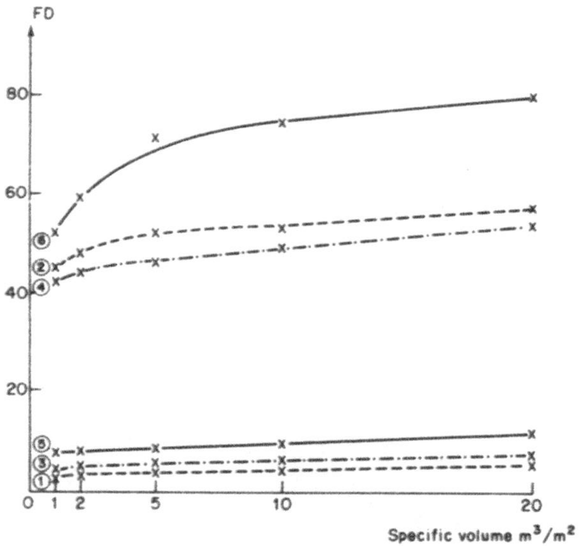

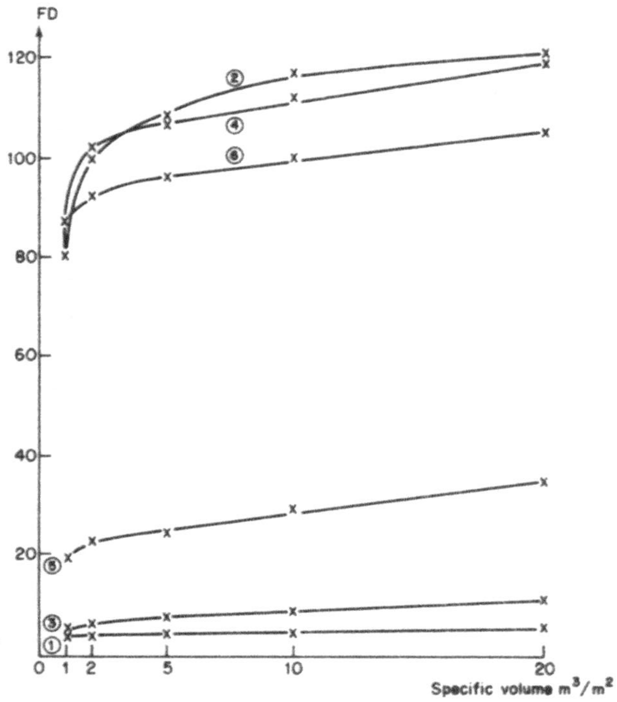

① 25 cm without gel ② 25 cm with gel ③ 10 cm without gel ④ 10 cm with gel ⑤ 5 cm without gel ⑥ 5 cm with gel

Fig. 13 - Decontamination of ferritic steel samples coated with hematite. Combination of chemical action with high-pressure water lance

EFFECT OF NOZZLE-TARGET DISTANCE P = 150 bars

Fig. 14 - Decontamination of ferritic steel samples coated with hematite. Combination of chemical action with high-pressure water lance

EFFECT OF NOZZLE-TARGET DISTANCE P = 600 bars

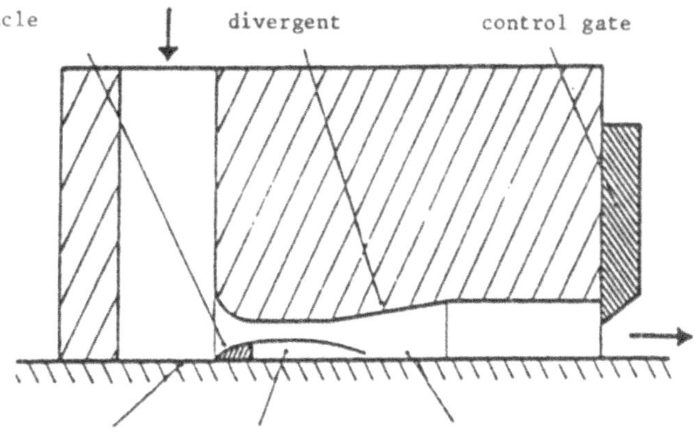

Fig. 15 - Cavitation erosion - cavitation device - Principle

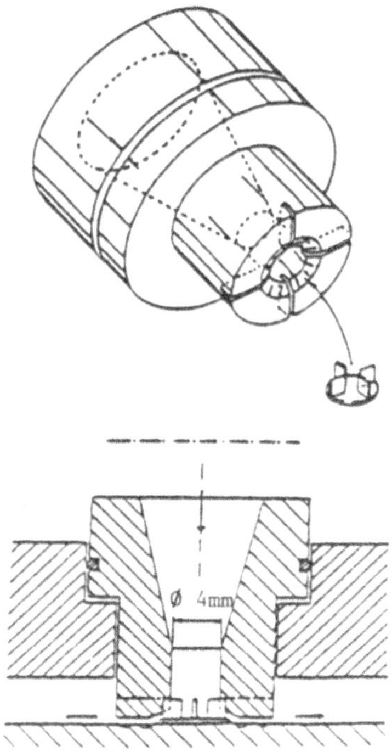

Fig. 16 - Radical flow cavitation device

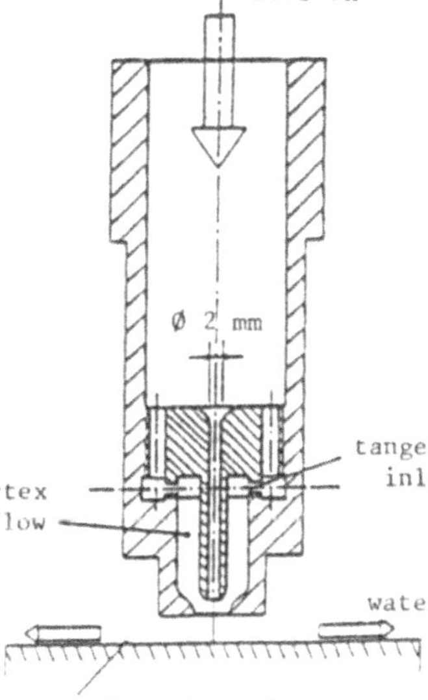

Fig. 17

FACTORS TO BE CONSIDERED IN DECIDING WHETHER TO DECONTAMINATE FOR UNRESTRICTED RELEASE

G. LOERCHER (NIS), A.M. CHAPUIS (CEA)
J. ESSMANN (PREAG)

Summary

In the event of the total dismantling of a nuclear power station, a decision has to be reached as to whether large masses of material should and/or can be decontaminated to such a degree that they can subsequently be recycled or taken to a public refuse dump without being subject to radiological protection regulations. Technical, legal and economic aspects have to be taken into account in such a decision. The estimated costs for the final storage of radioactive waste range from DM 4 000 to DM 6 200 per Mg of net mass, depending on the type of final storage. In the Federal Republic of Germany, 50 Mg of steel scrap and 30 Mg of brass have already been decontaminated and melted down as part of the re-equipping of Würgassen nuclear power station.

1. INTRODUCTION

The total dismantling of a decommissioned nuclear power station involves large quantities of contaminated plant components. Sufficient knowledge of possible decontamination processes and their costs is required in order to reach a reasoned decision as to whether the components should be decontaminated or disposed of as radioactive waste. The legal side (the licences necessary) also needs to be considered, as well as the recycling of radioactive plant components (e.g. recycling in the nuclear sector). The latter aspect, however is not covered here.

Decontamination, as used below, refers to the actual removal of decontamination or contaminants (the radionuclides) to a legally permissible level. Components decontaminated in this way are then no longer a radioactive substance and are therefore not subject to any radiological protection regulations and can be put to unrestricted use.

Unrestricted use (uneingeschränkte Verwendung) as used here is equivalent to the English "unrestricted release".

Extensive research and current projects under the Community's "Decommissioning of nuclear power plants" programme have shown that decontamination processes aiming at unrestricted use are still at an early stage of development and there is therefore a great reluctance to publish papers on the subject. No details of the cost of the decontamination processes could be obtained to date, mainly because large-scale application has not yet been adequately considered.

2. LEGAL ASPECTS

Decontamination for unrestricted use requires the setting of precise and clear limit values.

Such unequivocal limits must comply with existing legislation and be compatible with measurement technology and techniques.

In order to be able to decontaminate a component for unrestricted release, it is necessary to calculate or establish with certainty that the residual activity in that component will at most produce exposure consistent with the principle of "de minimis non curat lex" (the law does not concern itself with trifles), and to provide evidence to that effect to an authority. It is therefore understandable that there is a wish to see one or more overlapping and appropriate limit values laid down by law or regulation.

The Member States of the European Communities are covered by the Euratom basic safety standards /1/, which were drawn up on the basis of the Recommendations of the International Commission on Radiological Protection (ICRP Publication 26) /2/, and which are to be applied in all Member States. This should lead to standardization in 1984 of the basic principles applied in the Member States. There are three main types of limit values :

- dose or dose rate limits;
- activity limits;
- intake limits.

The dose or dose rate limits relate directly to that which is to be protected, in this case man. This means that the way in which doses or dose rates are calculted takes into account the actual effect on man. The calculations make due allowance for the type of radiation, the frequency of each type of radiation per decay event, effective quality factors and the characteristics of specific organs. The activity limit values in the Euratom basic safety standards and the legislation of Member States are based on such dose and/or dose rate limits.

Activity limit values are expressed in absolute activity, specific activity and area activity. Appropriate measuring devices and procedures and exact calibration make it easy to work with these limits. However, in the case of aggregate limits, i.e. when several different radionuclides are present, it is essential to incorporate weightings into the total calculation, since it is not the total number of decay events that is of primary importance, but the average energy released per decay event and the type of radiation.

Intake limit values are limits which specify the maximum permissible intake via air (inhalation), water and food (ingestion). The final portion of the exposure pathway, uptake and subsequent excretion by man, is considered in these limits. The "biological half-life", committed dose and behaviour in specific organs are all taken into account. Thus if it can be proved that uptake of a substance with a certain specific activity will never reach the ceiling values permitted by these limits (for general health reasons and as a result of retention and distribution mechanisms), the substance is be suitable for unrestricted use.

The individual limit values and the laws, regulations or guidelines in which they are found are given in /3/. When looked at more closely, the limits or derived limits in existence today, which are not expressly designed for application in the context of the unrestricted use of formerly radioactive components, leave a number of questions unanswered.

Two limit values are currently laid down in a number of licences in the Federal Republic of Germany, both of which are ceiling values. The first is a surface activity limit value of 0.37 Bq/cm^2 (without alpha emitters), and the second a specific activity limit value of some 3.7 Bq/g (valid for Co-60). No special radionuclide is applied in the surface limit, and only a weighting based solely on the broad distribution implied in the "free limit" is applied in the specific activity limit. Both limit values are aggregate limits, i.e. individual radionuclides must remain within the limits. The difficulties associated with these limits are too complex to be summarized. For full details refer to /3/.

Substantial quantities of materials have already been released for unrestricted use in the Federal Republic of Germany on the basis of these limits (cf. section 6).

In other countries of the European Communities, surface limit values exist in some cases which are higher (with or without alpha emitters) by a decimal power of 10. In France, a specific activity limit value of 74 Bq/g for individual cases has been approved.

3. CONTAMINATION

Contamination processes are extremely complex and require many more studies before they can be understood sufficiently accurately.

In addition to low phenomena involving stagnant water areas and speed distributions; chemical processes and conditions also play a role.

Because of nuclear transformation, a noble gas such as Xe-137, for example, can turn into an isotope of the element caesium, the strongest base-forming element. Finely distributed particles and water-soluble contaminants can escape with the slightest leakages and be carried through the air to be deposited as moisture. Size and chemical behaviour are then of paramount importance as regards the capacity to penetrate and depth of penetration through the decontamination coating into the concrete.

Chemical behaviour also plays a role in determining the liranness of deposits on steel components. The pressure and temperature of the medium, too, are important. Components that have been subjectd to higher pressurs and temperatures during operation will have contamination layers that are more difficult to remove than those in other areas.

Until now a distinction has been made between contamination at temperatures above 200° C and contamination at lower temperatures. Further research will show whether such a distinction is also necessary in the case of the hard decontamination planned prior to the unrestricted use of material as scrap.

The depth distribution of contamination generally displays an exponential kind of pattern, explained by conformity with the filter effect and the diffusion theory, and the gradual build-up of layers. Since contamination is not totally uniform during operation, i.e. exposure varies from one period of time to another, it is difficult to interpret information on contamination behaviour, even when sufficiently accurate measurements are available.

Nevertheless, studies of penetration or accumulation behaviour should not fail to take into account the decay patterns of individual radionuclides. Penetration or accumulation of activity occurs over a period of time. Deeper layers date back further than upper layers. With the passage of time, the different decay rates of the radionuclides alone will lead to different relative activity levels. This must not be interpreted as indicating different depths of penetration.

A "decontamination factor" is often quoted when means of decontamination are being considered and assessed. This is meaningful only if the contaminated part is practically identical in each test. The contamination must therefore be produced artificially under the same conditions, or a single part be cut into several pieces to which the various decontamination processes are then applied. However, "artificial contamination" cannot be directly compared with contamination which has been exposed to different pressures and temperatures over an extended period.

A high decontamination factor can naturally be reported in instances involving merely the successful removal of the first contamination layer, which does not adhere so strongly. Only the second or third step in the process will show the effectiveness of a process designed to remove practically all contaminants.

4. IDENTIFICATION OF DECONTAMINABLE COMPONENTS

Decontaminable components should be identified as early as possible in order to avoid wasting time and money on unnecessary decontamination work. Simple decontamination to remove weatly adhesive contamination is still useful because it reduces radiation exposure and facilitates subsequent handling. All other measures, which ultimately do not achieve the desired objective, merely cost time and money.

Greater experience with contamination, decontamination processes and handling and disposal of secondary waste will reduce the risks in decisionmaking. It is now possible to select a number of criteria on which decisions can be based.

One must first compare the two approaches, namely, decontamination for unrestricted use, or direct transfer to some type of final repository for radioactive materials.

The exposure of personnel and the respective costs of each approach must be considered. These are the principal criteria. However, there are other associated criteria which, in some cases, can lead to the exclusion of both approaches. These are :

- the type and degree of contamination;
- the geometry of the components;
- the mass of the components to be decontaminated.

We should like to make a few comments on these criteria.

It is increasingly possible to determine the type and degree of contamination. Particularly careful consideration must be given to whether components exposed for long periods to high pressures and temperatures (e.g. primary circuit) can be decontaminated, whereas components contaminated only by contact with air, whether through moisture or other factors, clearly are decontaminable. In between lies a grey area which needs further investigation before a proper decision can be taken, based on operating data and/or tests.

The geometry of the components plays a major role. For one thing, decontamination processes are mostly not able ot ensure even removal or cope with particular corners and cavities, which is precisely where substantially higher contamination can occur. And with complicated geometries, the measurements required to produce evidence that the limits have been observed can be carried out only with great difficulty or not at all.

The mass of components of the same type to be decontaminated, i.e. components which will be subjected to the same decontamination process, is important in two ways. First there is little likelyhood that very large masses will be transported to a final repository, simply because they contain a few grammes or kilogrammes of radioactive material. And, second, the decontamination process used must be suitable for such large masses. When selecting the process, one must ensure that the time needed for decontamination remains within acceptable limits - and this is possible - since personnel costs, a not unimportant factor, increase in proportion to the time needed, whereas the amounts received from the sale of decontaminated material are insignificant. It should be mentioned that the process need not in principle spare the material, on the contrary it can be quite aggressive, since it is not planned to re-use the material in its original form.

An idea of the mass of material that might need to be decontaminated is given by taking as an example two German nuclear power stations, one with a pressurized water reactor and one with a boiling water reactor.

An effort was made to make a distinction on the basis of the abovementioned criteria. The material was divided into three categories : non-decontaminable material, decontaminable material and material on which no decision can yet be taken.

Components which today are still very difficult to decontaminate or where decontamination is not viable can naturally be reclassified at a later date on the basis of experience and in the light of progress in decontamination techniques (Tables 1 and 2).

5. COST

Cost, as so often, is the decisive factor. Exposure, i.e. the collective dose, is of course important when the two approaches are compared. But since the collective dose can be kept low, at some expense, with appropriate measures, for example shielding and the use of remotelycontrolled systems, the necessity of such measures is the only overniching consideration, and this again affects the costs.

The costs for direct final storage are shown to provide a yardstick for the cost-effectiveness of the different decontamination processes.

The costs for direct final storage comprise :

- fragmentation costs : the components cut up on site must be further reduced to fit into drums or containers;
- conditioning costs : the fragmented components must be incorporated into concrete or other binding materials in the containers;
- transport costs : the containers must be shipped to the final repository (road and rail transport costs);
- final repository costs : the cost of constructing and maintaining the final repository must be compensated by charges. In the case of sea dumping, the loading and shipping costs are considered as final repository costs.

The costs shown in Tables 3 to 5 are costs for decontaminable components. They therefore relate only to components with a relatively low degree of contamination. These are the only costs that can be compared with the decontamination costs for unrestricted disposal. The latter too are subject to fluctuation as a result of new developments in conditioning and packaging or final storage. The estimated costs for the final respository, for example, have already risen and this will not necessarily be offset by the development of more suitable packaging. Decisions as to costeffectiveness can therefore only be made in individual, practical cases.

Certain comments must be made on the question of decontamination costs for unrestricted use.

The calculations must take into account all individual costs, including subsidiary costs. Specific costs must include the purchase, delivery and storage of the decontamination medium, and total consumption of the medium.

Since they must be undertaken with extreme care and generally require recording, measurements designed to serve as evidence of observance of limit values of whatever kind are very labour-intensive. Interim measurements are required to check the decontamination effect. According to experts, such measurements account for approximately one-third of total expenditure.

The amount, conditioning, handling, transport and final storage of secondary wastes must all be taken into account.

In France, an average of approximately FF 10/kg is considered to be a realistic cost for full decontamination, taking account also of decontaminability and the processes used, but no published material is available to back up this figure.

6. PRACTICAL EXECUTION OF FULL DECONTAMINATION

Non-destructive recycling of radioactive components was carried out in the Federal Republic of Germany at Würgassen nuclear power station as part of the re-equipping of the plant (1982/83). A total of approximately 900 Mg of contaminated metal scrap was removed for non-destructive recycling. The aim of this type of disposal is to keep to a minimum the amount of radioactive waste to be packaged and conditioned, placed in interim store and subsequently in a final repository. The fully decontaminated material was to be used as normal scrap. The entire 900 Mg of material were therefore transferred to the Karlsruhe and Jülich research centres for decontamination. The amounts are shown in Table 6.

Small metal parts (Ø < 50 mm) were fragmented at the plant, pelletized using a high-pressure pelletizer, packed in 200 litre drums and taken to the shipment hall at KWW for interim storage.

6.1. Shipment and documentation

Specially developed 10' steel and concrete containers, handled by crane and fork-lift truck, were used for shipment.

Components with a higher surface dose rate were transported in open-top concrete containers with a wall thickness of 120 mm. These containers too have protective coatings on the inside and outside which are easily decontaminable. The steel containers are 10'ISO containers with the same leaktinghtness and coating as conventional waste drums (strong industrial packaging). Open-top containers with and without doors were used as variants.

In order to avoid spreading contamination at the power station and decontamination plant, the ends of the pipes were sealed with ready-made caps and the individual components doubly wrapped in foil before being packed into the containers. All components to be packed into the containers were given a permanent identity code by the disassembler. The source (system), material, surface dose rate, weight of individual components and total weight, activity and surface dose rate of the container were entered on a material mouvement record card.

6.2. Limit values and decontamination processes

The aim of this type of disposal is to decontaminate to below the permitted limits as much material as possible for subsequent use as general scrap. The legally permissible limit values were 0.37 Bq/cm^2 averaged over 100 cm^2, and 10^{-4} times the free limits laid down in the German Radiological Protection Regulation /4/, giving 3.7 Bq/g in the case of Co-60, averaged over 10 kg. The activity in one shipment must not exceed the free limits by more than a factor of 10.

Some 50 Mg of steel scrap and 30 Mg of brass have been fully decontaminated and melted down so far. Various decontamination processes were used, depending on the type of material, contamination and geometry of the components. These were :

- high-pressure steam blasting;
- electropolishing;
- sand blasting;
- acid baths;
- mechanical processes such as sawing and cutting by abrasion:
- flame cutting

In most cases, two to three decontamination stages, sometimes of the same type, were necessary.

Non-decontaminable material, for example welds, was collected, fragmented, packed into 180 litre drums, pelletized with a high-pressure pelletizer, conditioned with the secondary waste in 200 litre drums and taken back to KWW for interim storage. On the basis of current experience, one can expect approximately 10 % of the material to be non-decontaminable.

6.3. Cost aspects of non-destructive recycling

The cost-effectiveness of non-destructive recycling compared with total removal of radioactive material stems from the following factors :

- no need for fragmentation of the material to suit the packging;
- no need to condition the fragmented material;
- no need for packaging materials;
- no need for interim storage of the waste containers at KWW or other interim stores;
- no need for final storage;
- sale as scrap possible.

Moreover, one cannot yet tell whether further costs will arise for final storage, for example at the Konrad pit in Germany, of the waste containers currently at the power station.

References

/1/ Council Directive of 15 July 1980 amending the Directives laying down the basic safety standards for the health protection of the general public and workers against the dangers of ionizing radiation. Directive 80/836 Euratom, Official Journal of the European Communities L 246, Vol. 23 (17 September 1980), ISSN 0376-9453

/2/ Recommendations of the International Commission of Radiological Protection, ICRP Publication 26 (17 January 1977).ISBN 3-437-10590-6

/3/ Lörcher, G.; Piel, W.
Dekontamination von Komponenten stillgelegter Kernkraftwerke Für die freie Beseitigung EUR 8704 DE 1983, ISBN 92-825-3904-0

/4/ Verordnung über den Schutz vor Schäden durch ionisierende Strahlen (strahlenschutzverordung- StrlSchV), 13 October 1976 (BGBl I S. 2905, ber. 1977, p.84 and p.269).

B W R - P L A N T		
Total mass of controlled area:		200.000 Mg
Mass of "free" building structure:		175.000 Mg
category	component/system	mass [Mg]
not deconta-minable	. total mass of activated components	1,100
	. pipes (partly)	2,700
	. heat-exchangers (some)	780
	. pumps, containers, valves (partly)	1,200
borderline cases	. spent fuel storage pit liner, storage racks	230
	. turbine	2,000
	. condenser	1,200
deconta-minable	. cables with holders	2,100
	. generator	1,000
	. containment	1,100
	. hoists, tools, electric motors control assembly drive	1,300
	. steel structure	2,200
	. vent system	1,250
	. pumps (rest) heat exchanger (rest) valves (rest)	500

TABLE 1 Boiling water reactor masses

P W R - P L A N T		
Total mass of controlled area		150,000 Mg
Mass of "free" building structure		136,000 Mg
category	**component/system**	**mass [Mg]**
not decontaminable	. total mass of activated components	1,100
	. pipes (primary loops)	2,300
	. heat-exchanger	1,200
	. pipes, containers valves, pumps (partly)	1,800
borderline case	. steam generator (secondary part)	600
	. spent fuel storage pitliner, storage racks	230
decontaminable	. containment	2,400
	. steel structure	1.200
	. cables with holders	1,500
	. hoists, motors, tools	800
	. vent system	190

TABLE 2 Pressurized water reactor masses

BASIC DATA

A) 200-L-DRUM

VOLUME	=	200 DM^3
DRUM MASS	=	40 KG
STEEL MASS	=	312 KG
CONCRETE MASS	=	368 KG (FILLING MASS)
FILLING FACTOR	=	0,2 VOL. STEEL/VOL. AVAILABLE

B) AVERAGE PERSONNEL COSTS PER HOUR : DM 90

FRAGMENTATION COST
(STORAGE IN DRUMS)

ONE NEEDS 10 MANHOURS PER MG OF MATERIAL

AND 90 DM PER HOUR PERSONNEL COSTS

THAT GIVES : 900 DM PER MG OF MATERIAL

MATERIAL COSTS OF 5 % HAVE TO BE ADDED

TOTAL 950 DM/MG

CONDITION COST

ONE NEEDS 4 MANHOURS PER MG OF MATERIAL

AT 90 DM PER HOUR PERSONNEL COSTS

PLUS MATERIAL COSTS OF 310 DM PER MG

PLUS DRUM COSTS, AT 3 PER MG, OF 945 DM

TOTAL : 1.600 DM/MG

TABLE 3 Fragmentation and conditioning costs

TRANSPORT COST

ONE DRUM HAS A MASS OF 720 KG (GROSS MASS)

THE MASS OF THE STEEL TO BE PACKED IS 312 KG (NET MASS)

THAT GIVES 2.3 Mg GROSS MASS PER Mg NET MASS

ROAD TRANSPORT COST PER 100 KM : DM 40 PER Mg GROSS MASS

RAIL TRANSPORT COST PER 100 KM : DM 30 PER Mg GROSS MASS

ASSUMPTION : 100 KM BY ROAD AND 400 KM BY MAIL

COST : DM 160/Mg GROSS MASS OR DM 370/MG NET MASS

DM 30/Mg TO BE ADDED FOR SURVEILLANCE

TOTAL : DM 400/Mg NET MASS

FINAL STORAGE COST

SEA DUMPING + EMBARKATION	= DM 945 PER Mg NET MASS
LAND BURIAL	= DM 1 400 PER Mg NET MASS
MINE (ASSE)	= DM 3 200 PER Mg NET MASS

TABLE 4 Transport and final storage costs

TOTAL COST (PER Mg NET MASS)

FRAGMENTATION COSTS	DM	950
CONDITIONING COSTS	DM	1 600
TRANSPORT COSTS	DM	400
FINAL STORAGE COSTS	DM	950; DM 1 400; DM 3 200 SEA DUMPING LAND BURIAL MINE (ASSE)

TOTAL : DM 4 000; DM 4 400; DM 6 200

TABLE 5 Total cost of direct removal

RE-EQUIPPING OF WUERGASSEN NUCLEAR POWER STATION

COMPONENTS REMOUED	Mg	MREM/H
RAIN STEAM PIPES, INCLUDING VALVES AND SUPPORT STRUCTURES	350	1500
FEEDWATER PIPES	100	1500
TURBINE COMPONENTS	100	50
CONDENSER PIPES	350	50
PLATE	50	50
	900	

TABLE 6 Masses and maximum surface dose rates

PARTICULAR ASPECTS OF POST-ACCIDENT PLANT DECONTAMINATION

D. HENNART * and M. LARAIA **

* : Centre d'Etude de l'Energie Nucléaire/Studiecentrum voor Kernenergie

** : Comitato Nazionale per la Ricerca e per lo Sviluppo dell' Energia Nucleare et delle Energie Alternative

Summary

Both authors each spent six months on the site of GPU's Three Mile Island Nuclear Generating Station, where a LOCA caused severe damage to the Unit 2 reactor on March 28, 1979. Working with the local NRC staff, they got insight in the decontamination works and the general evolution of the situation.

After the controlled release of the 1.6 PBq ^{85}Kr from the containment, the accident-generated water (2140 m^3 in the Auxiliary Building and 2600 m^3 in the Reactor Building) were decontaminated by filtration and ion exchange.

Extensive decontamination and dose reduction efforts were carried out in the Reactor Building and in the Auxiliary Building. Several decontamination techniques were tried ; emphasis was put on water flushing (hydrolance, rotating nozzle).

Inspections showed that most of the fuel was gone into a rubble bed, the upper level of which is about 1.5 m below the top of the undamaged core.

Reactor disassembly planning includes load test of the polar crane, reactor vessel head lift, plenum inspection and removal, and debris pickup with specially designed tools.

1. INTRODUCTION

On March 28, 1979 a LOCA at GPU's Three Mile Island Nuclear Generating Station Unit 2 near Harrisburg, Pennsylvania, USA, caused severe damage to the reactor core, and extensive contamination of the reactor building and the auxiliary and fuel handling building. However, only minute amounts of radioactivity were released to the environment and none was injured.

Both authors each spent a period of six months on the site of the damaged plant from October 1980 to April 1981 and from June to December 1983. They worked with the local NRC staff and obtained insight into the decontamination works and the general evolution of the situation.

Although financial and political problems nowadays outweigh by far technological aspects on the TMI scene, this contribution will essentially concentrate on the latter.

2. MANAGEMENT OF ACCIDENT-GENERATED GASES

At the time of the accident the charcoal filters in the ventilation of the Auxiliary and Fuel Handling Building proved ineffective and iodine release rates up to 1.9 GBq/day were registered. One month after the accident, a new ventilation system with four absorber trains was put into operation and the former stack was capped. This dropped the iodine release rate to 37 kBq/day, i.e. 0,05 % of the technical specification limit.

Anyway, some months after the accident natural decay has brought iodine concentrations to negligible values and attention was then focused on the ^{85}Kr-inventory which still precluded entry into the reactor building.

After evaluation of several management methods and increased perception of the hazard of keeping the plant in the then prevailing situation for a long time, the NRC and the plant owner finally agreed, that controlled venting of the containment atmosphere to the environment while monitoring the ^{85}Kr concentration in the reactor building and the meteorological conditions was the safest way to solve the problem.

The main purge, during which about 1.6 PBq of ^{85}Kr were released, was carried out from June 28 to July 11, 1980. It reduced the in-containment immersion dose by about 560 μGy/s* for the β contribution and 4.4 μSv/s* for the γ contribution, allowing a first reactor building entry on July 23, 1980.

Additional purges were made during subsequent months in order to vent the ^{85}Kr dissolved in the sump water and released as the main purge had disturbed the equilibrium. When in-containment entries became more frequent and sump water ^{85}Kr concentrations has decreased, containment was continuously vented.

The containment is still kept at a negative gauge pressure and the environmental radioiodine and ^{85}Kr concentration are regularly monitored at several locations around TMI.

3. MANAGEMENT OF ACCIDENT-GENERATED WATER

3.1. Position of the problem

Large quantities of contaminated water escaped from the Reactor Coolant System during the accident and were collected in the Reactor Building Sump, from which some was pumped over to tanks and sumps in the Auxiliary Building. That water, which contained particulate fuel debris and fission product concentrations up to 5.2 TBq/m^3, was the source of intense radiation fields and this prevented efficient decontamination and recovery of both buildings. Moreover, that volume of medium-level liquid waste constituted the most important environmental threat after completion of the containment purge, since long-term containment integrity cannot be guaranteed.

Therefore, accident-generated water nanagement was considered as a high priority task since the very beginning of the recovery planning.

3.2. Decontamination of the Auxiliary Building water : the EPICOR-II installation

Soon after the accident, the Auxiliary and Fuel Handling Building contained some 2140 m^3 of medium-level liquids with global activity concentrations ranging from 37 to 2200 GBq/m^3. The radionuclides inventory is given in table I.

An ion-exchange installation called EPICOR-II was designed to manage that water and constructed in the so-called Chemical Cleaning Building near the plant. Its main components are as follows (fig. 1) :

- a first -stage ion exchange vessel, filled with strong acid resin beads ;
- a second-stage ion exchange vessel, filled with a strong acid/weak base mixed resin bed ;
- a polishing ion exchange vessel with a classical mixed bed ;
- a clean water receiving tank to collect and monitor treated effluents ;
- an off-specification water tank to collect insufficiently decontamined effluents before recycling to the installation.

* : 200 rad/h, 1.6 rem/h

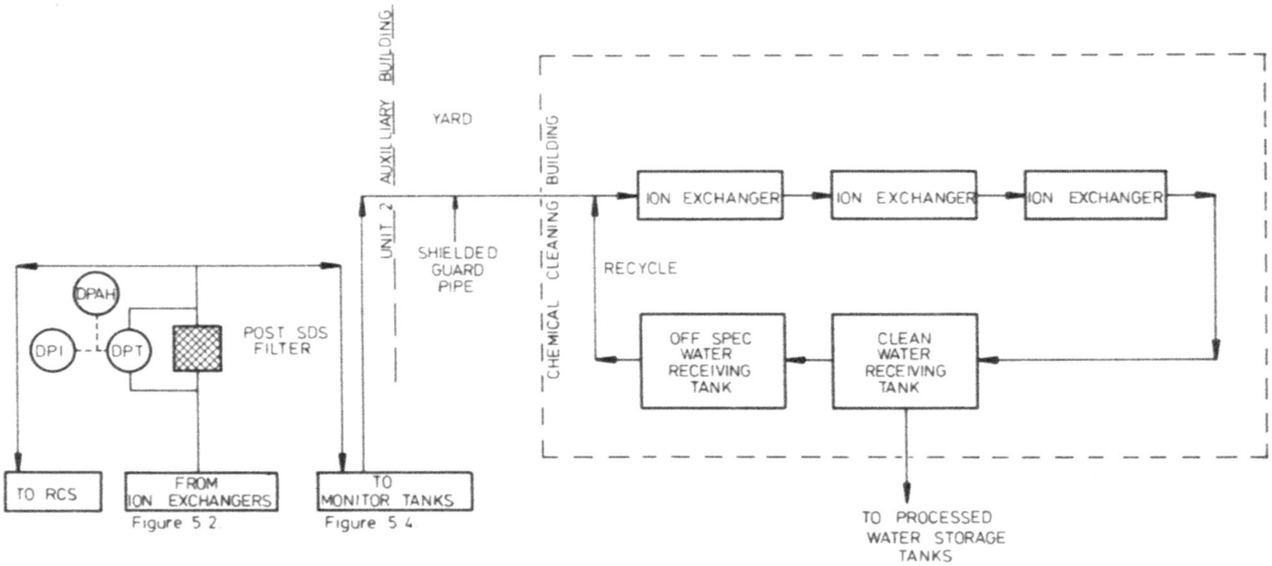

Figure 1. Liquid Flow Path of EPICOR II Processing System

The first stage vessel is intended to trap most of the cesium and strontium nuclides ; the second stage captures anion-bonded nuclides such as iodine and antimony compounds, and the third stage removes the last traces of the nuclides to yield the requested high decontamination factors.

Auxiliary building water treatment was carried out from October 1979 to December 1980 and caused a collective exposure of about 0.15 manSv to the operating personnel.

The treatment performances are given in table 1. The some 1.6 PBq of total activity were trapped in 50 high-activity first-stage and 22 second- and polishing stage vessels.

After interim storage, the high-activity vessels were transported to a DOE site in Idaho for R & D purposes while the other ones were disposed of by commercial shallow land burial. First stage vessels were vacuum-dewatered before transportation to limit hydrogen generation by radiolysis.

The processed water still contains tritium, which is not affected by the ion exchange process. It is stored onsite and used for decontamination works in the plant (see § 3.5).

Table 1 : EPICOR-II processing performances

Nuclide*	Auxiliary Building inventory before treatment (TBq)	Typical decontamination factor	Typical end concentration (kBq/m^3)
^{137}Cs	1250	4×10^6	146
^{134}Cs	227	4×10^6	27
^{90}Sr	69	10^5	35
^{3}H	19	1	8.65×10^6

* : ^{131}I had decayed to a negligible level when processing began.

3.3. Decontamination of the Reactor Building water : the submerged demineralizer system

Various causes prevented Reactor Building water processing before early 1982. At this time, the sump and basement contained about 2600 m^3 and steady leakage from the Reactor Coolant System at a rate of 25 l/h brought the level higher and higher, threatening some vital instruments.

The radioactivity concentration was in the range of 3.7 to 5.9 TBq/m^3, thus significantly higher than in the Auxiliary Building. The radiation intensity at the surface of the sump water was in the range of 8.6 μA/kg(1).

The water was also chemically impure : leakage from the Reactor Coolant System and criticality control considerations caused rather high concentrations of sodium (1100 ppm) and of boric acid (2000 ppm).

Due to the high radioactivity inventories and the wish to condition the waste by vitrification, mineral zeolites were preferred to the more radiosensitive organic ion exchangers for decontamination of that water. The so-called "Submerged Demineralizer System" (further referred to as SDS) was designed on that basis and constructed in the Fuel Handling Building, so using the ability of shielding the loaded ion exchange vessels with the spent fuel pool water.

The SDS consists of the following elements (fig. 2) :

- a floating pump in the Reactor Building basement ;

(1) : 120 R/h.

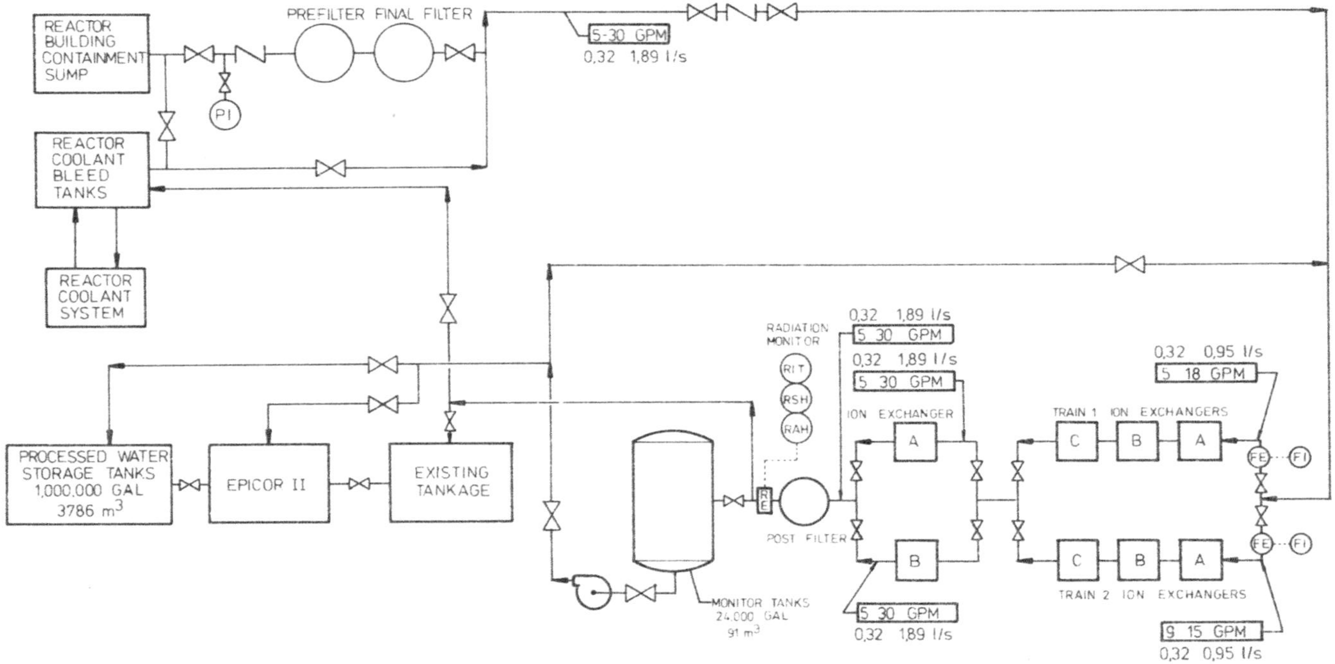

Figure 2. Submerged Demineralizer System Processing Flow Diagram

- a 416 m^3 collecting tank farm in an empty, concrete-shielded part of the fuel pool ;
- two cartridge or sand filters in series. They trap all particles with a size larger than 30 μm and so protect the zeolite beds against clogging and fouling ;
- two parallel trains of each three ion exchange vessels in series, followed by a backup of two vessels in parallel. The flow rate in each line is 315 cm^3/s (1.14 m^3/h). Some vessels may be bypassed if the influent activity is low ;
- a postfilter with a cellulose filter cartridge which removes all particles with a size larger than 3 μm and 98 % of those with a size larger than 0.45 μm ;
- two monitor tanks collecting the treated effluent before recycling to the SDS or polishing in the EPICOR-II installation (see § 3.2) according to the residual activity concentration.

Each ion exchange vessel contains 0.4 m^3 of a mixture of the chabazite type IE-95 and the synthetic Linde 4A zeolites, which are scavengers of cesium and strontium respectively. It can be loaded with 2.22 PBq of cesium and/or 222 TBq of strontium ; the radiation level at 6 m without shielding is then 30.1 μA/kg (1). Shielding is normally provided by a 5 m-deep water layer in the fuel pool.

The vessels are connected to the piping by means of quick disconnects allowing easy changeout. When the first vessel is loaded, it is brought to an interim storage location in the pool and the following vessels are shifted one place higher. A fresh vessel is installed at the fourth (back-up) place. This operating mode minimizes the waste volume to be conditioned while preventing activity leakage out of the system.

The treatment performances are given in table 2. The some 11.9 PBq of total activity were trapped in 9 zeolite vessels which were transported to the DOE site at Hanford for vitrification R & D purposes. They were vacuum dewatered and loaded with palladium catalyst before transportation to limit hydrogen generation by radiolysis. Polishing of the SDS-treated effluent in the EPICOR-II installation produced 31 ion-exchange resin vessels which were vacuum-dewatered and transported to a commercial shallow-land burial site.

Table 2 : SDS processing performances

Nuclide	Reactor building sump inventory before treatment (TBq)	Typical decontamination factors		Typical en concentrations (kBq/m^3)
		SDS	EPICOR-II	
^{137}Cs	10300	110000	4200	6.3
^{134}Cs	1160	110000	4200	0.7
^{90}Sr	426	390	680	422
^{3}H	74	1	1	5.4×10^6
^{125}Sb	0.93	1	5000	71

As in the case of the auxiliary building water, the processed effluents still contain tritium. They are stored onsite and used for decontamination works in the plant.

(1) : 420 R/h

3.4. Control of the Reactor Coolant System activity

The Reactor Coolant System contains about 333 m^3 ; at the beginning of 1982 it was still pressurized at 662 kPa and the temperature was in the range 27-39°C. The radioactivity concentration was about 1.2 TBq/m^3 ; the water chemistry is adjusted for corrosion protection and criticality control, so that the water contains 996 ppm sodium and at least 3500 ppm boric acid.

As the reactor coolant is in continuous contact with the fuel debris, complete decontamination is precluded by fission products leaching. However the activity level can be lowered so as to decrease the radiation intensity around some piping and equipment and use the reactor coolant water as a radiation shield in subsequent defueling activities.

As the core should not be uncovered, reactor coolant processing occurred by 150-200 m^3 batches, which were bled from the system while the same quantity of reactor coolant grade processed water was fed into it. The "A" and "C" Reactor Coolant Bleed Holdup Tanks were used as buffer storage for that purpose. The bled batch was processed through the SDS (see § 3.3.) and then recycled to the Reactor Coolant System without further polishing.

From May 1982 to July 1983, 13 such "feed and bleed" cycles were carried out and a total volume of 2230 m^3 of reactor coolant was processed. The decontamination factors were typically 11000 for cesium isotopes and 350 for strontium isotopes.

As the concentrations of the various nuclides in the reactor coolant is regularly monitored it soon appeared, that the strontium level did not drop due to leaching from the fuel debris and rather fast restoration of the equilibrium concentration. For cesium, the purification efforts were more successful as the target level of 3.7 GBq/m^3 was practically reached after the 13th cycle. However, when the Reactor Coolant System was depressurized and put into contact with the Reactor Building atmosphere, so shifting from reducing to oxidizing conditions, the cesium leach rate increased dramatically and the previous benefits of the "feed and bleed" campaign were lost in a few weeks. This evolution is depicted in table 3.

The precise leaching mechanism is unknown, but anyway this phenomenon is to be taken into account in further planning of the reactor disassembly and defueling activities.

Table 3 : Reactor coolant system activity evolution

Nuclide	Concentration (GBq/m^3)			
	Feb. 82	Jul. 83	Aug. 83	Dec. 83
^{137}Cs	518	8.5	16.3	255
^{90}Sr	481	93	104	333
^{134}Cs	56	0.52	1.0	13

3.5. Storage and disposal of accident-generated water

Up to now, the processed accident-generated water may not be disposed of, as the NRC recovery license and various political agreements forbid any release into the Susquehanna river.

Therefore, two storage tanks with a total capacity of 3786 m^3 were installed near the plant to collect the decontaminated water. The inventory is kept as low as possible by continuous recycling of that water for flushing and decontamination purposes in the plant and SDS/EPICOR-II processing of all effluents from in-plant cleanup operations. As of June 1983 those tanks contained 2830 m^3 of water, and another 800 m^3 was located in various buffer tanks of the SDS and EPICOR-II installations.

The radiological properties of the processed water are compared with the US drinking water norms in table 4. Taking into account that dilution with nonradioactive coolant water from various points on the island would give a dilution factor of about 3800, it can be seen, that the tritium concentration is the only actual concern for direct release.

Various disposal options are being evaluated ; they range from controlled release of the water to the river to concentration of the tritium and conditioning of the resulting high-tritiated water in concrete blocks to be buried in a desert area.

Table 4 : Radiological composition of processed water*

Nuclide	Concentration in processed water (Bq/m^3)	Concentration after cooling water dilution (Bq/m^3)	EPA 570 limits for drinking water (Bq/m^3)
^{3}H	5.37×10^{9}	1.41×10^{6}	790×10^{3}
^{90}Sr	422×10^{3}	111	296
^{137}Cs	6.3×10^{3}	1.7	7400
^{60}Co	$< 5.6 \times 10^{3}$	< 1.5	3700

4. DECONTAMINATION OF WALLS, FLOORS AND EQUIPMENT

4.1. Auxiliary Building

4.1.1. Walls and floor decontamination in the Auxiliary Building

The Auxiliary and Fuel Handling Building consists of 130 rooms, cubicles and access corridors which are more or less contaminated and contain several important radiation sources. After treatment of accident-generated water had made the building accessible, the walls of the gangways were decontaminated up to a height of 2.10 m and characterization works in individual cubicles were started to be followed by decontamination of concrete surfaces by processes such as scabbling and scarifying, typically removing 12 mm of material from the contaminated walls and floors before radiological survey and painting of the surface. Such works were carried out on the basement level to late September 1983 when lack of funding led the plant owner to temporarily suspend decontamination activities in the Auxiliary and Fuel Handling Building. The works were aimed at reducing the contamination levels from the original value (up to 1 MBq/m^2) to lower than 50 kBq/m^2.

In the highly contaminated Makeup and Purification Pumps cubicle on the Auxiliary Building basement, a remotely controlled 180 kg robot was used since mid-May 1983 for high-pressure flushing with water at 65 °C. The device worked trouble-free during about 6 weeks.

4.1.2. Makeup and Purification Demineralizers

The Makeup and Purification Demineralizers, which are part of the Chemical and Volume Control System, were loaded with essentially ^{137}Cs at con-

* : 1 Bq = 27.03×10^{-12} Ci ; 1 kBq = 27.03×10^{-9} Ci

centrations up to 560 TBq/m^3. Radiation intensities up to 82 μA/kg(1) at 0.3 m have been measured in one of both associated cublicles.

One of the vessels does not contain any more free water and due to radiolysis effect, the resin in it has shrunken to about one half of its original volume. The ion exchanger in the other vessel remained under water and withstood radiation much better.

Current management plans for those vessels call for cesium elution before removal of the solid phase. The resin would be contacted with a boric acid/sodium borate buffer (pH 7.6) to remove about half of the cesium contents ; the rest would be leached by means of a 1.0 M sodium hydroxyde solution. This process has been tested at laboratory scale on actual resin samples.

The effluent from the leaching process which should be carried out batchwise, would be processed through the SDS before storage as processed water. The zeolites used in the SDS have been found insensitive to the organic materials carried out from the radiolytically broken-down resins.

It is hoped that the ion-exchange material will meet the requirements for commercial shallow-land burial after completion of the cesium elution process.

4.2. Reactor Building

4.2.1. The Gross-Decontamination Experiment

After some preliminary tests in 1981 and as processing of the accident generated water in the Reactor Building neared completion, it was decided to go ahead with a so-called "Gross Decontamination Experiment" which was planned in late 1981 and carried out in March 1982. Various decontamination methods were tested on the polar crane, the operating (347') level, the entry (305') level, the refueling canal, the D-rings and missile shields and various bridges, hatches and stairwells.

Particular emphasis was put on flushing with water alone as this allowed reuse of processed accident-generated water stored onsite (see § 3.5) and produced effluents which could easily be processed through the SDS. On the other hand, chemical decontamination produces waste with high ionic concentrations which are incompatible with ion-exchange purification.

The following processes were tested as part of the Gross Decontamination Experiment :

- Low pressure flushing : hydrolance, 2.8 to 6.2 MPa, 3.4 to 5.7 m^3/h ;
- High pressure flushing : - hydrolance, 18-25 MPa, 2.3 m^3/h
  - skid-mounted spinning nozzles : 14-41 MPa, 1.1 to 5.7 m^3/h ;
- Strippable coating ;
- Chemical cleaning (with mechanical scrubber) : detergent solution phosphoric acid solution.

Post-experiment dose rates on the entry and operating levels are shown on figures 3 and 4. Decontamination of the operating level (347') was effective in reducing the smearable contamination levels by factors up to 17 on the floor and 13 on equipment, with a factor of 724 for the cable trays. The best flushing technique appeared to be the use of the skid-mounted spinning nozzle at 41 MPa and 4 m^3/h. Decontamination of the entry level (305') was rather ineffective due to incorrect operation of the spinning nozzle, use of primitive equipment to remove water after flushing, and inadequate contamination control practices leading to extensive recontamination by personnel crossing unflushed areas. Flushing of the polar crane

(1) : 1150 R/h

N°	γ	β
1	300 γ	190 β
2	200 γ	3.8Rβ
3	280 γ	266 β
4	180 γ	1.02Rβ
5	100 γ	190 β
6	500 γ	7.6Rβ
7	100 γ	950 β
8	150 γ	1.1Rβ

N°	γ	β
9	350 γ	266 β
10	150 γ	1.7Rβ
11	200 γ	3.04Rβ
12	500 γ	380 β
13	350 γ	3.9Rβ
14	280 γ	2.7Rβ
15	420 γ	304 β
16	200 γ	3.04Rβ

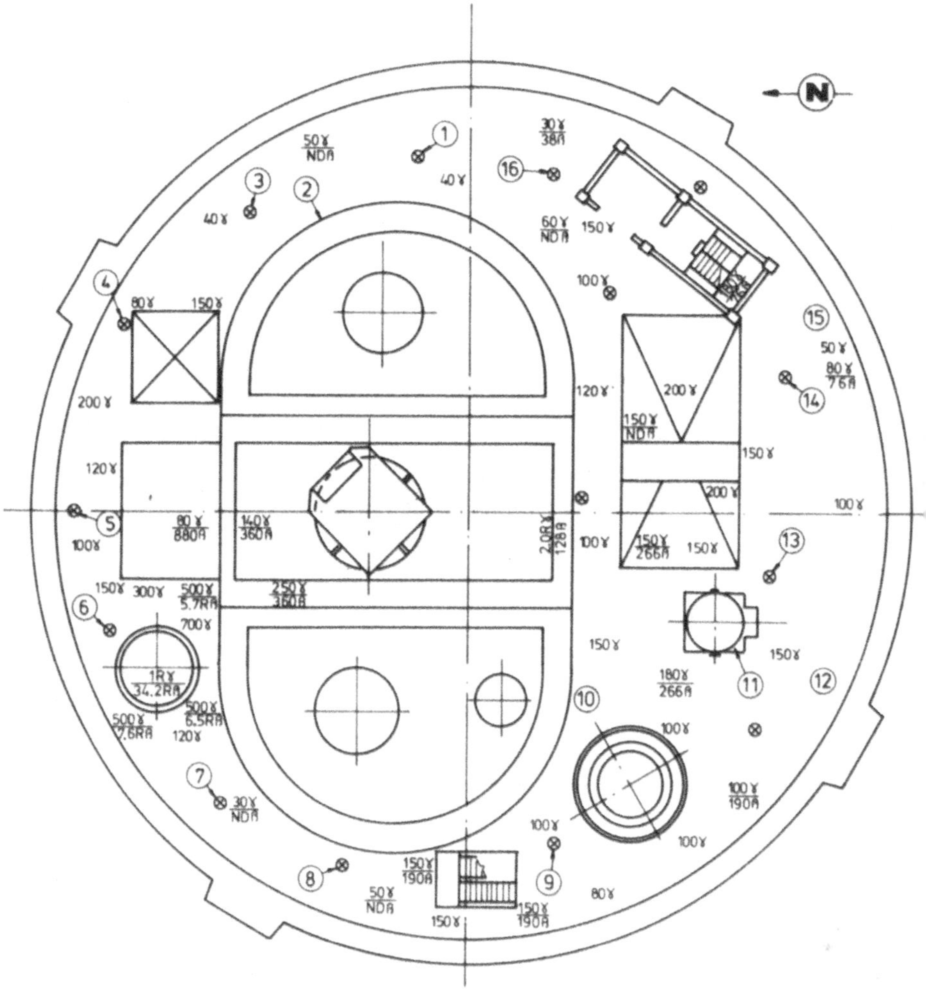

Figure 3. ELEVATION 347'- 6" RADIATION INTENSITY MAP.

N°	γ	β
1	1.0Rγ	1.9Rβ
2	600 γ	3.4Rβ
3	300γ	2.6Rβ
4	1.0Rγ	——
5	800	3.8Rβ
6	1.0Rγ	3.8Rβ
7	2.0Rγ	3.8Rβ
8	3.0 γ	3.42Rβ
9	1.5Rγ	5.7Rβ
10	3.0Rγ	——

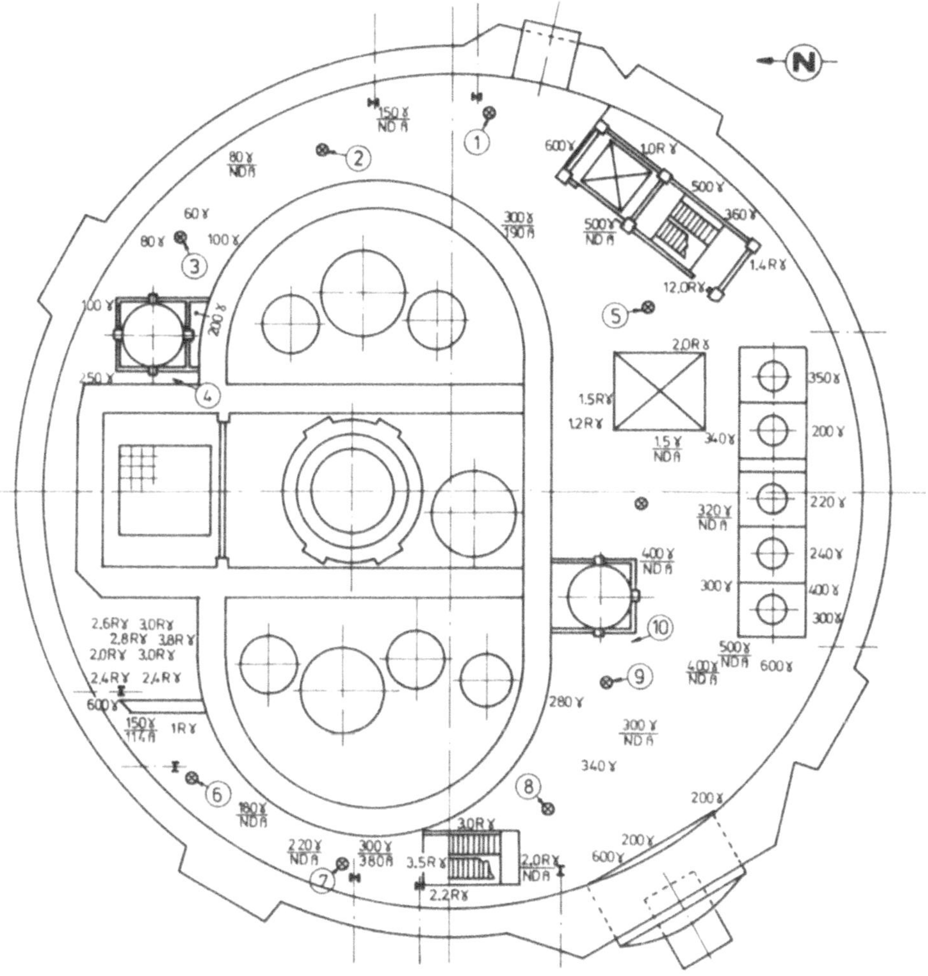

Figure 4 ELEVATION 305' RADIATION INTENSITY MAP.

was also unsuccessful because safety considerations and early refurbishing works (see § 5.1) hampered the decontamination. Low-pressure flushing of the D-ring walls and refueling canal succeeded in reducing the smearable contamination levels by factors up to 3.3 for ^{137}Cs and 27 for ^{90}Sr. Some typical data about the various processes are given in table 5.

Table 5 : Efficiency of various decontamination processes

Process		Waste production (l/m^2)	Contact RDF (★)	Swipe DF (★)	γ dose rate reduction (%)
Low pressure flushing		9.0	65.5	2.8	33
High-pressure flushing	spinning nozzle	26	29.8	69.9	23
	hydrolance (★★)	9.4	81.6	726	-
H_3PO_4 flushing		8.2	6.5	84.8	2
Strippable coating	floor		11.5	66.2	-
	equipment		9.7	1.6	-

(★) $\text{RDF} = \dfrac{\text{initial contact dose rate} - \text{final contact dose rate}}{\text{final contact dose rate}}$

$\text{DF} = \dfrac{\text{initial smearable contamination}}{\text{final smearable contamination}}$

(★★) on cable trays.

4.2.2. The Dose Reduction Program

Further decrease in the radiation dose rates was sought in the framework of the Dose Reduction Program, calling for flushing of local sources such as the air coolers and shielding or removal of some high-radiating components such as LOCA duets or enclosed stairwells. Emphasis was also put on removal of radioactive waste from the containment and good housekeeping practices.

After several months of efforts, it became clear, that the dose reduction objectives could not be met. Indeed, recontamination steadily occurred, because of the need to flush with borated water, causing some boric acid to become airborne and to carry activity throughout the building along the air recirculation paths.

Use of demineralized water for decontamination purposes is envisaged, yet it would make water management more intricate by adding a third water inventory (near Reactor Coolant Grade water and Accident-generated Processed Water).

It is also impossible to shield the important radiation source on the Reactor Building basement (282' elevation) : the walls and floor on that level remain heavily contaminated with accident-generated sludge. Plans to remove that waste are presently hampered by financial problems.

5. REACTOR CHARACTERIZATION

5.1. The Quick Look Experiment

On July 14,1982, the Reactor Coolant System was for the first time vented and partially drained down ; about 111 GBq of ^{85}Kr were released during that operation after the necessary measures had been taken to protect operating personnel.

On July 19, 1982, the leadscrew of the central control rod drive mechanism was removed and two days later, a closed circuit TV camera was lowered into the corresponding control rod guide tube.

The inspection revealed extensive collapse of the reactor core, which in this region had gone into a rubble bed, the upper level of which is about 1.5 m below the upper limit of the undamaged core. Water turbidity prevented further characterization of the debris in that area.

The experiment was repeated at the core periphery and at a point midway between the periphery and the core center. At that last point, the rubble level was also 1.5 m below the upper limit of the undamaged core ; however, intact pellets and springs could be seen at the top of the bed. Some rods protruded from the rubble and some pieces hung from the underside of the plenum assembly.

The rubble bed was probed with a metal rod ; the upper 35 cm appeared relatively soft.

5.2. The Underhead Characterization Program

On July 25, 1983, the Reactor Coolant System was depressurized again and partially drained, and various characterization tests were carried out.

First the centermost Control Rod Drive Mechanism was removed ; the deservice had a maximum contact radiation intensity of 1.4 µA/kg[(1)] which is about 2.5 times lower than the values measured for some of the leadscrews removed during the Quick Look experiment. The β dose rate at contact was about 833 µGy/s[(2)]. Contamination levels in the underhead space are estimated at 481 GBq/m^2 ^{137}Cs and 155 GBq/m^2 ^{90}Sr on vertical surfaces, and 3850 GBq/m^2 ^{137}Cs and 1240 GBq/m^2 ^{90}Sr on horizontal surfaces. The γ dose rates after head removal would be in the range of 14 µSv/s [(3)] at 5 m to 1.1 mSv/s [(4)] at 0.8 m. Such dose rates allow head lift without flooding of the refueling canal.

Two closed-circuit TV camera inspections of the plenum and of the core void were carried out during draindown of the Reactor Coolant System. The horizontal surfaces on the plenum appeared to be covered with a thin layer of sediments. The about 1 mm thick layer was sampled and tested for pyrophoricity ; the negative result allowed lowering of the water level down to 0.3 m below the upper plenum surface. The dried sediment probably contains some control rod alloy vaporized during the accident and recondensed there.

The examined leadscrew surfaces exhibited a wide variety of conditions from encrusted and scaly to clean and shiny.

The space between the lower plenum surface and the rubble bed was mapped by means of a computer-assisted ultrasonic device lowered to about 0.15 m above the top of the rubble bed. The expected mapping resolution is in the range of 1 cm.

The preliminary results confirm the information gained from the Quick Look Experiment. Rod pieces hanging at the plenum or protruding from the rubble bed are clearly visible. Very few intact fuel assemblies, if any, subsist at the periphery of the vessel.

Grab samples of the fuel debris have been taken at three locations and two different depths, yielding a total of six samples, which have been shipped to DOE facilities.

(1) : 20 R/h
(2) : 300 rad/h
(3) : 5 rem/h
(4) : 400 rem/h

Physicochemical and radiological characterization of those samples will allow design of defueling tools and accurate defueling planning.

As removal of the Reactor Vessel head can be carried out only if all Control Rod Drive Mechanism leadscrews are lifted to their parked position in the head service structure, it was tried to park four of them, so checking the operability of the drive mechanisms and getting insight into the dose commitment bound with that operation.

The four chosen leadscrews were successfully lifted at the beginning of October 1983, and the dose rate in the vicinity of the refueling canal seal plate went up from 280 to 830 nSv/s[(1)]. Radiation levels on the service structure platform did not increase significantly.

While bringing the leadscrews back to their original position in mid-October, it appeared, that one of them was stuck in the parked position.

6. REACTOR DISASSEMBLY

6.1. Polar Crane Issues

The main polar crane of the Reactor Building was rather extensively damaged during the accident and had to be refurbished ; those works lasted to early 1983 and were intended to allow limited use of the crane bridge, trolley and hoist. The electric bars around the building were not repaired and the crane is now powered by a central cable, so bridge rotation has to be limited to avoid excessive twisting of that feed line.

In March 1983, all crane functions had been tested for correct no-load working and GPU applied to the NRC for approval of load test and head lift procedures. Then three engineers who had worked at the design and survey of the refurbishment works released allegations concerning "unsafe practices" and "administrative irregularities" during those works, so triggering extensive NRC investigations, not only by the on-site staff, but also by investigators reporting directly to the Commissionners.

The so-called Head Indexing Fixture, a tripod assembly used to hang the Reactor Vessel head on the polar crane hook, was also extensively examined. Some welds were found undersized, and detailed stress calculations were performed by the plant designer to demonstrate, that the structure could be used safely in its actual state. The results were accepted by the NRC after critical review of the stress analysis report, on condition that three of the higher-stressed welds undergo nondestructive examination.

After 8 months of investigations, the NRC finally approved the polar crane load test procedure on November 18, 1983. However, due to lack of funds, the plant owner could not carry the test out at that time. Pending availability of fresh money and completion of preparatory works, the test is tentatively scheduled for April 1984.

The Reactor Vessel head could be lifted in September 1984, with more than 15 months delay on the initial schedule, after the two-step stud detensioning has been carried out.

6.2. Further planning

After head removal, the reactor will be fitted with a water-tight cylinder called "Internals Indexing Fixture" which will be filled with water for the purpose of radiation shielding. That water will be kept at a low activity level by use of the SDS.

The plenum will be examined extensively for eventual deformations ; in the meantime, that part of the spent fuel pool which is not occupied by the SDS will be refurbished and flooded along with the refueling canal. Those

(1) : 100 to 300 mrem/h

water bodies will provide sufficient shielding to remove and store the plenum assembly without excessive personnel exposure.

The water in the reactor (295 m^3), in the refueling canal (1226 m^3) and in the fuel pool (1665 m^3) will be filtered and deionized to keep turbidity and activity at low level and enable easy and safe defueling operations. Those should be carried-out by means of a wet vacuum cleaner operated remotely from the reactor service platform and fitted with various pick-up nozzles. The debris should be separated from the pumped water by means of knock-out canisters, hydrocyclones and backwashable sintered metal cartridge filters, and gathered in some 100 spent fuel canisters to be stored in the spent fuel pool.

7. CONCLUSION

Our stay at TMI and our work with the onsite NRC staff showed us, that major goals have been achieved during the 5 years following the March 28, 1979 accident. The most important of them are the decontamination of the accident-generated water by means of the EPICOR-II installation and the Submerged Demineralizer System.

Of course much cleanup work still remains to be carried out, but the experience gained up to now allows to anticipate that the subsequent operations, including reactor disassembly and defueling, can be accomplished safely, without endangering either the work force or the general public.

SESSION II : SUMMARY OF DISCUSSION

Paper : "Composition of contamination layers"

The Chairman remarked that in chapter 4.3. of the paper, it is mentioned that activity is found not only in the oxide deposits, but also in the base metal; this fact has not been noted in other publications. He asks if this is a peculiarity of the samples which have been examined or of the examination method, or if this penetration is a fact which is not generally examined; he would like to know more about the mechanisms underlying this penetration into the base metal. One of the co-authors, Mr. Ahlfänger (KW, Lingen) is convinced that this phenomenon, observed at KW Lingen, is generally present. He distinguishes a layer of up to 20 μm thickness which could be linked to deposition and corrosion processes. But ^{60}Co has been found in one sample at 90 μm distance from the surface, and this fact could be explained only by a diffusion process. The depth of penetration will depend on the type of steel and mainly on the grain size.

Paper : "Chemical and electrochemical decontamination"

A French participant asks for additional information on the gain in efficiency due to ultrasonic treatment when added to chemical decontamination, the ageing of the decontamination solutions or electrolytes or, otherwise expressed, the limit for re-cycling and re-use of these solutions, and the conditioning of the sludges obtained as a result of treatment of spent solutions; another question is whether conditioning of sludges leads to a preference for the electrolytical or the chemical decontamination methods. Mr. Pascali answers that tests to verify the influence of ultrasonic treatment have been made and the treatment has been found to be particularly efficient in removing loosely adherent oxide layers. In case of a combined treatment, for example with an acid treatment, the ultrasonic treatment diminishes the attack of the surface, and allows to reach a lower final activity of the sample. As far as the conditioning of waste is concerned, only small quantities have been treated up to now. The procedure is a precipitation with soda at a pH of 10-11; after this treatment, the solution is virtually clean and could be used infinitively. The final conditioning of the sludges will be studied later this year; the intention is to verify the possibility of conditioning by a cementation process. Mr. Gauchon, one of the co-authors, adds that during decontamination processes the metal content in the solution is steadily increasing. For electrolytical processes a limit of 50 to 100 g/1 is generally observed as reprocessing of solutions containing 100 g/1 of Fe is difficult to perform. By treating such a solution of concentrated phosporic acid, he observed that the quantity of secondary waste is as high as the initial quantity of electrolyte. For a solution of 2 mol/1 of sulphuric acid, 100 g of solid matter are obtained after precipitation. His experience in adding ultrasonic treatment is that decontamination efficiency is improved by a factor of 5 to 10, as compared to a simple chemical attack with re-circulation, when no oxide layer is present; this is thought to be due to a sort of cavitating effect created in the fluid. In some special cases, for example in removing oxides from a PWR primary loop sample surface, addition of ultrasonics is the only way for succeeding in removal. The usefulness of ultrasonics differs from case to case and a general statement on its efficiency cannot be made. Concerning the definite conditioning of sludges, all ways, like conditioning in concrete, bitumen, in thermosetting resins or other, are possible. The method to be chosen depends on the type and quality of the secondary waste generated. A pre-selected conditioning method will possibly determine the decontamination process to be applied. Mr. Lasch, one of the co-authors, adds some information to this important question. During the 100-ton decontamination test at Gundremmingen, about 800 m^3 of contaminated wastes for flushing arose. Treatment has been done by a chemical method, a precipitation - flocculation method, resulting in 1 ton of solid waste in total. Conditioning of the electrolytic solution for this test resulted in 2 tons of solid matter; as iron phosphate is a very heavy salt with a density higher than 2, the volume of this mass is about 1 m^3 equivalent to 5 drums of solid matter.

An FRG participant states first that ultrasonic decontamination will not allow treatment to unrestricted release, when penetration of activity in the base metal has occurred. In electro-polishing, for each geometry a copper cathode is needed which has to be fabricated for this purpose; he asks if the copper could be re-used. A further question on electro-polishing addresses the cost of decontaminating 1 kg of material to unrestricted release. Mr. Lasch answers that the copper cathode is universally used in tube-type parts and as it is a cathode it is not dissolved; consequently, it can be used infinitively. Mr. Stang (KRB, Gundremmingen) contributes the information on costs : decontamination by electro-polishing costs actually 2.50 DM per kg, but there is no charge for use of the restricted area included.

Mr. Pascali comes back to the effect of ultrasonic treatment. In general, time needed to reach a given cleaning effect is shortened, but the efficiency depends on the surface layer, and the nature of the chemical solution used. The mechanism of attack is different for solutions based on HCl, HF or HNO_3 : effectively, HCl solutions attack the surface progressively, the others attack on preferred lines. When preferred lines of attack are present, added ultrasonic treatment will evidently be more efficient, as this treatment allows an easier removal of the oxides.

Paper : "Decontamination of concrete, with special reference to flame scarfing"

A participant representing an international organization has noted that in the first slide Cesium has been shown to be 230 Bq/g, levelling off to 120 Bq/g at 4 to 5 mm depth. He asks for the thickness to be taken off to come lower than the 37 Bq/g limit mentioned elsewhere. Mr. Ebeling repeats that in the specific case of the KRB-A solid waste storage building, about 4 to 5 mm of concrete had to be removed in order to fall below the unrestricted release limit.

Paper : "Particular aspects of post-accident plant decontamination"

A U.S. participant adds a comment on the fate of the processed waste water. Three possibilities have been investigated : one is forced evaporation through the cooling house of the other unit, the second is open evaporation from a cooling pond on the site itself, and the third is direct discharge to the river. The National Council for Radiation Protection has just elaborated an evaluation of all three concepts and will probably recommend that it be discharged through the river during high-flows season in winter time. The resulting doses will be in the order of μSv to mSv. It has to be pointed out that those concentrations would be far less than concentrations that would normally be released during the operation of the plant.

SESSION III

DISMANTLING TECHNIQUES AND REMOTE OPERATION

Chairman : A. BUONO, ENEL

Secretary : R. BISCI, CEC, Directorate-General Science, Research and Development

Introduction by the session chairman

Dismantling techniques for steel components

Advanced cutting techniques : laser and fissuration cutting
1. Fissuration cutting

2. Laser beam cutting

Demolition of concrete structures

Remote operation in decommissioning

Summary of discussion

Technical panel : Remote operation in decommissioning

- Contributions by the Panel Members

- Summary of discussion

INTRODUCTION BY THE SESSION CHAIRMAN

A. BUONO
ENEL, Roma, Italy

The four papers given at this session summarize the studies, experiences and results that have emerged from nine contracts concluded between the Commission and firms or organizations in the Member States. The various activities have been grouped together to enable each author to discuss well-defined areas :

- traditional steel-cutting techniques of particular interest as regards certain aspects of their application in the nuclear field (e.g., production of powders and aerosols);
- non-traditional cutting-techniques for steel and other metallic materials, based on the use of laser or on the embrittlement of the material to be cut;
- cutting methods for concrete structures, such as the vessels of gas-graphite reactors or biological shields for water reactors, ranging from the use of traditional techniques involving pneumatic chisels, large diameter diamond-tipped circular saws, etc., to that of explosives.

A fourth paper, which concludes the third session, discusses the results of two studies of the use of remote-operation methods for dismantling operations.

DISMANTLING TECHNIQUES FOR STEEL COMPONENTS

F.W. BACH, University of Hannover
R. BISCI, CEC
B. BOEDEKER, Noell GmbH
J. KLARE, Salzgitter AG
H.G. KNACKSTEDT, Transnuklear GmbH
F. PIANA, Ansaldo Impianti

SUMMARY

In this research programme we looked out for possibilities to solve the dismantling problems in the reference BWR reactor with thermal cutting techniques.

It was shown that the stainless steel plated wall of the reactor pressure vessel with a thickness of 600 mm, can be dismantled by plasma-autogen-combination-cutting.

Further research work was done to find out the efficiency of oxygen lance piercing and powder flame cutting, specially at parts which cannot be done or only with great difficulties by flame cutting. It was carried out that the oxygen lance piercing gives the best results, when it is done by mankind.

A special dismantling problem was the recirculation inlet tube at the vessel wall with only a small clearance between tube and wall. It was solved by using the plasma arc cutting in air as well as under water.

The mastering of the dust and aerosol emission by application of the thermal cutting techniques is a serious problem for the future. The comparison of the dust and aerosol emission by plasma arc cutting in air and under water gave a remarkable reduction of the emissions by cutting under water.

SEGMENTING TECHNIQUES FOR STEEL COMPONENTS

Five research studies dealing with the segmentation of steel components were carried out under Project 3 'Dismantling techniques' of the research programme on the decommissioning of nuclear power plants sponsored by the European Community.

This report considers the following thermal cutting techniques :

- packed-oxygen-lance cutting;
- powder cutting;
- flame cutting;
- combined plasma-arc and flame cutting;
- plasma-arc cutting.

Plasma-arc cutting was tested both in air and under water.

The suitability of the techniques for the segmentation of steel components in nuclear power plants as shown by the test results is considered below.

PACKED-OXIGEN-LANCE AND POWDER CUTTING

The two techniques are based on the same principle.

Iron, in the form of rod in one case and powder in the other, is heated to ignition temperature in an oxygen flame and Burns up in an exothermic reaction at a temperature of approximately 2 500°C. This melts the material to be cut and the kinetic energy of the expanding gases and oxides expels it from the cut.

The process is suitable for cutting ferritic and austenitic steels, aluminium and copper alloys and concrete.

In order to test performance, tests were carried out with 1/4", 3/8" and 1/2" lances.

The best results in steel cutting were achieved by the 3/8" lance, since continuous cutting was impossible with the 1/4" lance, the flame dying repeatedly despite variation of the test conditions.

A mixture of 85% Fe/ 15% Al was used for powder cutting. The maximum depth of cut that could be obtained was 1 000 mm.

In a direct comparison, powder cutting proved superior to cutting with the packed oxygen lance, both in terms of cutting performance and the amount of slag and dust produced. A limitation of the packed lance is that the quality of the cut obtained depends to a large degree on the manual dexterity and experience of the operator. Remotely-controlled cutting using a packed lance is not currently considered promising.

However, the prospects for remotely controlled powder cutting are good.

Both techniques produce a high proportion of secondary waste, which gives rise to high filter loads when disposing of the activated or contaminated dusts and aerosols.

It would appear necessary to develop both techniques for under water application in order to minimize emissions.

FLAME CUTTING

Flame cutting has for decades been the state-of-the-art technique for use on materials of thicknesses up to 3 000 mm. Since it is suitable almost exclusively for the segmentation of ferritic steels, this technique can be employed only for cutting up the pressurizer relief tank and austenitically plated ferritic steels, when cutting is possible from the ferritic side. Tests have shown that, for example, a reactor pressure vessel head of a thickness up to 620 mm and with plating of 10 mm can be cut with this technique. The heat generated during the combustion of ferritic steel is sufficient to melt the austenitic plating.

COMBINED PLASMA-ARC AND FLAME CUTTING

The cutting of the reactor pressure vessel presents a particular problem. The pressure vessel has ferritic-steel walls whose thickness may be as much as 600 mm, with austenitic plating approximately 10 mm thick on the inside. Access when installed is possible only from the plated side.

A novel cutting technique was developed for pressure vessels (Fig. 1). The combination of a plasma-arc torch to melt the plating layer with a flame cutting torch to burn through the ferritic body permits thermal cutting in a flat position, in a vertical position (thicknesses up to 620 mm), and in an inclined position (thicknesses up to 300 mm) at a cutting speed of 60 mm/min. (Fig. 2). The test results are shown in Table 1.

This technique too has so far only been tested in and can be used only in air. Adaptation for underwater use would substantially reduce filter loads, as a result of the lower amounts of cutting wastes produced.

PLASMA-ARC CUTTING IN AIR AND UNDER WATER

Plasma-arc cutting is based on the disassociation and ionization of gases (e.g. oxygen) in an arc and their recombination on the plate, which is at a relatively low temperature compared with the plasma. Temperatures of approximately 8 000°C are generated in the plasma stream, thus allowing any material that conducts electricity to be cut.

The maximum thickness of plate that can be cut is 180 mm in air and 100 mm under water at the depth of 2 m at which the tests were carried out.

With plate of these gauges, however, special measures must be taken in order to prolong the service life of nozzles and electrodes.

Plasma-arc cutting is attractive by virtue of the relatively small size of the equipment, which allows cutting in cramped conditions, ease of use, good cutting quality and high cutting performance.

An additional advantage, particularly in underwater use, is that cutting is possible in all positions, even in an overhead position, This was demonstrated in tests on flat plates and on large-diameter pipes.

DUST AND AEROSOL MEASUREMENT

A feature of thermal cutting techniques compared with mechanical techniques is that the former generate substantial amounts of dusts and aerosols. Since, in practical application, the dusts and aerosols produced might be highly active, their safe capture and removal is of major importance.

The requirements the filter systems must meet are comparable to those in clean-room technology.

In order to be able to select appropriate filters, it is critically important to have knowledge of the type, amounts and particle size of the dusts and aerosols created during thermal cutting.

To that end, systematic tests were carried out both in air and under water with combined plasma-arc and flame cutting, as well as with plasma-arc cutting.

Cutting parameter lists for ferritic and austenitic steels designed to achieve minimum dust and aerosol emissions were drawn up, mainly on the basis of a direct comparison of plasma-arc cutting in air with plasma-arc cutting underwater. The test rig is shown in Fig. 3.

It was demonstrated that, when cutting the same material, underwater plasma-arc cutting reduces dust and aerosol emissions by at least a hundredfold (Figs. 4 and 5). Comparison of particle size distribution obtained with plasma-arc cutting in air and under water shows that the 0.2, m particles generated by the latter technique are approximately half the size of those obtained when cutting in air (Figs. 6 and 7). Generally speaking, it can be said that dust and aerosol emission diminishes the lower the power employed and the greater the workpiece velocity.

Measurements of the dust and aerosol emissions produced by combined plasma-arc and flame cutting showed a mean particle diameter of approximately 0.2 μm. A dust output of approximately 1 400 g per metre of cut length was measured when cutting components of a thickness of 610 mm with 10 mm of austenitic plating. The filter tests showed that sufficiently high retention rates exceeding 99.999% can be achieved with the filter systems selected. However, in no case so far have the service lives or dust storage capacities been adequate. With some types of filter, the permissible pressure differential was reached when 70 g of dust had been absorbed.

On-going research projects should therefore continue to deal with the problem of filter systems.

A further requirement for the development of underwater cutting techniques is the safe capture and storage of cutting waste dissolved in the water, present in the form of suspended matter and deposited on the edges and bottom of the tank. No research projects have yet tackled the question of the capture of this waste.

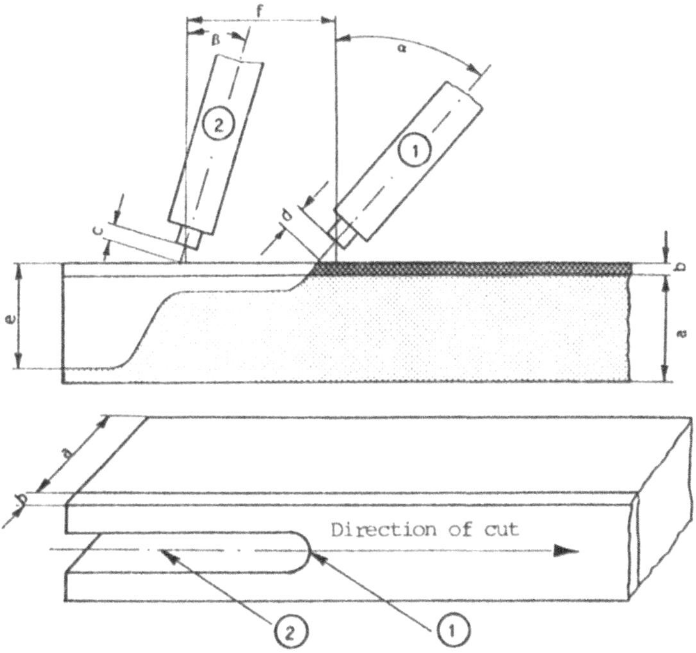

Fig. 1 : Schematic diagram of a combined plasma-arc and flame cutting torch (TW 424)

a) Thickness of the steel plate
b) Thickness of the chrome-nickel plating
c) Torch-to-workpiece distance (flame cutting torch)
d) Torch-to-workpiece distance (plasma-arc torch)
e) Depth of cut
f) Distance between torches
α Angle of the plasma-arc torch
β Angle of the flame-cutting torch

Constant values :

b = 10 mm
d To be adjusted in accordance with the manufacturer's instructions using a special measuring device.

Fig. 2
610 mm block cut using the combined plasma-arc and flame cutting technique

	UNIT	HORIZONTAL						VERTICAL INCLINED	
Cut number	----								
Thickness of the steel plate = a	mm	150	200	250	300	400	600	400	600
Diameter of nozzle (plasma-arc torch)	mm	4,5							
Angle of plasma-arc torch =	degrees	45				30			
Current	A	200							
Voltage	V	150				130			
Diamter of nozzle (flame cutting torch)		200-300				300-600			
Torch-to-work piece distance (flame cutting torche) = c	mm	8			12	37		35	
Angle of flame cutting torch =	degrees	5				0			
Fuel gas pressure	bars	0,22				1,1			
Heating oxygen pressure	bars	3,5				7			
Cutting oxygen pressure	bars	14	18	20		17			
Cutting speed	mm min	140	120	90	60				
Depth of cut = e	mm	160	210	260	310	410	610	410	610
Width of cut	mm							20	20
Distance between torches = f	mm	30				40		39	
Comments	----								

Table 1 : Cutting data for combined plasma-arc and flame cutting

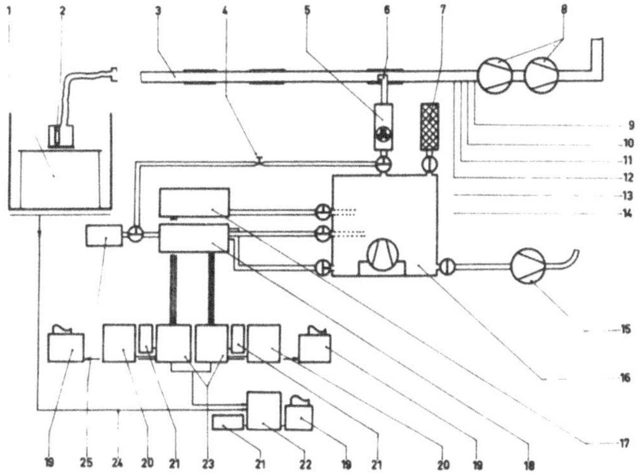

Fig. 3 : Dust and aerosol measuring system

1. Water tank
2. Torch
3. Exhaust air duct with steady-flow section
4. Damper
5. Neutralizer
6. Sampling device
7. Filter
8. Fan
9. Static pressure
10. Dynamic pressure
11. Condensation
12. Temperature
13. Atmospheric pressure
14. Room temperature
15. Atmospheric pressure
16. Dilution tank
17. Aerodynamic particle sizer (APs)
18. Electrical aerosol analyser (EAA)
19. Printer
20. Monitor
21. Diskette unit
22. Computer
23. Computer
24. Cutting data
25. Pump

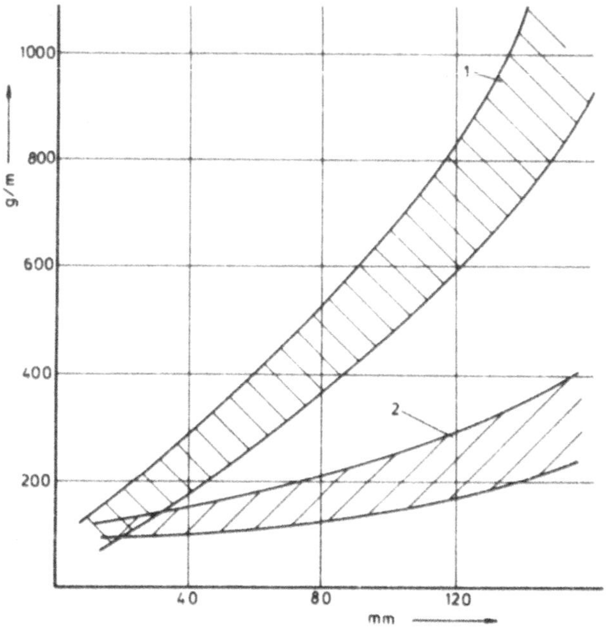

Fig. 4
Plasma-arc cutting in air
Dust generated as a function of plate thickness
(1) St 37
(2) CrNi

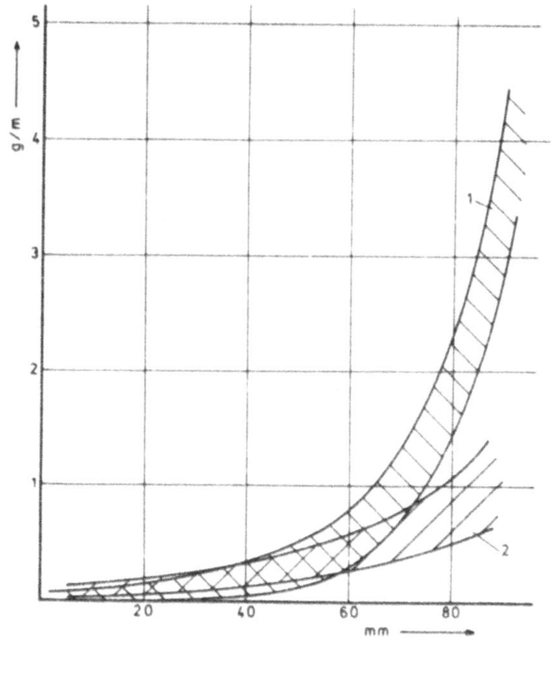

Fig. 5
Plasma-arc cutting under water
Dust generated as a function of plate thickness
(1) CrNi
(2) St 37

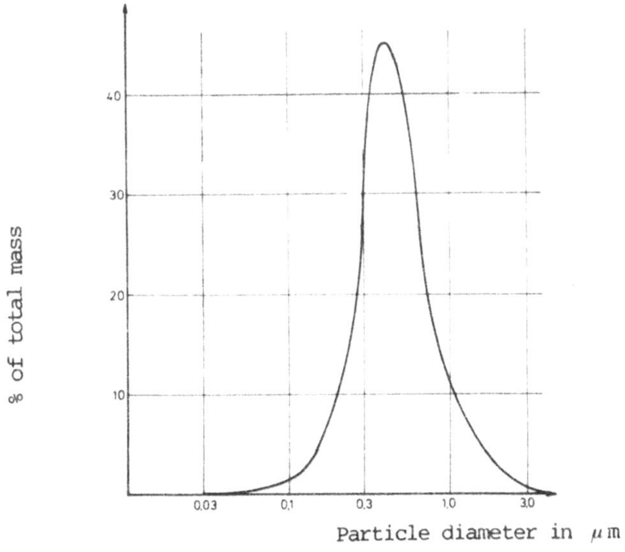

Fig. 6 : Dust mass distribution with plasma-arc cutting in air

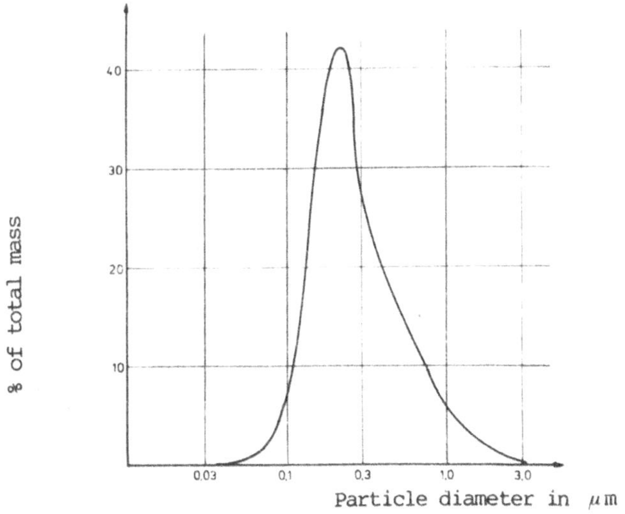

Fig. 7 : Dust mass distribution with plasma-arc cutting under water

ADVANCED CUTTING TECHNIQUES: LASER AND FISSURATION CUTTING

1. FISSURATION CUTTING

J. Perrin, G. Tanis (CEA)

Summary

The aim of this study is to develop a new method for cutting metal components of nuclear power plants, which produces virtually no secondary waste. In this method, a controlled intergranular fissure is produced in a heated area of the component by the addition of a molten material which gives rise to the formation of brittle compounds. With the presence of molten metal, the material or alloys in which we have tension stresses can have a brittle intergranular failure. Generally the failure occurs in a defined temperature range. The tension stresses are created by the thermal gradient induced during the local heating. At the same time, the molten metal is added.
Our work has been developped in two directions:

- Study on the maximum thickness (presently it is potentially possible to cut thick sheets up to 100 mm)
- Dismantling of nuclear parts as tubes, cans, hot all walls (we are on the biginning of the development in this field of work).

1.1. Description of the phenomenon

The fissuration process for austenitic steel sheets makes use of weakening by liquid metals.

The process is as follows:

- under the effect of stress, sliding occurs at the particle joints as the temperature is greater than T_f K/2, the temperature where this phenomenon appears in face-centred cubic structures (Fe γ);
- the liquid metal wetting the bottom of the fissure, decreases the energy of attraction between the atoms as has been shown by Stoloff.
  Under the effect of stress, these bonds break and the fissure progresses;
- without appearance of the fissure opening, the liquid metal is transported by capillary action to the bottom of the fissure, this enabling the rupture of the bonds between the atoms at the boundary of the fissure to continue.

For the opening to accur, stress must be applied whilst the wetting (fissuring) metal is still liquid.

Separation, in this process, is caused by thermal stress.

The essential characteristics of this fragile rupture are the following:

- sudden without significant lengthening of the test piece;

- essentially of the intergranular type, meaning the separation takes place along the particle joints;
- most often occurs within a well-defined temperature interval and it seems that outside this interval the material is not affected to any great extent;
- particularly concerns those structures of the face-centred cubic type.

According to our results, fatigue caused by liquid metals could appear when the steel is deformed by traction in a temperature region, such that the deformation is controled by - or causes - the intergranular sliding.

Intergranular sliding based on the movement of faults (very probably intergranular detachments) can only occur beyond a certain threshold level for the temperature (of the order of 600 to 700°C for steels). It necessarily leads to local concentrations of stress (triple joints, joint discontinuities) which, if not released by detachment in the matrix or by migration or recrystallisation from the joint, induce the intergranular ruptures.

When hot, we know that the possibilities of plastic release are significantly greater with ferritic centred cubic structures.

The intergranular fragility will therefore be much greater in austenitic face-centred cubic structures. These characteristics are in perfect agreement with our work: effect of the structure,temperature threshold level, intergranular separation.

This tendency towards intergranular separation is accentuated by the presence of certain liquid metals. If the liquid metal easily "wets" the material, all the fissures emerging at the surface can be stabilised and the critical number for sudden propagation should be lowered. This effect can be accentuated by a decrease in inter-atomic separating energy at the bottom of the fissure. Throughout this analysis, it can be seen that the low enthalpy values for the mixing of the material and the liquid metal must facilitate the sudden rupturing.

This is very probably the case for steel more easily than lead and cadmium or tin.

To summarise, we believe from our results and those of the literature that the sudden rupturing of steels in the presence of liquid metals induces simultaneously:

- specific deformation conditions of the materials which cause the intergranular sliding leading to separation along the particle joints. These conditions necessite a temperature threshold level according to the structure, the metallurgical state and the rate of deformation.
- a certain wetting ability of the liquid such that the liquid metal, by capillary action or surface diffusion, is permanently in contact with the bottom of the fissure, which then progresses under the effect of the stress applied. These conditions are favorised by raised temperature and stress.

1.2. Fissuration mechanism

The adaption of the fissuration cutting procedure to dismantling is based on work in which can be distinguished:

- the analysis of the thermal static field and criteria for the fissure propagation;
- the interpretation of the mechanics of the propagation and influence of structure geometry;
- the use of thermography and stress calculation.

a) Analysis of the thermal static field and criteria for fissure propagation

The temperature intervenes in two ways:

- by its level: it has a purely metallurgical action(energy of activation).
- by its distribution: it has a mechanical action.

As a result, the fissure accurs beyond a temperature threshold level in an area where the stress of expansion in positive therefore in traction (the central hot areas under the torch, being compressed).

b) Interpretation of the propagation mechanism and influence of the structure geometry

To make the fissuration continual, the torch must be moved in such a way as to always place the fissure point in the propagation area (which comes back to maintaining the conversion of part of the calorific energy into mechanical energy).

The two criteria for propagation and interpretation of the fissuration mechanism allow the case where the fissure progression could be problematic, to be anticipated.

Effectively, one can differentiate between:

- First criterium not satisfied: the propagation area does not exist. Possible causes:

a) Temperature lower than T_S threshold level
b) Tangential stress component too weak.
α-The temperature is too high in the centre inducing a drop in the mechanical traction characteristics of the material.
β-The stress field is released elsewhere than at the level of the fissure because of the presence of a nearby hole or free edge.
γ-Blocking of the deformation limiting the tangential stress: the case of transversal fissuration of a tube or the formation of a "rectangular opening".

- Second criterium not satisfied: the fissure is not in the propagation area.

Possible cause: a too rapid movement of the breach has caused "disengaging".

c) Use of thermography and stress calculation.

If the qualitative interpretation of the fissuration mechanics is satisfactory in its consistency with the observed facts, it will remain unsatisfactory, particularly when dealing with the cases numbered above.

In effect, the accessible experimental thermal field is not the only case: several parameters can modify its intensity and distribution in a way as to make fissuration become insensitive to the geometry of the structures encountered (position of the torches or inductors, electrical parameters, rate of movement). Once again, it is necessary to be able to quantify the different results. However:

- The fissuration, test only gives one set of information, yes/no
- The extensometry and the temperature measurement using thermocouples lead to considerable disadvantages.
- The modelisation using calculation of thermal phenomena in dynamic form is hasardous

We have resolved these experimental difficulties using thermography: effectively, it is possible by means of an infra-red camera to record thermal images.

At the rate of the standard video (20 ms per screen image)

These images, after digitalising and numerical treatment in a minicomputer, produce a true plan of the thermal field which can then be used in the determination of the stress field (structure calculation).

1.3. Cutting of large thicknesses

The stage visualized at the moment, is the fissuration cutting of austenitic stainless steels of 500 mm thickness and of carbon steels (type A 42) of 100 mm thickness.

To heat these plates of such a thickness, the use of T.I.G., rigidified T.I.G. and plasma arcs creates fusion baths much too wide and occasionally causes the projection of metal droplets. The electric arc no longer supplies rapid enough heating with regard to the problem of the molten region, to obtain a sufficient temperature on the other side of the sheet.

The various tests carried out lead to orientation towards induction heating.

1.3.1. Selection tests for the different heating methods

1.3.1.1. Fissuration head with the TIG torch

This head is fitted with a TIG torch of 800 A. For this intensity the torch must include a water-cooling system which is not easily acceptable in a hot cell.

The following defects were recorded:

- . too great a fusion of the metal
- . the risk of running depending on the position, and the risk of contamination of the cell by the particles.
- . too wide a heated area (poor power distribution)
- . presence of water-cooling systems.

1.3.1.2. Fissuration head with other torches

a) Rigidified Arc TIG: in spite of the effect of the concentration of the arc with this torch, the benefit is small with respect to the thickness to be cut.
b) Plasma Torch: this torch should heat a 100 mm plate without piercing the metal and without projection of metallic particles, however the tests are still to be carried out.

1.3.2. Induction heating tests

The following tests were carried out to test this heating method and adapt it to the fissuration technique:

- induction in the mass with the help of an aperiodic generator with frequencies from 25 to 230 kHz: fissuration obtained: 50 mm stainless steel.
- Induction with an undulating generator at a frequency of 3 to 10 kHz. Fissuration obtained: 60 mm on soft steel, fatiguing conditions were obtained on 100 mm of soft steel.
- Research and design of an inductors adapted to the generator, at the desired frequency and in the form of the area heated.
- Test on fissuring metal deposit in the heated area.

Conclusions

Using induction heating, it is not necessary to melt the metal. The attainment of 700 °C on the other side of the plate (temperature thereshold level) is sufficient.

The heating for cutting 50 mm thick stainless steel seems to be at frequencies between 10 and 145 kHz with a single spiral inducer "PIN" or "PANCAKE", and for an applied power of the order of 30 kW.

The heating for cutting 100 mm thick carbon steel seems to be possible using both techniques at 10 or 145 kHz.

The choice will be made depending on future tests both static and dynamic, which will take into account the magnetic field in the fissuring metal bath, the ease of introduction of the fissuring metal and the viability of the inductors, etc...

1.4. Hot cell dismantling

1.4.1. Cutting tests on an infinite sheet using special tool

The aim of this work was to cut a side of a cell in steel. Using specific apparatus enabling a series of small circular cuts to be made, a process has been developed. The apparatus comprises a torch support bench and lends a circular movement to the fissuration unit. The distribution of its cuts allows the side to be dismantled.

1.4.2. Tele-manipulated cutting tests

Although the PYE tele-manipulator used causes some difficulties in keeping the trajectories well-defined we have succeeded in marking horizontal and vertical cuts on flat sheets with the help of a fissuration head adapted for tele-manipulation.

These tests were carried out cylindrical rings and good results were obtained with vertical cutting.

Howeyer, the difficulties with the conservation of the trajectory due to this type of tele-manipulator indicate that the tests should be repeated with programmable apparatus.

1.5. Conclusions

The fissuration cutting process, creating no specific waste (shavings or projection of small particules) is an interesting method for dismantling.

It's use is anticipated in the case of a long series of cuts of a single object, using specific apparatus.

The tests, carried out on large thicknesses as well as various apparatus, are encouraging. The work needs to be persevered in order to popularise its use.

ADVANCED CUTTING TECHNIQUES: LASER AND FISSURATION CUTTING

2. LASER BEAM CUTTING

B. Migliorati (FIAT TTG), P. Gay (FIAT CRF)

Summary

Experimental tests have been performed using CO_2 laser with output power 1-15 kW to evaluate the effect of varying the following parameters: material (carbon steel Fe 42 C, stainless steel AISI 304, concrete), laser power, beam characteristis, work piece velocity, gas type and distribution on the laser interaction zone. In the case of concrete, drilling depths of 80 mm were obtained in a few seconds using a 10 kW laser beam. Moreover pieces of 160 mm were cut a 0.01 meters per minute. Results with carbon steel indicated maximum thicknesses of 110 mm, cutted at 0.01 meters per minute with 10 kW, depths about 20% lower were obtained with the AISI 304 stainless steel. A parallel investigation was aimed at characterizing particulate emission during the laser cutting process. At the end of the research it was possible to elaborate a preliminary proposal concerning a laser based dismantling system for the application to a typical Nuclear Power Station.

2.1. Introduction

A preliminary research of available literature, made in 1981, emphasized a substantial lack of information on laser cutting of steel at thicknesses more than a few centimeters and no specific information regarding concrete. In the following years further experimental data were published (1, 2). At the United Technology (UTRC) (1) nuclear fuel elements and 50 mm steel plates were cut. At the Bremen Institute (B.I.A.S.) (2) 35 mm cut depthswere achieved on carbon steel with a 4 kW laser source at 0.4 m/min.

Recently the spreading of industrial laser cutting systems was driven by the traditional oxyacetylene constructors of cutting devices.Laser system permitted both the acheivement of a better cut quality, expecially for few millimeters thick metal sheets,and the economical advantages of automatic processes.

At present, several constructorsoffer industrial laser sources with power range from 1 to 9 kW and a market survey (3) has predicted the construction of about 100 laser cutting systems for the present year. Many constructors offer laser with power equal or lower than 5 kW (Spectra Physics, Photon Sources, Coherent, etc...) but only a few are offering laser sources with power higher than 9 kW (FERRANTI, UTRC) more suited for cutting high thicknesses of metals and concrete.

In the present paper, an experimental research that was carried out to understand the capability of laser technology for cutting the high steel and concrete thicknesses required by the nuclear power plant dismantling, is described.

2.2. Laser cutting: experimental results

All experimental tests were carried out at the FIAT Research Center with an AVCO 15 kW laser source. The laser beam was focalized to a one millimeter diameter spot using a Cassagrain telescope with an f number of 7 (radio between focal length and beam input diameter). The high power densities on the workpiece ($> 10^6$ W/cm^2) produce high surface temperatures. Due to these temperatures chemical reactions, material melting and evaporation were induced during laser cutting process.

The typical laser cutting parameters investigated were:

- the laser power with a variation from 2 to 14 kW;
- the laser beam cross-section and focal depth using telescope with different f number;
- the distance between the optical plane and the work piece surface;
- the gasdynamic flow regime in the laser material interaction zone: the type of gas, the nozzle shape and position, the inlet gas pressure and mass flow ;
- the work piece speed was changed from about one centimeter to one meter per minute according to the material type and thickness.

Optimal parameters were found and then kept constant during various tests. Oxygen gas was blown towards the melted zone with a tubular nozzle with a 1.5 mm diameter. The nozzle tip distance from the workpiece surface was 5 mm and the angle between nozzle and beam axis 15°. The gas mass flow was changed between 6 and 8 cubic meters per hour with an inlet pressure variation from 7 to 8 bar. The incident laser power was changed from 2 to 14 kW (specific power density at focal plane of 1 kW/mm^2 and 7 kW/mm^2); the larger amount of cutting tests was made at 10 kW in order to be able to refer the results with laser sources available on the present market.

The focal plane distance from the workpiece surface was about - 1/3 of the plate thickness for steel and - 5 mm for concrete. After a first series of cuts on several materials, tests were carried out on carbon steel (Fe42C) and stainless steel (AISI 304) plates and on concrete blocks (200x200x200 and 150x150x150 mm); these blocks were either steel reinforced or not. Some cutting tests were also done on superimposed pipe bundles of Inconel 600 (diameters Ø 21 and 27 mm).

A summary of results is described in Fig. 1, 2 and 3. In Fig. 1, cut depth on concrete versus workpiece speed is reported for a 10 kW laser power. For decreasing speeds a relevant depth increase up to 160 mm at speed of 0.01 m/min is shown. Steel reinforced concrete resulted easier to be cut due to exothermal reactions developed with the reinforcement material. Drillingdephts up to 80 mm were achieved with repeated, five seconds long, pulses of 10 kW laser power.

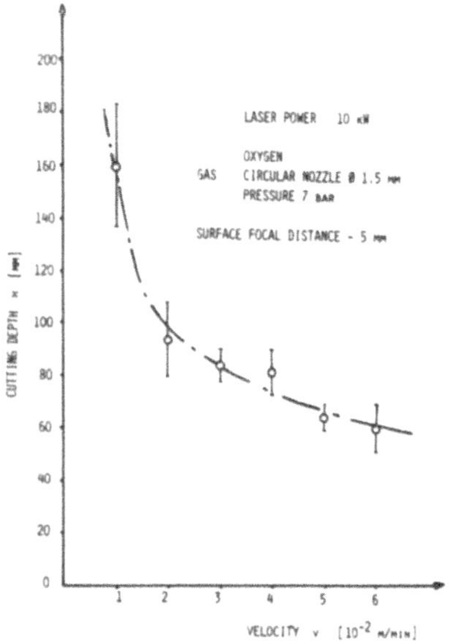

Fig.1 Cutting depth in concrete versus velocity

Depth limitations were related to the difficulty of expulsion of molten slag. Hole diameters were limited between 5 and 8 mm and therefore resulting holes are not easly suited for explosive charge positioning.

Several cuts were carried out on carbon steel and stainless steel by changing laser power and workpiece speed. In Fig. 2 cutting depths on carbon steel (Fe 42C) are reported versus workpiece speed with a 10 kW costant power.

A maximum cut depth of 110 mm at 0.01 m/min was achieved. Fig. 3 shows a comparison of the cutting depth on carbon steel and stainless steel for different laser power at a constant speed of 0.2 m/min.

For carbon steel cut depths about 20% higher are shown.

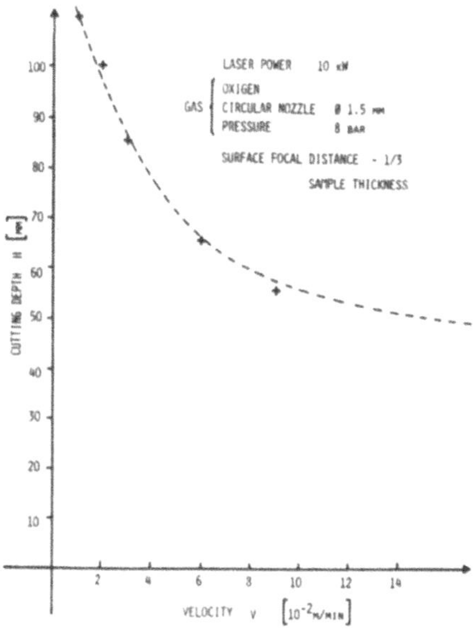

Fig. 2 Cutting depth in the carbon steel (Fe 42C) versus velocity

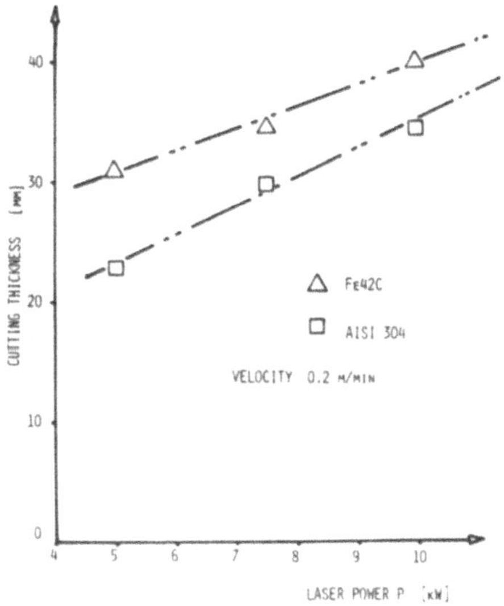

Fig. 3 Comparison of the cutting thickness between carbon steel (Fe 42C) and Stainless Steel (AISI 304) versus Laser power

2.3. Analysis of laser cutting emission products

The laser cutting interaction zone is characterized either by material temperatures higher than the melting and the evaporation point either by turbolent flow. Therefore during cutting processes; expecially due to presence of oxigen, there is a generation of particles with different diameters some of these particles are injurious to health.

Then for a thoroughly evaluation of the laser cutting techniques for contaminated materials three series of measurements were carried out to understand emission products.

The first set of measurements based on isokinetic sampling, were aimed at finding the optical properties of type particles, the second were based on a global weighting for mass balance based on isokinetic sampling were aimed at finding aerodynamic and diffusive properties of the particles.

All experiments were carried out on the same container containing the moving workpieces. As seen on Fig. 4 the container has a window for laser beam entrance and an output manifold connected with the exhaust system.

Fig. 4 Experimental set-up for aerosol evaluation; the cutting container and the sampling section are visible.

During the first tests an isokinetic emission sample was taken on the exhaust pipe and in the meantime gas velocity and pressure were measured.

To analyse the samples the following devices were used:

- centrifugal separator for particle selection;
- a particle counting system based on optical scattering to evaluate the granulometric distribution;
- a Beta ray balance to measure the particle mass retained by filters;
- analyzers for organic vapor and for nitrogen oxides.

Emission volumetric flow was 1350 m3/h and dust mass flow varied from 20/30 g/h for steel to a maximum of 500 g/h for concrete. Particle components were steel oxides for steel cuts and silicates for concrete.

Granulometric distribution measurements showed more than 60% of particles between 0.3 and 0.5 μm. Organic vapor concentrations were less than 0.16 ppm and nitrogen oxides total emission was less than 2g/h.

During the second set of measurements quantitative weight determinations of volatile particles produced during concrete laser cutting process were carried out. The measuring method was based on the difference between concrete weight variation after cutting and the weight of large particles collected on the walls of the box. For a standard cutting condition (10kW 0.02 m/min) the average value of 1850 g/h resulted nearly three times higher than that of earlier experiments.The discrepancy was explained with the limitations of the first measurement method at high ($>$ 5 μm) and low ($<$ 0.3μm) particle diameters.

The aim of the third set of experiments was to evaluate particle aerodynamic and diffusion properties. Particles were collected both by sedimentation/thermophoresis techniques and by isokinetic sampling. Transmission electon mycroscopic analysis on collected particles have shown dimensions lower than one mycron ($\sim$ 0.01 μm) and the formation of aggregates at one meter distance from the laser interaction zone.

The following instrumentation was used for analysis of the sampled aerosol:

- Various type of impactors (mini SIERRA, Andersen MARK I)
- Inertial spectrometer
- Electrostatic precipitator
- Diffusion battery
- Total filter

Emission volumetric flow rate was about 1000 m^3/h: and the dust mass flow varied from 18-55 g/h for steel to 480 g/h for concrete.

2.4. Proposal for laser application to the decommissioning of nuclear power plants

The test results summarized in the above chapters of this report outline the characteristics of the laser cutting technique as may be applied to the decommissioning of nuclear power plants and lead to enhance the reliance on the use of this tecnology.

This positive preliminary conclusion is supported by the following considerations.

1. The steel cutting capability of the laser beam (thickness not less than 110 mm) is suitable for dismantling a substantial portion of nuclear Plant Components
2. The concrete cutting capability (thickness not less than 180 mm) increases as more as the steel bars content is increased. The laser capability to dismantle concrete components must be more deeply investigated for large thicknesses. This could be performed in a possible connuation of the present research program.
3. The laser beam can be remotely controlled. The up-to-date laser application tecnology comprehends various robot mounted lasers.
   The possibilities of "robotization" are more effective for the laser in comparison with other cutting techniques.
   On the basis of the above considerations a proposal has been sketched for decommissioning a generic PWR plant.

The proposal relies on the assumption that the plant is to be submitted to "Stage 3 decommissioning" and that the wastes produced (including dismantled components) should be finally conditioned (e.g. by cementation) in the Reactor Containment Building. The decommissioning is performed in dry conditions.

Inside the Containment Building two dismantling cells should be set up (see Fig. 5):

- One, stationary, is located inside the Refueling Pool (empty of water) for dismantling of the main components;
- The other one is installed on the polar crane and is provided with telescopic arm and movable laser head,for reaching any position inside the Containment Bulding.

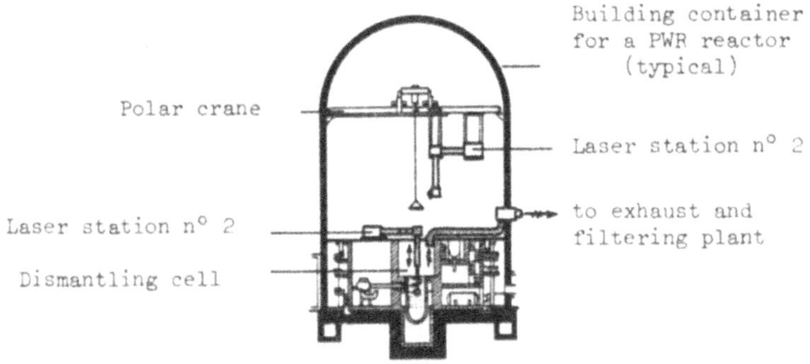

Fig. 5 Decommissioning of a typical PWR. Laser cutting cells

The two cells are remotely controlled and human direct intervention is reduced to a minimum.

During operation of the stationary cell the refueling pool is covered in order to reduce contamination problems on the structure inside the containment.

After cutting, the pieces produced are collected inside a standard container and then conditioned for final disposal. (e.g. by cementation)

In this way the Containment Building provides the maximum of confinement of the radioactive contamination and the waste containers need no additional conditioning.

The meaning of the above proposal is primarily to allow envisaging the advantages that could be connected with the use of the laser beam in decommissioning Nuclear Power Plants.

The technical definition of a laser cutting system for decommissioning must be postponed to a specific detailed study, that could begin in the next future.

REFERENCES

1. J.P. Carstens. High Power CO_2 Laser Materials Processing Technology at United Technologies Research Center - Paper presented at International Workshop organized by Verein Deutscher Ingenieure (VDI) Technologie Zentrum on "Material Processing with CO_2 High - Powers Lasers "held at Stuttgart, Germany, April 1-2 1982
2. W. Jüptner, R. Rothe, K. Teske."Thermal cutting of metals plates with laser beams". Paper presented at 3 rd International colloquium on welding and melting by electrons and laser beams. Lyon 5-9 September 1983
3. C.B. Hitz. The Laser Marketplace "1984". Laser & Applications January 1984

DEMOLITION OF CONCRETE STRUCTURES

I.Ll. DAVIES - Taylor Woodrow Construction Limited
A.R. GREGORY - Central Electricity Generating Board
H.G. KNACKSTEDT - Transnuklear GmbH

Summary

Many methods have been developed for the demolition of conventional concrete structures but when the special requirements associated with the demolition of structures in nuclear facilities are imposed considerable revision in working practices is essential. In fact, when these requirements are introduced some of these conventional techniques are completely inappropriate and all other methods need considerable development.
This paper outlines work which has been undertaken under the auspices of and with support from the Commission of the European Community in their research programme on the decommissioning of nuclear facilities. It considers the methods which appear to be most promising and reviews them under three main headings:-
i) mechanical systems
ii) chemical techniques
iii) thermal methods.
Rubble, slag, dust or smoke arising from the various cutting methods is of enormous importance because it is this active waste material which presents the final problem of disposal. Activity of concrete is low, but the reinforcing presents difficulties. It is important that cutting methods should not increase the quantity of radioactive materials arising from the demolition process.
It may be concluded from the work completed that it will be possible to remove activated concrete from nuclear structures in a way which ensures that radioactive materials cannot escape into the environment. Further developments are required to bring procedures to a state where they may be completely acceptable to regulating authorities.

1. INTRODUCTION

During the operation of nuclear power systems concrete and steel in those structures adjacent to the reactor will become radioactive due to the neutron irradiation to which the material is subjected. Thus, in the biological shields around all light water reactor and the early gas cooled reactor systems and in the prestressed concrete pressure vessels (PCPV) of the later gas cooled systems, considerable quantities of the concrete and reinforcement become radioactive and demand that great care be used in the demolition of such structures.

It is evident that removal of the activated material must precede the general demolition of the heavy concrete structures which form the biological shields and PCPV so that it is necessary to demolish nuclear structures from the inside outwards.

To study the problems associated with the controlled removal of activated material from nuclear facilities the CEC established, as part of their research programme on decommissioning such systems, a series of programmes aimed at studying the various techniques which might be used for the removal of activated material. These programmes included:-

i) a study of a large circular saw with a diamond tipped blade capable of cutting into R.C. to a depth of 1m (probable depth of activation)

ii) controlled use of explosives to remove active concrete and where necessary drill suitable holes for the placing of charges to optimise performance

iii) the use of various high temperature thermal sources (thermic lances, powder assisted flames and lasers) to cut and remove active concrete.

This paper outlines the results of the work completed in the first five year research programme, it draws some tentative conclusions and attempts to identify the future developments required for the solution of problems.

2. THE PROBLEM

Techniques used in the demolition of conventional R.C. and P.S.C. structures are crude, unrefined and aimed at rapid and inexpensive site clearance. Such methods are ill suited, as currently practiced, for the dismantling of nuclear facilities. Nonetheless, these approaches offer the most effective way forward for developing appropriate methods to achieve the closeness of control which will be essential in removing nuclear structures. The major problem is to refine these existing techniques to match the requirements for such activities.

The biological shield of a typical light water reactor frequently forms a shielding pool around the reactor vessel cap and comprises a set of reinforced or prestressed concrete walls forming an open topped hollow cylinder approximately 10m internal diameter and up to 15m high; the pool may form an enlargement at the top of this cylinder. This structure also provides the support structure for the reactor system and is generally in excess of 1m thick. The biological shield for typical gas cooled reactors (e.g. Magnox) is a closed, right hollow cylinder of reinforced concrete approximately 25m high and 20m internal diameter. The walls are a minimum of 1.8m thick and have reinforcing adjacent to both the internal and the external surfaces.

Two basic forms of structure have been used for PCPV. One contains a single large cavity about 25m internal diameter and 25m high within which the reactor and steam raising units are positioned with the reactor being separated from the boilers by an internal non structural shield and separation wall. In the other form (the multicavity vessel) the reactor is housed within a central close fitting cavity about 10m internal diameter by 10m high and the boilers in a set of planitary penetrations approximately 6m internal diameter positioned in the thickened walls around the reactor cavity. In both systems the cavity(s) is lined with steel plate to provide the gas tightness of the pressure circuit.

In all biological shields and PCPV the thickness of the activated layer of concrete varies but is approximately 1m. This is defined by a neutron induced activity level of 10^{-11} Ci/g, a notional level for unrestricted disposal of active wastes. In a single cavity PCPV the activation may be substantially contained within the non structural shielding wall separating the reactor from the boilers if the system is specifically designed for this; some activity may be induced in the main structure along "shine paths".

Many methods exist for the demolition of conventional RC and PSC structures similar to the biological shields and PCPV in nuclear facilities although some concern is expressed with respect to the demolition of prestressed concrete systems with bonded tendons. Mechanical methods include the use of hydraulic breaker, burster and wedge systems. The use of explosives with or without stitch drilling is commonplace in quarrying - a not dissimilar problem - whilst thermic lance and flame cutting techniques are also included in the tools for demolition of heavy concrete structures. These latter methods tend however, to be self inhibiting as the thickness of concrete is increased due to resolidification of the molten slag before it escapes from the cut. Heavy wrecking balls have been used for structures up to 1m thick and a machine with crushing jaws is available for walls up to 0.9m thick although this requires that the top or edge of the wall be available for the initial bite. Notwithstanding that these methods exist, such practices for the demolition of heavy RC and PSC structures are far from being well established.

When such methods are considered for the demolition of nuclear facilities it is found that the following additional special requirements of:-

i) complete retention of all radioactive materials
ii) remote working for all operations in the radioactive environment
iii) minimum increase in quantity of active materials

considerably exacerbate the problems of demolition, require radical revision of working practices and eliminate totally some of the conventional techniques because they are found to be completely inappropriate. It is found that, with the depth to which material becomes activated in a biological shield, for instance, constituting over half the shield wall thickness, the demolition procedures required are those with which active material can be cut away under very precise control to ensure that damage is not inflicted upon the non active exterior shell of the structure which is needed to provide structural support and containment of radioactive materials. A somewhat less demanding requirement may exist in LWR systems which are always housed within a containment structure, but in this case the use of careful demolition practices will ensure that radioactive dusts are contained within the minimum volume so that large areas of the surfaces of other structures do not become contaminated with these radioactive dusts etc.

The dose rates which will be experienced within the biological shield and PCPV structures soon after reactor shut down and before the activated layer is removed will be far too high to permit manual operation of the dismantling equipment. Two options exist, therefore, for this situation:-

either i) the whole of the system is allowed to stand for a period of time so that activity levels may decay sufficiently for "hands on" working to be accomplished without exposing workers to unacceptably high radiation doses. Delay times of the order of 100 years are required for this reduction to enable the main isotope, Co_{60}, to decay sufficiently to allow access. After the Co_{60} hazard is removed other isotopes with much longer half lives take over so that further delay yields insignificant reduction in activity levels. Even following such a delay great care will have to be used and special protective measures will be needed to prevent inhalation or ingestion of dust.

or ii) all equipment will have to be operated remotely. This will demand that all such equipment must be capable of remote operation reliably for long periods with little "hands on" maintenance.

It is perhaps worth noting that in a multicavity PCPV or Magnox biological shield there will be up to 8000t of active material to be removed without any release to the environment. Although time and speed of operation will not be critical factors in such demolition work it must always be remembered that any programme will become increasingly costly with increased duration.

The third special requirement of minimum increase in the quantity of active materials arising from these demolition processes demands close attention to such operations as drilling and cutting with tools requiring an abundance of cooling water. The need to keep the total quantity of such water to a minimum will undoubtedly place heavy demands for an adequate recirculation system with extensive filtration and cleaning. Additionally, although material in the operating machines will not become active the production of smoke, or dust will undoubtedly cause contamination of machine surfaces with radioactive materials so that extensive decontamination operations will be required before machines can be salvaged; alternatively the machine itself becomes added radioactive material requiring disposal.

3. MECHANICAL SYSTEMS

3.1 Hydraulic Breaker

The hydraulic breaker is a mechanical impact device similar in concept to the pneumatic road drill but driven hydraulically instead of pneumatically. It is currently widely used for breaking stone and concrete in the mining, construction and demolition industries. The principle by which the breaker works is that pressurised hydraulic fluid causes a piston to reciprocate inside a stationary cylinder. At the end of each power stroke the piston impacts with the tool which carries the impact energy to the concrete to be broken.

Breakers are available with impact energy of up to 2500J per blow and tools 135mm in diameter, enabling them to cut through the steel reinforcing bars as well as removing the concrete. Impact rate can be up to 450 impacts per minute, giving an energy transfer of up to 33kW. Such a unit weighs about 1.6t and can be mounted on a hydraulic excavator weighing 15t to provide a versatile tool for removal of the inner face of the shield. It would be capable of cutting any arrangement of reinforcing bars found in the shield as well a the light steel angle in the ring beam. If necessary an oxygen torch mounted on a manipulator arm could be used to cut particularly awkward steel sections. Debris would be allowed to drop into the bottom of the biological shield as it was produced.

Tests have been conducted using a breaker of 2,000J per blow which had no difficulty cutting 25mm reinforcement and removed concrete at $2m^3$/hour. Spoil size varied but was mainly of less than 150mm diameter with pieces up to 600mm diameter. With the more powerful breaker and an experienced operator higher removal rates would be achieved, possibly $4m^3$/hour. At this rate 900 working hours would be required, about 6 months, to remove the active layer from a Magnox biological shield.

Water sprays may be necessary to control the airborne dust and increase visibility for remote control. The noise level was 108 decibels at 6.5m from the tool in the tests, which is unlikely to cause any problems. Vibration of the structure may need to be monitored to guard against resonance within it.

The versatility of the tool makes it particularly suitable for the removal of concrete and reinforcement around penetrations and where activation has penetrated to a depth greater than average. The breaker would be mounted on a 'universal joint' so that it may be tilted in any direction. This would enable it to leave a reasonably smooth surface without damaging the remaining concrete.

Hydraulic power for the breaker comes from the pump on the excavator, and is controlled by valves. The whole machine can be easily adapted to remote control by umbilical cable or radio control. No major modifications would be required to a standard machine except that the diesel engine used to drive the hydraulic pump would be replaced by an electric motor. Alternatively the hydraulic pump could be located outside the biological shield with fluid supply through high pressure hoses. A number of proven designs are available requiring little maintenance apart from lubrication and an occasional tool change.

Similar pneumatic breakers are available. They are lighter and deliver more blows at lower energy per blow, which is considered less suitable for breaking reinforced concrete.

3.2 Concrete Saws

Concrete can be sawn using large circular saws similar to those used for cutting stone in quarries. Mobile concrete saws are used for sawing slots in motorway slabs and cutting openings in walls and floors. No mobile saw was available to cut 1m deep into concrete so such a saw was developed under contract for CEGB and the CEC. The blade 2.5m diameter and tipped with diamond abrasive has cut a smooth slot 13mm wide and 1m deep. It was mounted in a carriage on the end of a crane jib and driven by a 75kW hydraulic motor.

It is envisaged that vertical and horizontal cuts could be made in the face of the biological shield at 1m spacing and $1m^3$ blocks of concrete then broken out with wedges or jacks, making use of the weakness of concrete in tension. Trials with the largest currently available wall saw of 2m diameter on a reinforced concrete wall showed that the saw was capable of cutting the steel reinforcement, but that progress when cutting steel was relatively slow. As the reinforcement in the biological shield is mainly within 150mm of the surface it was proposed to cut through the reinforcement first with a smaller saw, followed by a larger blade cutting to the full depth in concrete only. However contrary to initial expectations, it was not necessary to use a smaller diameter blade to cut the reinforcement followed by a 2.5m blade to produce a 1m deep slot. The 2.5m blade was plunged into the reinforced concrete to 1m deep and cut the concrete and reinforcement at a traverse rate of 160mm/min.

Development of the saw enabled cutting rates of $10m^2$/hour to be achieved, giving a removal rate of up to $7m^3$/hour. The blade will have to be changed and retipped when it becomes worn, each blade is capable of cutting at least $200m^2$. 3% of the material, is removed as a water borne slurry and the blocks could be cut to any convenient size, simplifying the problem of material removal. Water is used for cooling and lubrication of the cut face and the removal of debris from the cut. The problems of airbone dust are thus eliminated, but the water must be collected for decanting and recycling. Vibration is much less than for the hydraulic breaker.

The saw must be placed and manoeuvred with greater precision than is required for the breaker, but this is achievable with minor modifications to available machines and the special cradle for holding the saw in position on the wall. The present design would be unsuitable for cutting the top of the biological shield from inside.

In order to break the blocks out a light hammering is required on wide-faced wedges in one of the saw cuts to force the cut faces apart. Such a force can be easily achieved with a hydraulic wedge. The concrete at the back of the block will then fail allowing the block to come free from the wall and be drawn out for removal by a crane. Remote control would be slightly more complicated than for the breaker because of the greater degree of control required. A separarate manipulator would need to be developed for the jack or burster.

As with all methods involving cracking, if any piece of reinforcement passes through the cracking plane then the block cannot be removed as the reinforcement will not be broken. Subsequent cutting of the reinforcement is not easy as it first has to be located and then access gained to it. This could limit the application of the method around penetrations. Weaknesses in the concrete could cause cracks to run in other than the required direction, in particular the possibility of cracks running into or through the outer part of the shield must be guarded against.

3.3 Stitch Drilling and Bursting

This method is similar to sawing but instead of cutting the slots with a saw they are produced by drilling a series of holes side-by-side with a diamond abrasive tipped drill. The holes may just intersect or have small amounts of concrete left between them. They would be drilled with a hollow drill to minimise the quantity of material removed; typically a drill would cut an annulus 200mm diameter with a cut width of only 4mm. The core would then be broken out once the drill had reached the required depth. Once all the holes had been drilled and the cores removed a hydraulic burster could be inserted into one of the holes and expanded until the block was broken away.

Two types of burster are available, wedge and piston. The wedge type burster consists of a hydraulic cylinder with a piston rod having a wedge-shaped end arranged to expand two feathers. The burster is inserted into the hole and the cylinder pressurised. The wedge forces the feathers apart splitting the concrete. The piston type consists of a hydraulic cylinder with a number of small pistons acting in a radial direction which move out when pressure is applied, again splitting the concrete.

The drills and bursters that would be required have already been proved in commercial use but manipulators would have to be developed for them. The method would be slower than sawing and more complicated as more operations are involved. Similarly, remote control would become more complicated. The method is more versatile than sawing, being able to cut out irregular shaped pieces if required. As the drills are smaller than the saws they cut less volume of concrete before needing to be changed.

It would be possible to drill holes further apart and make more use of the burster to crack the concrete, providing the drill holes were placed to cut all the reinforcement.

The method is quiet and produces little vibration and airborne dust is prevented by water lubrication and cooling. There are no significant advantages over the saw and the method is slower and more complicated.

4. CHEMICAL TECHNIQUES

The chemical techniques which have been considered for use in the controlled cutting of activated materials include:-

i) The use of explosives for drilling and cutting (shaped

charges) and for material removal.

ii) The use of slurries which expand on setting.

iii) The use of chemically aggressive reagents (strong acids).

4.1 Cutting with Explosives

Explosives have been used extensively for structural demolition and material quarrying over a very long period. In general the requirements have been for rapid site clearance and material loosening although occasions have occurred where close control has been a requirement but rarely, if ever, has it been necessary to contain the dust which may arise from such operations. Where military demolition has been required to produce barriers an overkill situation has been utilised to ensure that critical members in a structure are completely cut and collapse occurs.

For the demolition of nuclear structures, in which activation of the materials of construction (concrete, steel reinforcing and prestressing steels) has occurred to a depth which may be more than half the thickness of the structural component, removal of radioactive material must be effected without damage to the non-active material which will be required to form the containing structure for dust and rubble, etc. This will mean that the activated material will have to be removed in thin layers to ensure no venting cracks develop. To determine exactly the number and thickness of the layers required to remove a total of 1m of reinforced concrete from a structure 2m thick a series of tests was undertaken as part of the CEC programme of research on decommissioning, support being given by the Central Electricity Generating Board, the Department of the Environment and Hunting Engineering Limited. The tests also sought to determine the most effective explosive charge weight and consider how best explosive charges might be positioned to obtain maximum material removal with minimum residual effects from blast and debris scatter.

4.1.1 Modelling

To facilitate the experimental work, tests were performed with models at a linear scale of one quarter. A dimensional analysis, which extended previous similar analyses for:-

a. the response of structures to blast from very large weapons,

b. the response of structures to impact loads.

was performed for non-dimensional parameters associated with detonation and shock wave effects. This indicated that "replica scaling" could be applied and that, so long as the constitutive relationships for the materials of construction and the explosive used in the charges were identical, the structural behaviour would scale linearly if the ratio of charge weight was varied as the cube of the ratio of linear dimensions. Table 1 lists the various parameters and shows how each will scale in "replica modelling".

As part of the experimental programme tests were performed on reinforced concrete discs at both "full scale" and "¼ scale" in which the material charactersitics and dimensional details were homologeous. Comparisons were made of the cratering behaviour of charges buried at scaled depths in the concrete using a charge weight ratio of 64:1 (i.e. 4^3:1). These tests confirmed the analysis and demonstrated that such procedures could be adopted and that model results showed close similitude to full scale behaviour.

4.1.2 Depth of Burial and Charge Weight

A series of tests was performed on model targets in which the depth of burial was increased for a single weight of explosive. Careful measurement of the crater dimensions was made and observations made of the discharge of material from the craters formed. As the depth of burial was increased the volume of material ejected, as indicated by the crater diameter, increased almost linearly until the charge was insufficiently powerful to produce a crater and a camouflet was formed. The maximum depth of burial at which a crater was formed was little affected by the strength of the concrete or whether there was any reinforcing adjacent to the surface being cratered. Some tests were performed with charges of various weights up to 2½ times greater than the charge weight used previously. As would be expected using the cube root scaling law the crater dimensions were only increased by up to 35%.

4.1.3 Charge Spacing and Patterns

Tests were performed in which a number of charges were placed at the optimum depth of burial but at various separation distances to assess the most effective array of charges which might be used in stripping concrete in layers to form a composite crater with the flatest bottom. Various separation and positioning arrays were compared and it was found that an array based on an equilateral triangular grid gave most promising results. The bottom of the composite crater was formed with a series of conical peaks which developed at the centroid of each triangle. A separation distance of about three times the depth of burial produced the most effective results.

For the removal of the second layer it was found that the new charges should be placed in holes drilled at the centroids of the original equilateral triangular pattern but the depth of burial had to be adjusted to ensure that cratering occurred; charges for this second layer could not be buried at double the original depth but at the original depth below the peak of the "cones" which had been formed. It was found that to remove concrete to the full depth four layers of charges were required.

To complete this set of tests a closed right cylinder of reinforced concrete was constructed and the inside stripped away in layers. Various small changes in charge positioning, layout pattern and total charge weight were made in a series of tests in which the thickness of the wall of the structure was reduced to one half ot its original dimension. These tests essentially confirmed all the preliminary studies but the advent of some external cracking of the wall of the structure illustrated the need that such stripping be undertaken with extreme caution. Although cracking occurred it was of a very minor form and as far as could be observed did not allow venting from inside.

4.1.4 Shaped Charges for Drilling and Cutting

In all the tests performed charges were placed in holes drilled mechanically. In the demolition of nuclear structure this would mean drilling into active concrete. For drilling from the inside the operation would of necessity be by remote control and would involve an increae in the quantity of radioactive material because water cooling of the bit would be required. Drilling from outside would allow "hands-on" working but some radioactive material would be washed out and would need special treatment. Shaped charges are available for drilling and Hunting Engineering Limited - one of the collaborating groups in this research - developed a small shaped charge which drilled a hole suitable for use in the removal of activated concrete. Such charges can be remotely placed and supported inside the structures to be decommissioned using a simple feed and support system.

Another shaped charge was tested for linear cutting in an effort to divide the internal face by a grid of cuts which severed the reinforcing. In this instance it was found that the weight of the charge required to achieved a 1m long cut was far too large for firing inside a closed vessel similar to a biological shield. The pressures developed within the cavity would be very difficult to contain. This research was therefore discontinued after a very few tests.

The presence of reinforcing adjacent to the face being stripped was found to little effect the results obtained from charges buried within the concrete. In firings against such a reinforced face the cover was stripped off and the reinforcing was exposed so that it could readily be cut with a gas flame or mechanical croppers. Behind the reinforcing a crater with only slightly reduced dimensions was formed although in this case the rubble was retained by the reinforcing.

4.2 Expanding Slurries

There is on the market one slurry with the trade name of "Bristar" which when allowed to set expands. Placed in a hole in rock or plain concrete the expansion produces large forces which breaks and cracks the host material very much as do some of the mechanical bursting systems.

Tests performed with "Bristar" confirmed that manufacturers claims. It disrupted plain concrete blocks effectively, with the formation of very little dust and without the development of other disrupting effects. It was found, however, that with a concrete structure of finite thickness cracks could propagate through the thickness so that in a biological shield external cracks might develop. The introduction of reinforcement into the concrete test sample significantly inhibited crack formation and propagation. The extent of cracking decreased as the quantity of reinforcing steel increased and where the quantity and position of the reinforcement was similar to that which has been included in biological shields from Magnox stations little cracking was observed. Where cracks could be seen these extended only as far as the nearest reinforcing rods and it appeared that the presence of reinforcing completely inhibited the expansion of the solidifying slurry.

4.3 Aggressive Chemicals

Assessments were made to determine whether concrete might be removed using aggressive chemical reagents such as strong acids. It rapidly became apparent that the quantities of reagents required would be extremely large, the breakdown processes rather slow, the equipment required to maintain the circulation and application (spraying) of the acids would be highly specialised and the hazards involved in the processes unacceptably high. In addition, it was evident that the products of the reactions would need extensive washing and other treatment before they could be placed in disposal casks and solidified for final disposal. Such washings would add considerably to the total quantity of active and contaminated wastes needing eventual disposal.

5. THERMAL METHODS

Various thermal methods for dismantling radioactive concrete structures have been examined. These have included the cutting of concrete using:-

i) oxygen lance methods
ii) powder assisted flames
iii) laser techniques.

The efficiency of oxygen lances of various sizes was assessed with regards to the amount of slag (secondary waste) formed, rate of lance burn-off and oxygen consumption. Powder assisted flames were assessed against normal and high density concretes and lasers were tested to determine cutting rates, depth of cut and slag behaviour at a constant laser power of about 10kW.

5.1 Oxygen Lance Methods (Thermic Lance)

A thermic lance works on the principle that when carbon steels are heated to about 900°C they will burn readily when supplied with oxygen. The reaction raises the temperature to about 3500°C which is sufficient to melt the silica in concrete to produce a low viscosity slag.

The lance consists of a carbon steel tube containing a bundle of steel rods, the tube being inserted into a holder through which oxygen is supplied via a control valve. Once the end of the tube has been preheated, the oxygen supply is turned on and the high temperature self-sustaining action takes place at the tip of the tube gradually consuming it. A 17m od lance is commonly used and produces a 25 to 35mm diameter hole.

As well as the slag, the reaction produces large quantities of toxic smoke and gas. When the lance is used to form holes, the slag is ejected from the hole by the excess oxygen and gas from the reaction. This is not the case when a slot is being formed which can lead to slag solidifying in the cut making cutting of blocks diffucult.

5.1.1 Testing with the oxygen-lance

There are two types of thermal cutting with the oxygen-lance : perforation cutting and bevel cutting. The purpose of the tests was to compare these two methods to see which produced less slag and therefore kept secondary waste to a minimum:-

a) Perforation cutting
In perforation cutting, severance is achieved by piercing lines of holes from top to bottom or horizontally. The holes are lanced either immediately adjacent to each other or at intervals, leaving concrete between them.

b) Bevel cutting
Bevel cutting is carried out from top to bottom at an angle of approximately 45°. Unlike perforation cutting, no individual holes are made but the concrete is "shaved" off.
The methods were compared by cutting a ring of heavy concrete with a diameter of 2.5m, a cross-sectional area of $0.3m^2$ and a density of 4000 kg/m^3. The test parameters recorded were: lance consumption, amount of slag and cutting time. The results are given in the following Table:-

Cutting method		Perforation	Bevel
Oxygen pressure	bar	8	8
Lance consumption	m	42.76	37.86
Amount of slag	kg	178.800	147.90
Time	min	52	54

Result

Bevel cutting produces 20% less slag than perforation cutting with 13% lower lance consumption, the cutting time being 4% longer. This result indicates that bevel cutting is preferable to perforation cutting.

5.1.2 Testing to Find the Optimum Lance Diameter

Holes were pierced in heavy concrete and normal concrete and oxygen consumption, lance burn-off, cutting speed and amount of slag were measured.

The following tables give the results for dependence of the above parameters on oxygen pressure when 10mm and 12.5mm lances were used:-

10mm lance

O_2 pressure	bar	6	8	10	12
O_2 consumption	l/min	378	490	635	728
Lance burn-off	cm/min	81	106	112.5	132
Cutting speed	cm/min	11.5	15	19.5	23
Amount of slag	kg	2.16	2.44	3.1	3.7

12.5mm lance

O_2 pressure	bar	6	8	10	12
O_2 consumption	l/min	590	730	855	935
Lance burn-off	cm/min	73.5	91	105	128
Cutting speed	cm/min	14.4	14.6	15.8	20.4
Amount of slag	kg	2.81	3.41	3.76	4.90

Result

The result shows that for use on heavy aggregate concrete in nuclear power plants the 10mm lance is to be preferred to the 12.5mm one for the following reasons:

1. Oxygen consumption as a function of oxygen pressure for the 10mm lance is appreciably lower than in that for the 12.5mm lance.
2. For all oxygen pressures used the 10mm lance produced much less slag than the 12.5mm lance.
3. Cutting speeds are higher with the 10mm lance, which means shorter times and therefore a reduction in the dose rate for operating staff.

In the two previous tests, it was shown that with oxygen lances the cutting speed increased and the amount of slag decreased with decreasing lance diameter.

To check the continuation of this trend a 6mm lance was introduced into the test series. The results are given below:-

6mm lance

O_2 pressure	bar	6	8	10	12
O_2 consumption	l/min	176	234	320	550
Lance burn-off	cm/min	75	77	78	152
Cutting speed	cm/min	9	10.5	15.2	26.5
Amount of slag	kg	0.84	1.2	1.51	2.84

Result

Lance burn-off increases linearly between 6 and 10 bar and then rises very steeply as the pressure is increased to 12 bar. On average, lance burn-off is 12% less for the 10mm lance. Cutting speed is 15% less and oxygen consumption is reduced by 45%. The amount of slag and therefore of secondary waste, is reduced by 47% for a loss of cutting efficiency of only 15%.

For use in nuclear power plants therefore, where reducing the quantity of secondary waste is of paramount importance, the 6mm packed lance is to be recommended for cutting concrete.

Assessment of the oxygen lance method

During work with these lances, the following significant disadvantages for the cutting of radioactive components were found:

1. Burning time of each lance too short (7 min).
2. Lances extinguished at the connecting sockets owing to

discontinuity of the core wires.

3. Lances extinguished during cutting as a result of too rapid burn-off of the core wires inside.

These problems mean that personnel are exposed to a higher radiation dose since they have to spend more time in the vicinity of active components.

The applicability of the method in the nuclear field is further limited by the fact that the lances cannot be remotely controlled. Optimum parting-off is ensured only by manual and therefore more flexible guidance of the lances.

Oxygen lances are suitable for cutting active components in nuclear power plants when dose rate and local circumstances permit working close to the components to be cut.

Owing to the considerable weight of each lance, safe and reliable handling is possible only when no more than three lances are screwed together.

Since this method produces large quantities of dust, the air in the cutting room has to be changed at least 10 times every hour. The cutting room, therefore, has to be kept as small as possible to avoid the need for an excessively large filter system.

5.2 Powder Assisted Flames

Flame cutting with powder assist uses a similar reaction but in a somewhat modified way. A powdered mixture of aluminium and iron is carried into a gas flame in a jet of oxgen where the reaction temperature again reaches about 3500°C. The jet reacts with the reinforcement and concrete in much the same way as does the thermic lance but in this case the cutting head cannot be fed into the hole as it is being drilled and in consequence appears to be less efficient for drilling blind holes than when an oxygen lance is used.

5.2.1 Testing Powder Assisted Flame Cutting

The purpose of the test was to find the maximum cutting depth attainable in heavy concrete (density 4000 kg/m^3) and normal concrete (density 2200 kg/m^3) with the concrete fusion cutting machine (B 80 cutting machine made by Messer Griesheim GmbH, Frankfurt/Main).

Cutting parameters to achieve maximum depth were as follows:-

Cutting oxygen pressure	:	12.0 bar
Heating oxygen pressure	:	2.2 bar
Powder pressure (FeAl)	:	1.2 bar = 1700 g/min
Fuel gas (C_3H_8)	:	0.6 bar
O_2 consumption	:	1670.0 l/min

Result

The tests show no difference in the maximum cutting depth attainable in heavy concrete and normal concrete, which in both cases is about 1 metre.

5.3 Comparison of Powder Assisted Flames with Oxygen Lance Methods

5.3.1 Comparison With Heavy Concrete Ring

For the direct comparison of the efficiency of powder assisted flame and oxygen lance cutting, three sets of cuts were made in a heavy concrete ring. The time required and the amount of slag produced per cut were measured and are shown below.

Oxygen lance

		Perforation cutting	Bevel cutting
Amount of slag	kg	178.01	147.90
Time	min	52	54

Powder assisted flames

Amount of slag	kg	81.64
Time	min	33

If this amount of slag is compared with the average amount of 163 kg produced when the packed lance is used, it can be seen that the amount of slag produced by powder assisted flames is reduced by about 50% and about 38% of the time is saved.

5.3.2 Comparison With Normal Concrete

This test was carried out in order to make a direct comparison of the packed lance (10mm) and the hand-held powder assisted flame torch in use on a normal concrete.

The test specimen was a channel section concrete slab 970mm wide, 280mm long and 120mm thick.

Cutting parameters of the two methods were:-

10mm oxygen lance		Manual powder assisted flames	
Oxygen pressure	: 10 bar	Cutting oxygen pressure	: 6 bar
Lance consumption	: 12.45 m	C_3H_8 fuel gas	: 0.5 bar
Oxygen consumption	: 600 l/min	FeAl powder pressure	: 0.4 bar
Amount of slag	: 20.94 kg	Amount of slag	: 24 kg
Cutting time	: 32 min	Cutting time	: 37 min

Result

These tests showed that the feed motor must always be used in the cutting of concrete when using powder assisted flames since any uncontrolled movement with the torch causes the molten concrete to solidify. The result of the comparison shows an increase in cutting time by about 14% and an increase in the amount of slag produced by about 13% for the powder assisted flame over the oxygen lance.

5.4 Lasers

Lasers with sufficient power output can, when properly focussed, inject sufficient energy into a material placed at the focal point to raise the temperature to the melting point of the silica contained in concrete, and will cut, by melting, reinforced concrete. It has been found that a laser with an output of 9.5 kW is able to drill a hole in concrete 80mm deep at 1.6cm/min. It has not the power to cut through a 200mm thickness of concrete.

5.5 Advantages and Disadvantages of Thermal Methods

Thermic lances and powder assisted flames will cut both concrete and steel at about the same speed and will cut any shape of surface. In operation there is no vibration and little or no noise from any of the thermal techniques while material more than 25mm from the cut or hole is undamaged by the high temperature reactions.

All thermal cutting methods produce a large quantity of smoke which will spread contamination unless it can be contained effectively, cooled and filtered. This smoke is formed of submicron sized particles which present significant problems in retention. The lance is consumed rapidly and feeding such a heavy component requires considerable deftness. Both the lance and the assisted flame methods use large volumes of oxygen whilst the overall efficiency of lasers is approximately 8% so that a 10kW output calls for some 125kW power to drive the system.

6. CONCLUSIONS

The three projects supported by the CEC on the demolition of concrete structures have shown that it will be possible to remove activated materials from the inside of biological shields and pcpvs and that no damage need be inflicted upon the non active concrete. The studies have indicated the techniques which appear to be most promising, the rates at which cutting may be achieved and the limitations and disadvantages of each technique.

The large diameter saw worked extremely well and it was found that breakout of the cut material could be achieved successfully. Problems needing attention appear to be the need for cooling water and the quantity of water involved in the cutting, the requirement that a heavy rigid support machine will need to be introduced into the shield or pressure vessel structure to manipulate the mechanism which weighs about 2½tons and the difficulty that can be foreseen with cutting in corners and around heavily reinforced lined penetrations.

Explosive techniques for drilling and cratering have been shown to be effective and capable of close control in the removal of layers of activated concrete. The presence of reinforcing has not been found to be inhibiting and the development of a small shaped hcarge has indicated that holes suitable for the positioning of cratering charges may be drilled with ease. Although the method does not produce secondary waste materials there is an emotive reluctance to the use of explosives on a nuclear site. Quite clearly the use of explosives can only be undertaken by a highly qualified and experienced expert who will have to work to closely controlled instructions. One major problem will undoubtedly arise from the possibility of secondary damage breaking the containment structures but with due attention to the maximum total quantity of explosive fired at any one time this problem can be eliminated. The major disadvantage lies in the fact that explosions produce noise, vibration and quite sever shock effects for which working practices are of paramount importance.

Thermal methods are found to be feasible but pose a number of problems which make them unattractive. Control of the cutting procedures is very sensitive and the manipulation of lances and torches is found to be critical for continuous cutting. Extinguishing of the flame by non continuous lance construction and slag formation and solidification can occur. The rate at which materials are used tend to make the operations expensive. Smoke produced in the cutting contains a high loading of sub micron particles which are difficult to extract to prevent environmental pollution.

Nonetheless, at this time a hydraulic breaker, mounted on a suitable machine is probably the best method available for removing the activated layer of concrete from biological shield or PCPV structures. It could use equipment that is proved to be reliable under site conditions and is known to need little maintenance although major modification to facilitate remote control will be required. For its operation reasonable visibility will be required but the dust arising will cause some problems with visual observations using closed circuit television systems. The large saw and the use of explosives have attractions but do require some further development.

TABLE 1 - SCALING RELATIONSHIPS ASSOCIATED WITH SHORT DURATION LOADS

Parameter	Dimensions	Full Scale Value	General Model Value	Model value using REPLICA modelling	Typical Quantities
Length	L	d	$s_L d$	sd	Missile dimensions Target dimensions Geometry
Mass*	M	m	$s_m m^*$	$s^3 m$	Missile mass Target weight
Time	T	t	$s_t t$	st	Duration of Load Natural period of target
Density	ML^{-3}	ρ	$s_\rho \rho$	ρ	Densities of materials in structure and missile Density behind shock
Strain	—	ε	$s_\varepsilon \varepsilon$	ε	
Stress	$ML^{-1}T^{-2}$	σ	$s_\sigma \sigma$	σ	Includes pressure and material moduli
Velocity*	LT^{-1}	v	$s_v v^*$	v	Missile velocity Stress wave velocities Ejecta velocity Shock and sound velocities
Acceleration	LT^{-2}	a	$s_a a$	$s^{-1} a$	(Except gravity)
Rate of Strain	T^{-1}	r	$s_r r$	$s^{-1} r$	
Force*	MLT^{-2}	w	$s_w w^*$	$s^2 w$	Applied load Structural reactions
Energy*	ML^2T^{-2}	E	$s_E E^*$	$s^3 E$	Kinetic energy of missile Energy in explosive

The scale factor s is defined as $s = f_m/f_p$

Parameters marked * are not independent quantities

REMOTE OPERATION IN DECOMMISSIONING.

L.da Costa (ACEC), F.De Buck (ACEC), M.Leconte (ACEC)
E.Kirchner (GST), P.Drücke (GST), N.H.E. Larcombe (U.Warwick)
and J.R. Halsall (U.Warwick).

Summary :
In order to reduce radiation exposure and occupational hazards during decommissioning tasks in nuclear facilities, and to allow intervention at an early stage after shutdown and in situation of accident, remotely controlled and, during execution of specific operations,autonomous systems are needed.
This paper presents the results of two research programs on the subject of remote operation and robotics.
One of these programs, "Review of Systems for Remotely Controlled Decommissioning Operations", conducted in close cooperation by GST and ACEC, was aimed at analyzing the tasks arising, at the review of applicable existing equipment and first conceptual studies of needed remotely operated equipment.
Based on the evaluation of tasks and existing equipment, recommendation for future research and development was made.
The second program, "Possible Advances in Remotely Controlled Operations in Hostile Environments", conducted by the University of Warwick, concentrates mainly on advanced control and data transmission systems.
This study reviews the present state-of-the-art in remote controlled robotics (REMCON) and gives total systems predictions for possible future applications within the nuclear industry.
In decommissioning tasks, the main fields of application to be considered are : monitoring by visual inspection and radiation or position measurement, decontamination of assemblies or of work space, dismantling of metal or concrete structures, and possibly treatment and transport of resulting debris.
It is clear that in terms of human life and health, the use of remote control or robotic devices in nuclear industry in general, and in decommissioning tasks in particular, is a necessity.
Results of both studies confirm that most of the technologies needed to develop future REMCON systems exist as of today. In fact, with the exception of control aspects, the technology requirements are not excessive.
The main problem lies in the necessity for a totaly new systems approach to be developed. An approach involving the marriage of two high level technologies, namely robotics and nuclear.
In other words, to start a coherent program of development for REMCON, we need to foresee in sufficient detail both the application problem areas and the robotics potentials and problems.

1. INTRODUCTION.

Decommissioning, be it immediate or deferred, is widely accepted as the safest way to dispose of old and obsolete nuclear facilities. Some actual decommissioning work has already been performed worldwide, and concerns mainly research and pilot installations. Due to the usually short working periods, and to the nature of these facilities, radioactive contamination and activation levels were found to be reduced and limited to rather well known confinements. It was therefore possible to carry out these operations using mainly hands-on techniques, while keeping overall radiation exposure at low levels.

In the case of standard commercial power plants and reprocessing facilities, long operating periods are expected thus leading to much higher radiation levels and huge quantities of contaminated components and structures.

This aspect is of even greater importance while dealing with emergency situations, such as accidents, where whole containments have been exposed to leakeages of primary loop fluids.

In both cases, it is clear that decommissioning will take years, during which time we will be dealing with levels of radiations and waste quantities still unheard of. Increasing social concern and protection legislation excludes the use of past hands-on techniques.

We are therefore left with no choice but to develop new remote operation capabilities encompassing all the aspects of decommissioning.

The use of remote control or robotic devices within the nuclear industry will not be a matter of choice - the problems which have been generated already in terms of waste handling and decommissioning dictate their use - and the potential for incident which must always be present serves to underline that dictat.

2. ANALYSIS OF DECOMMISSIONING TASKS.

Analysis of tasks to be performed while decommissioning show that we are confronted with specific, one-off, long-sequence and multi-technique tasks demanding increased levels of adaptability, flexibility and above all control capacity.

It is therefore necessary to accurately determine conditions and requirements under which remote controlled systems have to perform, in order to derive the resultant technical specifications.

Decommissioning of nuclear facilities involves a large number of operations or tasks which can be structured in a variety of ways. We can for instance adopt a plant specific approach, draw an exhaustive liste of operations, classify them according to expected man-rem exposure or any other criteria, and derive concepts and specifications for remote controlled systems.

Another option, is to adopt a task specific view and establish a classification based on function, from which we could then expect to derive a number of generalized concepts adapted to the time and environment-induced conditions of the individual plant.

The first approach leads to plant specific equipment, making a generalized concept unlikely. Nevertheless, for some tasks this may be the only possible solution for instance while dealing with dismantling of reactor primary loop.

A survey of existing or under development dismantling equipment shows that special purpose devices for cutting reactor vessels have or are being developed mainly in conjunction with the primary loop constructor.

The second approach leads to a better understanding of the variables involved and will result in the development of more efficient economical and user-oriented equipment. This approach will also allow a rather straightforward extension from preplanned dismantling schedules to accident-induced plant interventions.

Based on this approach, tasks arising in decommissioning can be classified into four categories :
- survey and monitoring
- decontamination
- dismantling of components and structures
- transport, treatment and conditioning of resultant waste.

This is only one of many possible classifications and it is open to discussion. It is based upon well understood and indentifiable functions which constitutes and obvious asset when time cames for conceptual design.

Under each heading comes a variety of tasks which have to a great extent been identified in this study.

Further in depth analysis of all tasks and related techniques led to a classification based on the need for and the usefulness of remote operation. Certain interventions can be done only by remote operation - strong radiation fields, heavy or oversized components - while for other tasks, remote handling will allow reduction of individual exposure and, as is the case with treatment, packaging and transport of waste, reduction of collective exposure of a group of workers at reasonably low cost.

This classification led to the identification of a number of needs in terms of systems functions, such as monitoring, appliance carrier, heavy duty handling and transport.

A specification of performances for each corresponding system was evolved, leading to a conceptual design of the most promising systems.

A major problem facing this approach is that availability of knowledge is often restricted. This is particularly true of fuel reprocessing plants which have clear security risks and of research establishments which are not usually exclusively concerned with civil R & D. In many cases it is not a simple matter for an evaluator to gain entry to assess the potential for robotic operation within a facility. Furthermore details of plant and processes are often themselves restricted limiting even paper studies.

3. CONCEPTS.

Though conducted in parallel and independently, both study contracts arrive at identical conclusions concerning future programs of development.

Most promising systems identified are :
- development of a family of small and medium scale mobile manipulators using advanced control techniques.
- development of small mobile monitor units operating in largely autonomous fashion.
- development of handling and transport machines for standardised biological and nuclear waste containers.
- development of general purpose control workstation units incorporation appropriate ergonomic features, and of data and power transmission systems.

Proposed concepts are still rather vague and considerable work will have to be done to implement actual working systems in a real environment.

Most of needed techniques exist today. Control aspects is the exception to the rule, since important developments are needed to achieve desired levels of system autonomy.

Systems identified have a very wide-based function definition. Proposed development is along lines of increased modularity in order to ease task-extension adaptability. Features such as propulsion, manipulator kinematics, energy sources, data transmission, control modes and so on are recurrent and form the basis of most remote controlled systems.

We can envisage R & D work along these lines, since there is no such thing as "robotics and remote control" industry. What we have are specialist manufacturers of equipment.

The aim of such a program would be to integrate existing knowledge, and help create new one in order to lay the bases of a viable remote control industry.

Existing nuclear facilities are complex mazes of concrete structures, piping, reservoirs, wiring etc. No provision has been made to ease access into critical areas. On the contrary, labyrinths of neutron absorbing concrete act as biological shielding. Sometimes access is difficult to man, let alone remote controled vehicles.

In parallel with the development of new remote operated systems, we should start planning ahead and incorporate into the construction of future plants standardized features that would make possible to achieve higher levels of remote operation in routine inspection and monitoring, in plant maintenance, and above all, in accident managment situations.

Such breakthrough would be profitable both in terms of human health and operating costs.

Better standards of inspection made possible by almost continuous monitoring of systems status will help optimize maintenance and will contribute to more effective decision-making processes.

Already some developments exist in this area such as computer controlled obstacle avoidance systems for use in fuel handling machines, reactor vessel ultrasonic seam weld inspection machines, in-service reactor containment inspection vehicles, to name a few.

Plant operators are still reluctant regarding use of such first generation robots, specialy so as many never go past the prototype stage. But, as technology develops and constructors acquire much needed experience, we will begin to appreciate the full potential of remote operation in nuclear industry.

4. CONTROL ASPECTS.

In order to clarify the terminology employed, we propose the following definition from the University of Warwick.

ROBOT - a system capable of non-trivial mechanical movement through an environment known only in part to the system and at most partly controllable, such motions made being by intent and being influenced by observation of the environment either by the system control or through the agency of one or more operators external to the robot itself.

A robot system designed to be largely directed by an external operator while having the features above will be refined to as a REMCON system (REMote CONtrol).

This definition is quite different from what is accepted in the industrial context. But so are needs and performance expectations.

Industrial robots usually have a fixed location, limited reach and process-specific performances with severe requirements for accuracy and repeatability. Mechanicaly speaking they are rigid and relatively heavy machines with very unfavorable weight-to-payload ratios. Apart some more complex laboratory applications they have little adaptability and need only a limited range of sensors and remote control interfaces. Their sofware is based on a number of fixed routines. These are machines made to evolve on a rather deterministic environment with little or no capacity to adapt to changing conditions.

Decommissioning operations are the exact opposite. Here robots have to work in individually structured plants, executing unrepetitive tasks in a constantly changing environment.

New advanced remote control developments are needed in the sense of our proposed ROBOT definition.

A major theme of the form of advanced remote control is the provision of protective autonomy within the remote system. By making use of obstacle detecting sensors it is possible to devise a control system which reduces the risk of inadvertent impact with the environment. A number of algorithms exist that elaborate obstacle avoidance stategies though the majority is applicable only in 2-D situations. To effect this protection it is necessary to translate the operator's control input in the perceived context of the sensor data. In this way the operator's controls do not act as demand input but rather as request inputs.

There will be times when the operator wishes to override the autonomous protection, for instance when using a vehicle to push something. This would be an intentional operation and would require the operator to indicate to the system that contact is intentional. If however the operator is reversing a vehicle or retracting a manipulator his attention is usually on an immediate task which is in the forward direction. As in the operation of conventional lifting vehicles accidents occur due to the operator's inattention to the reverse direction. In the use of protective autonomy the system itself would restrict the motion and would in doing so exhibit a difference between the operator's command input and the actual action taken which is detectable and would be relayed back to the operator in the form of a graduated alarm signal.

Further forms of autonomy can greatly assist the operator.

The remote operation of a remote manipulator of more than 4 degrees of freedom cannot usually be done smoothly unless very specific control input devices are made. When the manipulator does not possess anthropomorphic form, as is the case with the multi-link manipulators being developed for reactor maintenance and inspection, unaided control of more than 4 degrees of freedom becomes practically impossible. In cases such as these it is possible to reduce the operational demand to the control of the tip of the system alone. In order to make this possible it is necessary to interpose a level of transformational autonomy which transforms the operators requests for tip motion into the distributed control of the manipulator. When the manipulator is in free space this is known as the inverse kinematic transformation and it is fairly well understood how this can be effected for redundant linkages. In the case of the manipulators to be used in nuclear applications the space in which the manipulator is to work is far from free. It is however possible to use either sensor data, such as proximity measurement, or data derived from a computer model of the environment with a transformation process to ensure that not only is the system tip moved in such a way as to best approximate the operator's wishes but is also held clear of obstructions.

This form of transformational autonomy is applicable to mobile systems as well and is of particular interest in the case of the mobile manipulator where the degrees of freedom of vehicle movement are added to the manipulator's degrees of freedom.

In many remote control applications there will be actions which are routine. These may be such functions as the'journey to work' of a mobile system or the deployment and retraction of a manipulator. Given the protective autonomy described above many of these functions may be conducted autonomously by the system and would be described as sub-function autonomy. In any such autonomous action the operator would be able to interrupt the action and to take over direct control. The action to be taken on such interruption will not always be a stoppage of the system. In some circumstances, such as the transport of liquids in open vessels, impulsive arrest is no longer the safest action and the system must reflect this fact by stopping in a safe manner.

In some applications the sub-function autonomy will be able to perform a task better than a direct operator control could do. In remote manipulation using proximity and tactile feedback it is possible to program a system to have greater dexterity in grasping and lifting arbitrary loads than is possible by using direct control even with force feedback. Effectively the sensor data can be more rapidly used and the necessary corrections be made by making the function autonomous.

Clearly when sensor feedback is used quite complex sub-function autonomy can be given. From the University of Warwick's experience with autonomous vehicles they have shown that even the negociation of heavily obstructed environments can be executed autonomously using relatively simple tactics. When a human operator is at hand to 'advice' simple tactics a fully strategic control is possible without recourse to expensive computer controls.

All the processes mentioned above serve to simplify the operator's task. It must be remembered that in tasks such as reactor decommissioning the operator will not be involved for a few hours, as is the case with most existing remote control applications, but may be involved in three or four hour continuous operating shifts over months or years. This means that we must pay much greater attention to the ergonomics of the operator's workstation. Many existant remote control systems have not been designed with these factors in mind, indeed many should be described only as experimental systems. In practice they are difficult to use. This stems not from the design of the effector mechanisms themselves, which often show considerable mechanical ingenuity, but from the operational difficulties arising from the control station layout and operation and the lack of sufficient panoramic awareness. It is for this reason that a number of prototype remote control devices are currently gathering dust rather than being used.

Figures 7 - 8 - 9 show the succesive development of remote control towards the form described. To distinguish this advanced form of remote control from more elementary systems the name REMCON is proposed.

5. CONCLUSION.

From the results of the two study contracts we can conclude that REMCON operation in the nuclear industry is at its infancy.

Some laboratories have been working on this subject for long periods of time, and have accordingly developed extremely valuable know-how. Some very promising REMCON systems have been proposed, or are under development.

Unfortunatly, existing experience within the nuclear industry on REMCON manipulation exists in a very restricted context.

Actual operating experience probably exists most fully within the marine technology environment where applications present greater control problems as they are not directly linked to a fixed reference, but have the considerable advantage of operating in a generally less obstructed environment.

Increased mobility is of utmost importance in future remote operated systems. Existing systems are often fixed location ones and few examples of applicable REMCON vehicles exist.

Problem lies not with the existing mechanical and control systems, but on the fact that first concepts are not followed by sustained development efforts.

Some "hurried-up" vehicles have even disappointed potential users, their lack of development and reliability being the source of considerable problems.

Other developments, despite proving successful, were designed to tackle such low frequency tasks that they could not be maintained.

This lack of sustained working experience prevents a proper assessment of existing short falls from being made.

In order to develop a coherent R&D program it is necessary in the first place to identify existing application problem areas. This is the example case where we should avoid looking for a problem to fit an existing solution.

Inproved relations between plant operators and the REMCON industry are needed in order to make existing knowledge more widely available. This very same plant operators are often reluctant regarding the usefulness of existing or proposed systems, mainly because they have not been directly involved in their development.

Future development is most needed regarding control aspects, although subjects such as manipulator kinematics and dynamics, power supply, information transmission systems and ergonomic design will have to be carefully looked at.

These are only a few of the technologies involved, but it is clear that the success of REMCON depends on our aptitude to integrate them into a coherent working system.

Fig. 1

Hands-on decontamination of cell interiors at EUROCHEMIC - MOL using high-pressure water jets.

Fig. 2

Hands-on cutting of loading tube of a dissolver at EUROCHEMIC - MOL.

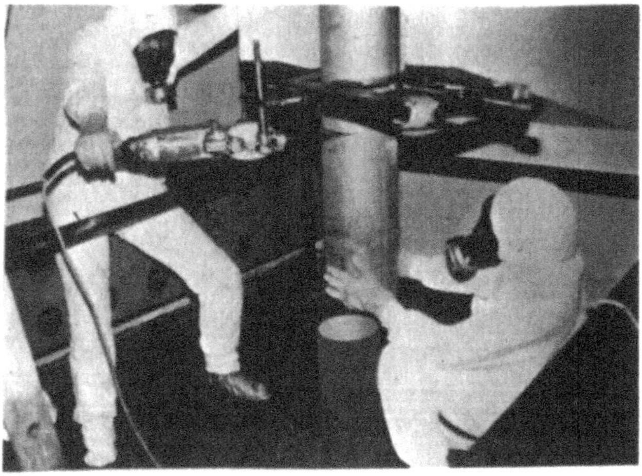

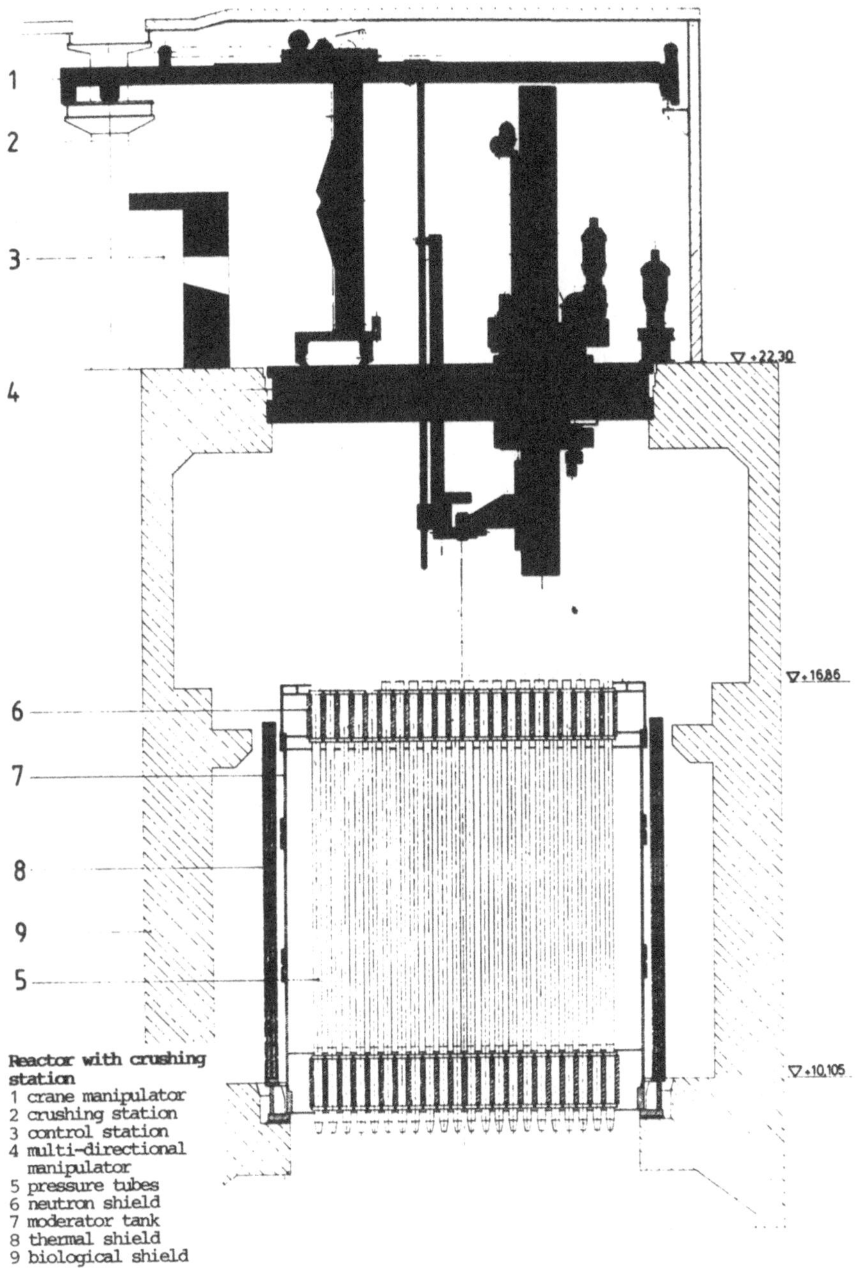

Fig.3 : Custom developed NIEDERAICHBACH reactor dismantling equipment - NOELL.

Fig. 4

"PIADE" - CEA
Porteur d'intervention adapté au démantèlement.
General view.
Dismantling system specially developed for use in decommissioning of nuclear facilities.

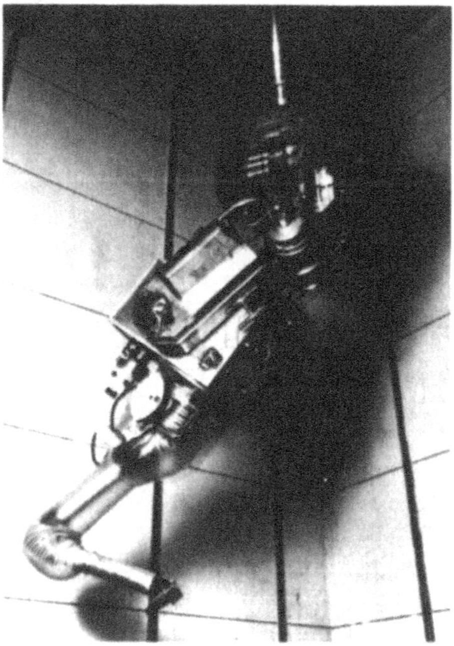

Fig. 5

"PIADE" - CEA
Close-up view of standard manipulator MA-23M

Fig.6 : Remote operated in-service reactor
inspection, monitoring and repair vehicle - ACEC.

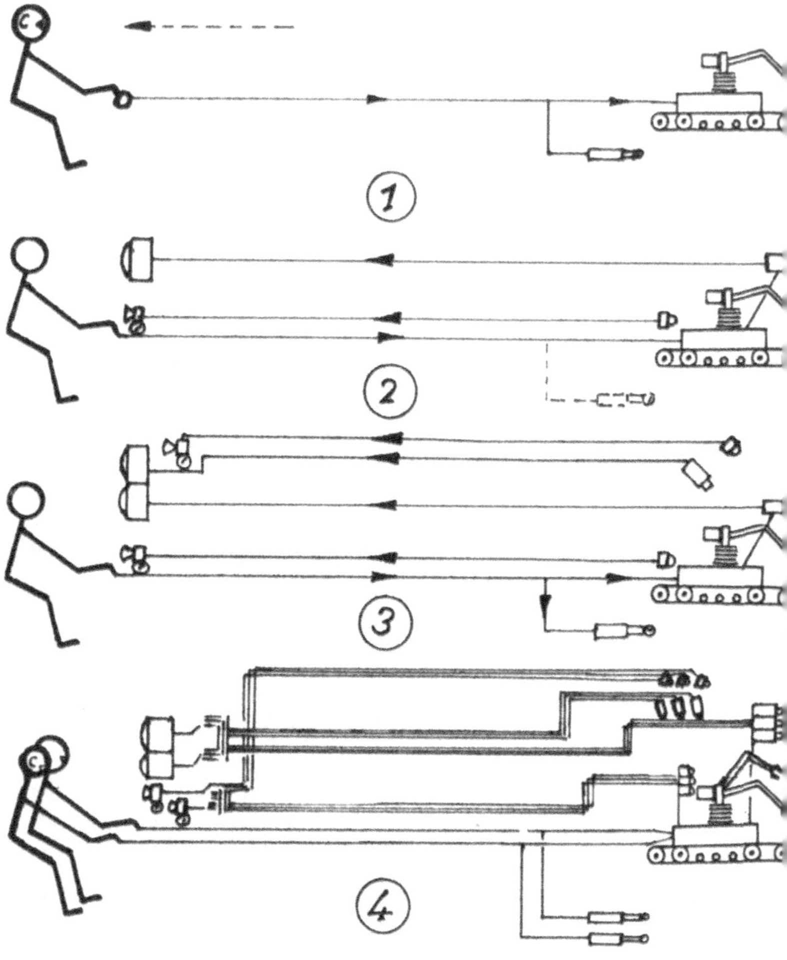

Fig. 7 : Elementary Remote Control.

1. Control by direct vision and sound, both main and environment control.
2. Simple main system CCTV with auxiliary acoustic and/or instrumentation, environment control usually only when view of main.
3. Main system plus environment CCTV and auxiliary sensors.
4. Multi-operator multiplexed multi-channel main and environment CCTV and auxiliary sensors.

8.

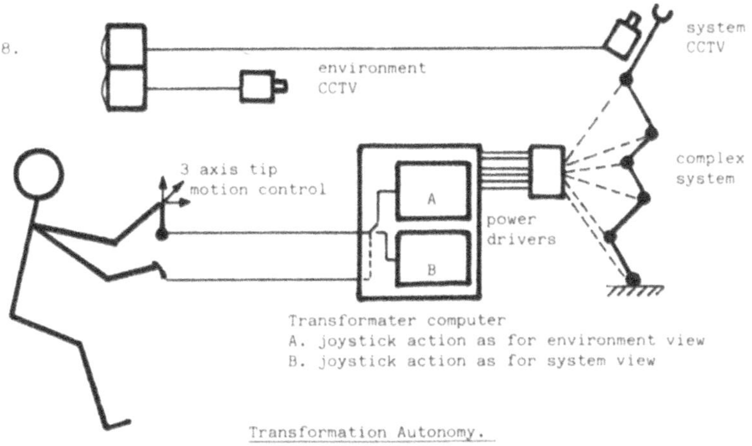

Transformation Autonomy.

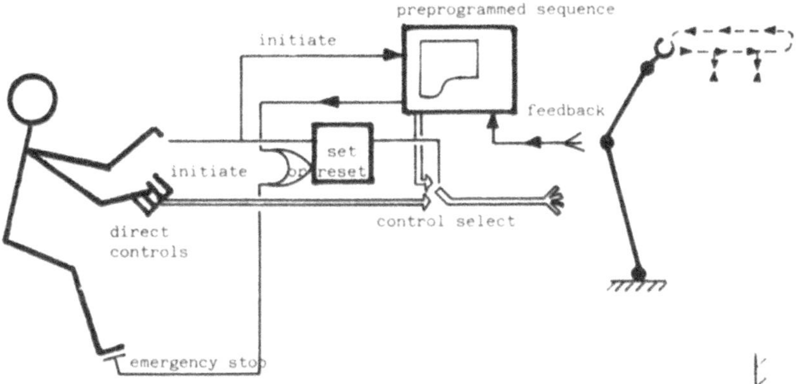

Sub-function Autonomy.

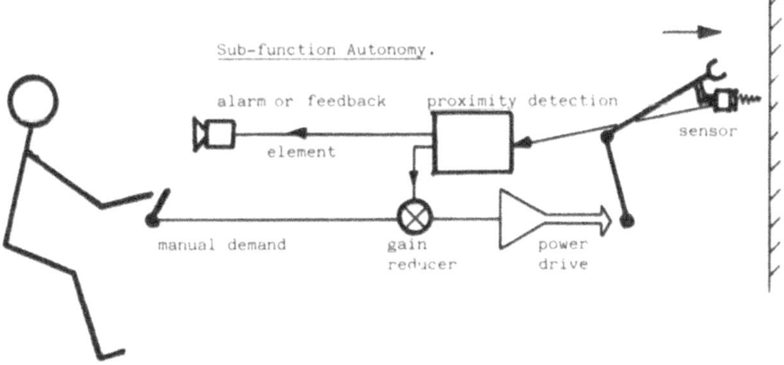

Protective Autonomy.

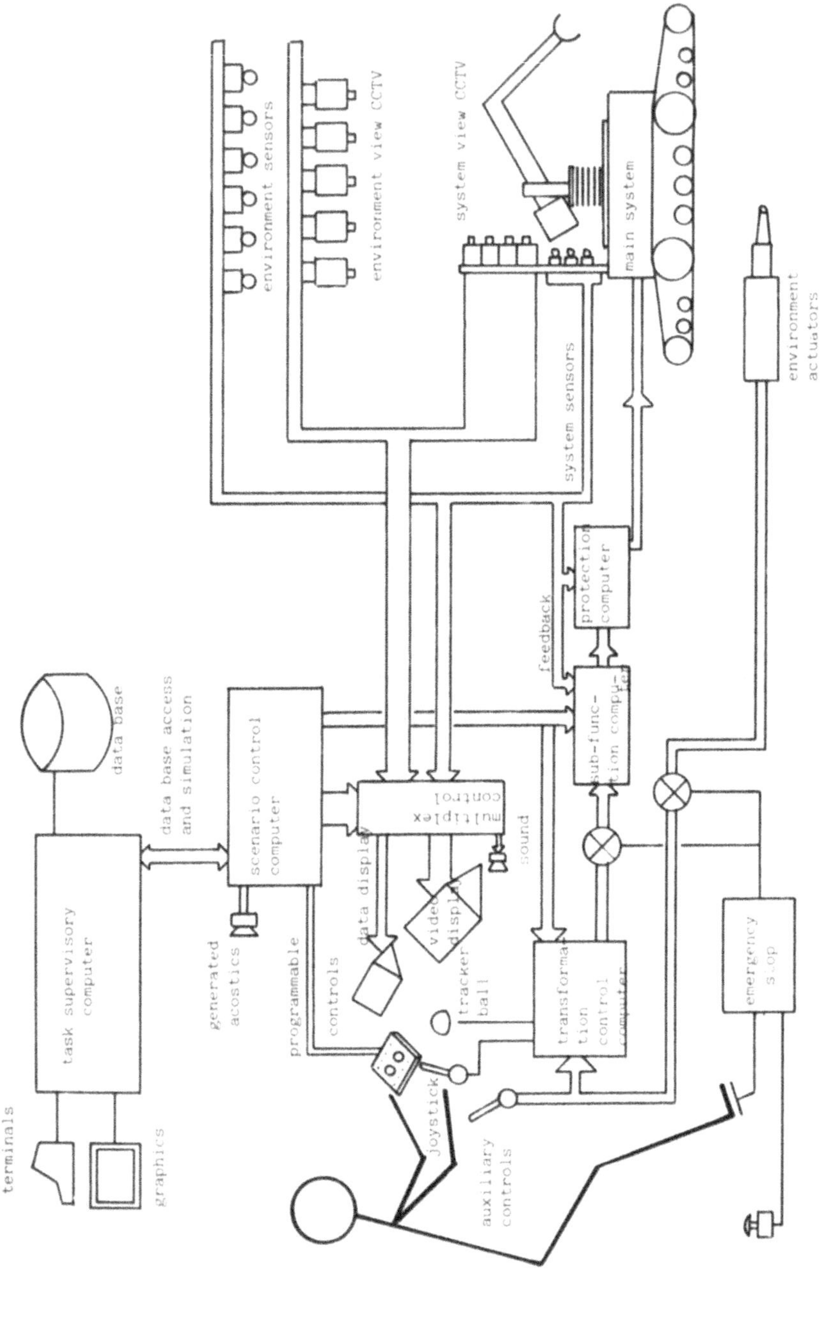
environment sensors
environment view CCTV
system view CCTV
main system
system sensors
environment actuators
data base
data base access
and simulation
scenario control
computer
multiplex
control
feedback
protection
computer
sub-func-
tion compu-
ter
task supervisory
computer
terminals
graphics
generated
acostics
programmable
controls
data display
video
display
sound
tracker
ball
transforma-
tion
control
computer
joystick
auxiliary
controls
emergency
stop

SESSION III : SUMMARY OF DISCUSSION

Paper : "Advanced cutting techniques : laser and fissuration cutting"

A U.K. participant asks, with regard to fissuration cutting, the reason for preference of copper as a more useful metal than, perhaps, cadmium, tin or lead - for example its melting point, or its reaction with the steel. He would appreciate some comments on the suitability of the metal that generates the fissures. One of the co-authors, Mr. Tanis, explains that copper had been chosen instead of tin or other materials for its low vapour pressure and easy penetration into the grain boundaries of steel material, thus facilitating the fissuration process.

A U.K. participant, recalling that a laser with an output energy of about 10 kW had been used, asks for the power requested in terms of input energy. Mr. Manassero (Fiat CRF) replies that, in order to have an output energy of about 15 kW in the laser beam (AVCO-Eyerett type), it is necessary to have about 180 kW of electrical energy because the gain of laser beam is very low.

Paper : "Demolition of concrete structures"

An FRG participant asks for the life-time and the costs of diamond-tipped saws and for experience obtained in cutting reinforced concrete horizontally and vertically. Mr. Gregory, one of the co-authors, answers that the cost of the blade equipped with the diamond tips was about 5,000 pounds. The cost of development of the whole machine turned out to be about 150,000 pounds. The blade has been about half used in these trials for this particular development and there is still life left in the blade and indeed it is possible now, having spoken to the people who manufacture these things, that the life be extended by making the tips thicker and by choosing a different type of mix between the diamond grit and the copper binder.

An FRG participant comments on the disadvantage of this type of saw for not being able to cut into the corners. He asks, regarding demolition by explosives, what type of firing had been used, the cascade firing or the instantaneous firing. Mr. Davies replies that one part of the test was aimed at checking whether cascade or instantaneous firing of all the charges together should be used. In the cascade firings, with a delay of the order of 20 milliseconds, a totally surprising misfire occurred. The last charge of a series of 6 misfired. In fact, the shock wave arriving from the first of the firings had disengaged the detonator so that it fired but not in intimate contact with the explosive. After that, it was decided to fire all charges simultaneously in order to make sure that this problem of misfires be excluded.

The Chairman comments that each of the methods discussed in the first three papers has its own well-defined field of application and a narrower optimum field. In his view, their fields of application overlap to the extent that they can be used in conjunction to dismantle all the systems and structures of a nuclear power plant.

Mr. Davies adds some information about the small shaped charge, which has been developed. Weighing 4 kg, it could be compared to a conventional bee-hive. It will drill a hole 1000 mm deep, by about 100 mm diameter. As compared to the small charges of 200 g, the charge is 20 times greater in mass, drilling a hole 1.6 times deeper and 3.3 times greater in diameter.

Paper : "Remote operation in decommissioning"

A U.K. participant comments that the presentation has correctly concentrated on control systems but another subject of interest is heavy duty handling; he would like to ask what the state-of-the-art today in heavy duty handling is, particularly in terms of weight-lifting capability. Mr. Da Costa indicates that some developments for heavy-weight handling equipment in decommissioning have been done, an example being the special machine built for decommissioning the Niederaichbach reactor in the FRG. To his knowledge, there does not exist a machine in nuclear industry handling more than 1 ton, apart from those used in normal operation in the plant.

The co-author, Ms. Larcombe, adds that although there are no heavy duty manipulators or mobile manipulators developed for the nuclear industry there are, however, heavy duty manipulators with capabilities up to 2 metric tons as standard products. There are some special forms of manipulator for the forging industry which go up to much larger weights, but they are very specific and could not be treated as general purpose manipulators. There also is heavy duty lifting equipment made for the mining industry and these are all sources of external equipment which may be brought in to assist in the decommissioning tasks.

Mr. Viecenz (Noell, Würzburg) adds more information on the earlier mentioned KKN equipment. The manipulator for the Niederaichbach plant is designed for 3 metric tons lifting and handling capacity of mechanical equipment to cut pieces and handle them out of the reactor core in order to bring them to the upper floor. So it is a special application for the nuclear industry.

TECHNICAL PANEL

REMOTE OPERATION IN DECOMMISSIONING

Chairman

A. CREGUT
Chef de Département, Chef de l'Unité Centrale de Déclassement des Installations Nucléaires, Centre d'Etudes Nucléaires de la Vallée du Rhône, Site de Marcoule.

Secretary

R. BISCI
CEC, Directorate-General Science, Research and Development

Members

A. BARKER
Assistant Chief Engineer, Nuclear Plant Design Office, Directorate of Engineering, UKAEA Risley, Warrington.

M. DEVRESSE
Chef du Département Marketing, Stratégique et Robotique de la Division Nucléaire des Ateliers de Constructions Electriques de Charleroi.

M.H.E. LARCOMBE
Computer Scientist, Department of Computer Science, University of Warwick, Coventry.

J. VERTUT
Adjoint au Chef de Département de Protection Technique, Institut de Protection et de Sûreté, Centre d'Etudes Nucléaires de Fontenay-aux-Roses.

W. ZIMMERMANN
Hauptabteilung Ingenieurtechnik, Kernforschungszentrum Karlsruhe.

CONTRIBUTIONS BY THE PANEL MEMBERS

Introduction by the Chairman

I would like to share a secret with you by admitting that I am not a specialist in remote control : if I have the privilege of chairing this round table, it is, I think, essentially because, like all dismantlers, I ask the remote control specialists and even the robotics experts a lot of questions.

Dismantlers have in fact many options available to them and have, in particular, choices to make between the remote-handling and remotely-controlled devices they will use.

I hope this round table will provide at least some valuable partial answers to the two questions which, in my view, are important when one is in charge of a decommissioning and dismantling operation.

First, when work is to be done in the presence of ionizing radiation, is it necessary to have a machine with dexterity, i.e. a machine that can replace man, be as agile as the human hand and give the operator all the feedback necessary for his work, namely sight, the sense of touch, etc. ?

In other words, the question that must be asked is how far should the sophistication of remotely-controlled devices be developed, or would it be preferable instead to have a set of far simpler and more rugged tools, each capable of carrying out the specific task for which it is designed ?

Secondly, once a tool or a manipulator has been chosen, to what extent should such a remote-handling device be assisted ? Let me explain myself : is it necessary for the operator to act directly on the equipment or should the equipment itself be fitted with devices that make it possible to receive sensations, accelerate movements started by the operator and even, where appropriate, complete them itself, etc. ? In other words, the question I am asking is : once a given device or piece of equipment has been chosen, is it necessary for that equipment to be computer assisted ?

I would now like to propose to you that our round table be conducted as follows :

- Mr. Zimmermann from the Federal Republic of Germany will give an account of the state of the art in his country today, on the basis of actual experience;
- Mr. Barker from the United Kingdom will give us a progress report on studies in the UK and the project that is in hand on the dismantling of the WAGR reactor;
- Mr. Devresse will describe the remote manipulation and control methods that have been used in Belgium for inspections and tasks in hostile environments;
- Mr. Vertut will describe in broad terms the French programme for developing remotely controlled and remote-handling equipment;
- Ms. Larcombe from the United Kingdom will put forward a few futuristic ideas on the possible subsequent directions our research could take.

W. ZIMMERMANN

The Nuclear Research Centre of Karlsruhe (KfK) has been involved in remote operations for activated components since the mid sixties. This task resulted out of the operation of the research reactor FR2 for maintenance and repair work. The manipulator vehicle MF3, which must be well known, was developed at KfK, who are also developing a complete remote operated maintenance system for a reprocessing plant.

Since the mid seventies KfK hass been involved in the decommissioning of the reactor FR2 and the Nuclear Power Plant Niederaichbach (KKN). For the FR2 it is intended to dismantle the reactor block and its internals only. For the KKN decommissioning to stage 3 is intended. The remote handling facilities for dismantling the activitated parts of these two reactors have been developed. While the decommissioning of the FR2 is delayed for the moment, we hope to get the licence for decommissioning the KKN at the end of this year.

For dismantling the KKN three manipulators (Kranmanipulator, Ringmanipulator and Trennmanipulator) are under construction.

A central cutting device is being developed for the dismantling of the reactor block of FR 2. All documents required for the licensing procedure are written.

Experience in remote operation was obtained during the exchange of the core tank and the thermal shield of the FR2 reactor, as well as during an incore repair operation of an experimental channel of the core tank.

A. BARKER

I will first mention the latest state of design of the machine for use in the demolition of the Windscale gas cooled reactors. I can then reply to Mr. Crégut.

The initial design state following a concept study is now commencing at Stracham and Henshaw, one of the UK engineering firms specializing in the design and manufacture of materials handling equipment and nuclear fuel charge machines. To those of you who attended the conference at Windscale last year, it will be appreciated from the slide illustration that the concept represents second thoughts. An inverted rigid mast supports a manipulator and cutting tool. Although consideration has been given to extension of the mast by hydraulic means it has been decided to extend the length as work proceeds by the addition of sections. However, for daily maintenance, an elevating platform permits the withdrawal of the manipulator and cutting tool to the upper region of the machine. Serious consideration is being given to the use of an industrial manipulator already in use for automobile engine assembly. The advantages arising are low cost and proven reliability.

In reply to Mr. Crégut's question, first, the manipulator possesses the maximum number of degrees of freedom. I think this is essential. One tool only is supported but there is the means for rapid tool changing in the maintenance facility. Secondly, I believe that the maximum assistance must be given to the operator particularly with regard to vision. We are considering both closed-circuit television and fibreoptics. However, in case fume and smoke interfere with vision a teach and repeat facility on the manipulator is being provided to pre-programme an operation under conditions of clear visibility.

In reply to a question from the floor regarding reduction of dose to operators by virtue of the addition of computer controlled operation the operator of the machine now in design for the decommissioning of WAGR will be sited within a control room shielded to a standard of 0.25 rem per hour i.e. a standard not requiring control of personnel access. The addition of computer controlled operation will not reduce operator dose during normal operation. However, dose uptake to maintenance staff is of concern. Use of programmed self-maintenance of the machine, if this is possible might be advantageous however it is difficult to judge what dose uptake might arise during un-scheduled repair; the best safeguard against this, in my opinion, is good basic mechanical design and freedom, as far as possible, from complication in control equipment.

M. DEVRESSE

ACEC Nuclear division experience in the field of remote operation started in fact in 1964, when the first European commercial PWR SENA was shut down after 1 year of operation due to failure of the support structures of the thermal shield. We were obliged to design and use special tools remotely controlled to be able to remove the thermal shield from the lower reactor internals. Special inspection, underwater cutting tools and milling tools were used.

After 1 year of design and repair work, SENA plant was again put at full power on the grid. Since we were convinced that Nuclear Robotics could solve problems encountered during plant operation and plant dismantling operations, and mainly save man-rem exposure, we have decided to develop a mobile autonomous robot for inspection and intervention in nuclear reactor buildings. It would either be teleoperated for unstructured tasks to be executed, or used like a robot for repetitive inspection tasks. It is equipped with two light master/slave telemanipulators with feedback mounted through an interface system on the carrier chassis. The information transmission is made through a radio link.

J. VERTUT

To avoid terminological difficulties, the terms recognized in France should be explained:

Télémanipulation (remote manipulation): dexterity with merely a fixed manipulation;

Télémanutention (remote handling) : handling at a distance (and heavy handling operations without dexterity);

Téléopération (remote control) : dexterity with mobility (vehicle) and generally a transporter. The vehicle provides the agility and the transporter the posture. The author recently proposed the term "télétrans" in order to extend the semantic field to convey the idea of the role played by transport and transfers (in a confined and/or shielded environment) in remote control.

This demonstrates the physically three-dimensional nature of remote control :

- an axis for payload, the latter ranging between 0.5 kg and several tonnes : as the payload increases, there is a gradual transition from dexterity to handling. A new trend is to broaden the concept of dexterity, while retaining the same positional accuracy and/or effort, to include heavy loads. Dexterity then belongs to the field of control and mechanics;
- an axis for reach which is associated with the dextrous arm (in combination with the transporter and the mobility of the vehicle);
- an axis for operating distance, which covers large distances in new space and submarine applications and in the nuclear field (for work in accident conditions).

Decommissioning, like maintenance, corresponds more often to the case where reach and distance are identical, or at least similar.

In reactor maintenance (and, to a certain extent, decommissioning) however, the distance to be covered in order to reach the task location is the source of acute difficulties for vehicles.

These considerations show how the approach adopted in the "nuclear robotics" sector is similar to that taken in "robotics in hostile environments", which is incorporated at the CEA in the slightly wider field of "service" (i.e. non-production) robotics.

The corresponding economic approach is similar to an insurance that makes it necessary to invest in safety margins, i.e. the value of flexibility in the event of difficulties. A plant which is equipped for remote maintenance is therefore already well prepared for remote working in the event of an incident or accident (as in the case of TMI) and lends itself all the more to decommissioning.

Why use the word robotics here and not merely remote control? Because the advent of systems, automation and information technology in computer-assisted remote control places the latter in the field of robotics.

Industrial robotics deals with the problem of substitution. Service robotics aims at doing the same thing, when the tasks are not repetitive; man remains in control but is assisted by the computer for the physical actions, the presentation of the information, the monitoring of the system, safety and the real-time preparation of the task sequence.

The current French service robotics programme covers all these aspects.

M.H.E. LARCOMBE

We must remember that in the crucial stages decommissioning is an act of demolition rather than one of disassembly. As with any form of demolition, a considerable degree of disorder will be created, whether we wish it or not. To manage this disorder any decommissioning process must have access to handling systems of great flexibility - a flexibility which will often be partly satisfied by mobility of part of the handling mechanism.

Virtually all the crucial phases of decommissioning will have to be effected by remote control. This remote control will not be conducted in the well ordered world typified by conventional case operations. It will be in a world of potentially unexpected disorder. Furthermore the operations will be carried out in typically 3-hour-sessions within 8-hour-shifts over periods of months or even years.

Our ideas of conventional remote control are often simplistic, based on systems where direct and wide angle viewing of the operation is possible. In practice, we will be operating through the medium of TV image feedback, where the field of view is of the order of $\pm$ 10°. If we imagine the problem of driving a conventional automobile with such a limitation on the field of view we see that we would expect to have accidents rather too frequently for comfort.

Within the study conducted by our Robot Laboratory for the Commission we introduced the artificial word REMCON to indicate that the advanced remote control techniques now possible are very different in concept to conventional remote control. We introduced this word to avoid the use of the heavily abused word "robot" in contexts such as "remotely commanded robot system".

In a REMCON system the control involves a high degree of system intelligence. This is typified by the three autonomies introduced in our joint paper "Remote operation in Decommissioning" and also by those discussed by Mr. J. Vertut on this panel.

Here the system, through the virtue of multi-sensor environment detection, can be given the ability to perform certain functions autonomously or to change the apparent degree of freedom of control to the operator (such as occurs in the protection and tracking autonomies). There are also functions, such as the control of multi-link manipulators (which involved the discipline of many degrees of freedom) beyond unaided human control, but which may be readily effected by computer aid. This aid may come from sensory feedback or from computer modelling or both system and environment (including C.A.D. forms of spatial model).

The most useable sensor feedbacks are those of system disposition, tactile and force sensing, short range proximity detection and longer ranging systems (such as sonar). All these sensors are readily and simply integrated within the computer control and the computer power requirements are very small by to-day's criteria. A system using perhaps four or five small microprocessor systems being typical.

To ensure efficient operation of such REMCON operations, which may well involve simultaneous control of two or more systems by multiple operators, much greater care must be exercised on the ergonomic aspects of control workstations. In particular the presentation of information to the operator must be readily comprehensible not only to the individual operator but also to co-operators and to supervisory staff. The information will be comprised of both TV images and computer generated displays, including computer graphics (possibly derived from the C.A.D. databases mentioned above).

All the component parts for the construction of REMCON systems exist. Within the criteria of to-day's technology we can describe these components as simple and, most importantly, robust. The systems described have been developed already to a high degree of readiness for application not only in Reactor Decommissioning but also in other similar hostile environment conditions.

TECHNICAL PANEL ON REMOTE OPERATION IN DECOMMISSIONING : SUMMARY OF DISCUSSION

The Chairman, upon conclusion of the presentations by the panel members, points out that the participants, after having been informed on what is presently going on, what is in project and what could be designed in the future, should raise a number of questions after this rapid travel in time.

A U.K. participant comments that a full range of options to solve the problems of decommissioning has been described. He gives a note of caution that it should not be attempted to design the universal decommissioning machine. He thinks the simplest answer to solve a given problem should be used, rather than opting for a more universal solution. This does not exclude the need for the much more complicated options that have been described.

The Chairman confirms that in any case the simplest solution has to be applied, but in situations with difficult problems to solve it is important to have a whole technology of remote operation, as well as remote transportation and remote manipulation, available.

An FRG participant asks - as there is clearly a relationship between the dose rate to the personnel and the use of robots - how much funding should be committed to obtain which dose reduction to the personnel.

Mr. Vertut answers that remote operation begins when it is definitively impossible to perform an operation directly. This case exists clearly in part of the decommissioning tasks and in a large part of nuclear maintenance tasks. In intermediate cases, which appear in decommissioning, in maintenance and in accidental situations, the time necessary to complete the task becomes very important, and applying simple tools may take a long time. Consequently, the economical analysis of an investment has to consider time, including extra time for unexpected difficulties during execution. This security margin may be the real justification for a more versatile machine but not for the universal machine.

Mr. Barker adds that in the concept he had described, the operator will sit behind shielding and there will be no need for control of access to that area. Introducing complexity in the machine will not materially help to reduce the dose rate to the operator when things are operating normally. What cannot be predicted is the dose rate that the maintenance staff will have during non-routine maintenance or repair. The computer complexity is associated with vision; it allows help to the operator in seeing what is going on or in giving him the option of practicing a run without the presence of either fume or smoke; afterwards, the computer facility will go over that run and do the actual operation.

The Chairman, in completing the comments, considers that there are two answers to the question which, taken in a simplified manner, concerns the cost of the man-rem :

- the answer of the designer in charge, whose tendency would be to try to obtain a maximum of financial efficiency, without taking any risk at the level of radiation exposure of the workers;
- the answer of nuclear engineers in charge who know that, in the interest of everybody, funds have to be invested without immediate or short-term return, in order to advance in science and technology as well as to supply more performant systems.

A French participant doubts that an intervention vehicle like the one presented by Mr. Devresse would have a large field of application in a PWR.

He thinks that in most cases, specialized equipment would be needed. To carry out maintenance or decommissioning operations, access will be one of the main problems. He adds that an additional difficulty is that engineering drawings do not always reproduce the "as-built" situation. Photogrammetry or other techniques coul be one of the means to allow checking the exact configuration of the working area.

Mr. Devresse confirms that the vehicle has effectively limitations in its means for intervention. Rooms housing components like steam generators, primary pumps and the vessel are so crowded that, for example, between the protective wall and the outer surface of the steam generator, a man could not pass. On the contrary, during operation of the reactor, the vehicle is able to take over regular inspection tasks inside the reactor building to verify essentially the behaviour of the secondary equipments. There is a multitude of pipings, valves and systems in accessible pathways. To sum up, this vehicle is devoted to inspection with a capability of intervention limited to systems and auxiliary loops which are accessible. In case of an incident leaving the reactor building inaccessible for a longer period, such a vehicle could perform relatively rapid inspection tasks, allowing the operator of the plant to establish a diagnostic of the state of the reactor building.

A French participant means that a group of people from different fields interested in remote handling, from designers to operators, should set up scenarios together that would need remote operation.

The Chairman agrees to the merits of cooperation, but supposes that this discussion would be more general in its need for the inclusion of robotics. He concludes the discussion by expressing his opinion, that in the present situation an effort has to be made in order to meet requirements in dismantling operations, decontamination, management of waste, in developing tools, systems and equipment for remote operation.

SESSION IV

MANAGEMENT OF DECOMMISSIONING WASTE

Chairman : J. MARSHALL, UK-Doe

Secretary : K. PFLUGRAD, CEC, Directorate-General Science, Research and Development

Introduction by the Session Chairman

Immobilization of contamination on metals and concrete

Treatment of metal waste by melting

Management and disposal of graphite waste

Large transport containers for decommissioning waste

Summary of discussion

Technical Panel : Requirements and possibilities for the disposal of decommissioning waste in the Community Countries

- Contributions by the Panel Members
- Summary of discussion

INTRODUCTION BY THE SESSION CHAIRMAN

J. MARSHALL
UK-DOE, London, United Kingdom

At this conference, and elsewhere, a great deal of effort is being given to the problems of decontaminating and dismantling nuclear facilities. However, when the facility has been dismantled the radioactive waste will need to be disposed of.

The study of radioactive waste disposal is a major topic in its own right but it is important that decommissioners do not forget the problems of waste disposal. For while the only difference between decommissioning waste and operational waste is in the quantity and the rate of delivering, we must carry out research to assess the best radiological and economic means of handling and disposing of this waste.

The papers in this section cover every aspect of these problems, immobilization, re-cycling, radiological assessment of disposal options and design of transport containers. This shows the degree of interest and experience which exists to solve these problems within the CEC.

IMMOBILIZATION OF CONTAMINATION ON METALS AND CONCRETE

C. DE TASSIGNY (CEA), I.L.L. DAVIES (TAYLOR WOODROW)
and G. BRAMBILLA (NUCLECO)

Summary

The dismantling of nuclear plants produces large quantities of concrete dust and metal fragments. Suitable coatings are necessary to fix surface contamination and limit the diffusion of radionucleides during transport and temporary storage. Three comparative studies have been performed by NUCLECO (I), CEA (F) and TAYLOR WOODROW CONSTRUCTION LTD. (GB).
The metal coatings were chosen from a selection of two component polymers : epoxy and elastomeric polyurethane resins. A layer of epoxy resin 150 um thick will effectively fix the contamination. The polyurethane resins can be projected to a thickness of at least 5 mm, which is necessary to ensure sufficient mechanical protection. Weakly active metallic waste resulting from the dismantling of nuclear reactors has been coated in this manner.
Sodium silicate combines with the concrete dust to give a compound which is hard and stable. The most effective composition contains a $SiO_2:Na_2O$ ratio of 2,85:1 with a binding efficiency of 32 to 35 %. The dust is collected in a conveyor system and transformed into pellets.

1. INTRODUCTION

The dismantling of nuclear reactors produces quantities of concrete and metal with different degrees of contamination. Suitable coating techniques are required to solve the following problems :

- fixing of the contamination and reduction of diffusion,
- prevention of metal corrosion,
- reduction of dust formation during the demolition and transport of the concrete.

Several initial techniques have been investigated :

- NUCLECO (I) and CEA (F) use polymers to fix the contamination and reduce diffusion.
- TAYLOR WOODROW CONSTRUCTION LTD. (GB) use sodium silicate to bind the dust in pellet form.

The studies were undertaken in a comparative manner with the following aims :

- to study the coatings available commercially which are easily applied and where possible, inexpensive.
- to evaluate the quality of the coating by means of tests for chemical and mechanical resistance, contamination fixing and behaviour under radiation.
- to study the design of equipment for the application of the chosen coating.

2. APPLICATION OF EPOXY RESINS FOR CONTAMINATION FIXING. WORK UNDERTAKEN BY NUCLECO.

2.1 Choice of coating.

Due to the presence of radiation and contamination, consideration must be given during the dismantling operations of nuclear plants to the surfaces to be coated and the particular application conditions.

The investigations were restricted from the beginning to painting products and were directed towards those products available on the commercial market which can be applied to all types of support (steel, cement, previously painted materials) with different surface conditions.

Generally all painting procedures presuppose suitable preparation of the support (degreasing, sand-blasting, cleaning) which, for the considered case, is not obviously possible for all the contaminated material.

These considerations led in practice to restrict the number of suitable products to the group of two-component epoxy paints (resin and hardener).

Theoretically, the epoxy systems have the following features :

- minimum shrinkage during hardening ;
- no secondary reaction products ;
- very good mechanical characteristics ;
- resistance to chemical agents ;
- very good adhesive properties ;
- good electrical characteristics.

Table I gives the trade names and types of epoxy paints recommended by the manufacturers.

TABLE I - Paints for the containment of contamination.

MANUFACTURER	PAINT	TYPE	RECOMMENDED CYCLE Steel	Cement
MAX.MEYER-DUCO Fombio MI	TITANIA 5.23.E.0080	Epoxy anticorrosive zinc phosphate primer.	1X	
	TITANIA 5.28.E.	Thick epoxy paint	1X	2X
IMPERITALIA Turin TO	FONKOR 1444/AS/N	Epoxy anticorrosive zinc phosphate primer.	1X	
	KORALL SPE/AS/N	Thick epoxy protecting paint.	1X	2X
	KOTIXOR CRE/S	Epoxy protecting covering and finishing coat, easy to decontaminate.		
PARAMATTI Turin TO	ETICOAT BIT A.S. LBV Series	Epoxy-bituminous pigmented paint, coverable with further paint.	2X	2X

These paints contain solvents. The use of epoxy paints without solvents would be safer during application as they would be classed as non-flammable products. However, the non-flammable epoxy paints have been rejected for other reasons : the available commercial products only give good adherance on iron sanded surfaces.

2.2 Properties of Epoxy Paints.

The epoxy paints from table I were characterised from the conventional viewpoint (mechanical and chemical resistance) by testing adhesion; porosity; marking ability; behaviour on direct and indirect impact; bending on a conicle spindle; behaviour in salt fog, moisture, de-ionised water, SO_2 atmosphere and when dipped in acidulated and alkalised water.

The knowledge of these properties is useful to evaluate compatibility of the paints with the supports and to foresee their effective lifespan.

Among the paints chosen, TITANIA EP.23 was also characterised from a nuclear point of view, determining its ability to retain the contamination, its resistance to radiation and its decontamination properties.

The possibility of preventing contamination diffusion (barrier effect) was evaluated by measurement of activity released on dry smearing and on water leaching. While in the first case the release of contamination is not significant, in the second case it starts only some days after immersion.

The resistance to radiation was determined in the ENEA laboratories in Casaccia (Rome) using a ^{60}Co source with a real exposure time for the samples of 1000 hours for an absorbed dose of 600 Mrad. The irradiated surfaces of the samples were compared directly with those of reference samples. The variations in decontamination were also compared.

From this series of tests it can be concluded that the resistance to radiation of the paints under test is good up to the level of the absorbed dose. Decontaminability of a paint is determined by means of the percentage residual activity after one or several contamination-decontamination cycles. The corresponding index is poor for the irradiated samples or controls.

The indications drawn from these characterisation tests on epoxy paint applied to different supports can be summarized as follows :

- NUCLECO recommend the application of two successive coats of paint (an anti-corrosive primer and one protective coat on iron, two protective coats on cement) to give a total thickness of at least 150 to 160 μm.
- the effective lifespan of the coating depends on the nature and condition of the support surface. It can be approximately evaluated as follows in table II :

TABLE II - Effective Lifespan of Coatings.

SUPPORT	COATING EFFICIENCY
Previously painted materials.	5 years
Cement	5 years
Carbon steel, slightly rusted.	2 to 3 years
Carbon steel, very rusted.	1 year

As it is possible to re-paint over an epoxy paint when the indicated timescales have come to an end, NUCLECO suggest maintenance consisting of a partial or total re-painting of the surface to be protected, which could double the lifespan of the coating.

Table III summarizes the properties of epoxy paints :

TABLE III - Properties of Epoxy Paints.

PROPERTY		BEHAVIOUR
Resistance to corrosion.		very good
Resistance to radiation.		quite good
Decontaminability		poor
Containment of contamination	dry conditions	very good
	wet conditions	very good
Duration		1 to 5 years

2.3 Equipment and Handling.

For the application of epoxy paints in a contaminated environment, the "airless" system proves to be the most suitable method because of the following advantages :

- the absence of air jets which could diffuse surface contamination in the environment ;
- possibility of operation with long pipes (up to 50 to 100 m) between the pump and the paint tank ; the tank can thus be kept out of the contaminated area.

For operation in highly radioactive areas, it is possible to adapt manipulator robots already used in industry which reproduce the complex movement of the painter's hand. Such robots should be modified and mounted on a remote-controlled mobile vehicle.

2.4 Conclusion.

The following remarks can be made in conclusion :

- During the dismantling of nuclear plants, commercially made epoxy resins can be advantageously used for the containment of material contamination. The general properties of epoxy paints are summarised in table III (characterisation tests).
- The chosen epoxy resins contain flammable solvents and their application in a nuclear environment requires safety measures which could limit their field of use.
- The coatings used are temporary. An effective lifespan of 1 to 5 years was determined and depends on the type of material and surface conditions.
- The epoxy coatings could only have a protective function against contamination diffusion during dismantling operations, transport or temporary storing of materials.

3. APPLICATION OF EPOXY PAINTS AND POLYURETHANE COATINGS FOR CONTAMINATION FIXING. WORK UNDERTAKEN BY CEA (CENG. Grenoble).

3.1 EPOXY PAINTS.

The same general and technical reasons led to the choice of epoxy paints by the CEA. The suggested formulations are nevertheless different as they were chosen as being suitable for an electrostatic application method. Two types of coating to be applied in this manner were studied :

- thin-layer epoxy coating. The trade reference for this paint is : Blancomme red-brown 03 FT 1/SE/ES. It contains 12 % solvent and polymerisation takes place at ambient temperature. It is in the form of a liquid pigmented paint with an application time in the absence of dust of one hour at 20°C. This requires a drying time of more than 10 minutes between each application.
- thick-layer epoxy coating. The trade reference for this paint is : Blancomme 95KT 17 red-brown. In powdered form (without solvent), this type of paint is formulated to give layers of 1 to 2 mm thickness with electrostatic application. Polymerisation takes place at 200°C after a heating and baking process in a specially made tunnel, which lasts approximately 20 minutes per layer.

3.1.1 Electrostatic Application.

Two different spray-guns were used :

- SAMES STATAIR MRV 96 for the liquid paint,
- SAMES STAJER JR 62 for the powder coating.

The high voltage (90 kV) applied to the nozzle of the spray-gun electrically alters the paint particles which converge on the piled-up metallic waste. This method reduces the loss of paint during application and avoids the dispersion of the polymer : 95 to 98 % of the liquid epoxy paint was shown to collect on the material to be painted.

The application tests with powdered epoxy coating revealed difficulties in industrial application in a nuclear environment. The rest of the work therefore only dealt with the liquid epoxy paint.

3.1.2 Properties of epoxy paints.

Two parameters condition the use of these coatings in the nuclear industry :

- the fixing of the contamination,
- the diffusion of the contamination.

Contamination Fixing.

The contamination fixing factor is measured with a technique currently used in radiation protection : dry or moist rubbing is carried out on a surface of 100 cm^2 (10 cm X 10 cm) defined in the contaminated area.

Steel plates are contaminated by dipping in a solution containing ^{60}Co and ^{137}Cs and then dried. The test consists of three rubbings :

- an initial rubbing on a non-painted surface giving an activity Ao,
- after application of the first layer of paint, a second rubbing on another surface of 100 cm^2 giving an activity A_1,
- after application of a second layer of paint, a third rubbing gives an activity A_2.

The relationships Ao/A_1 and Ao/A_2 give the value of the fixing factor for the coating.

The tests were carried out with epoxy paint Blancomme 03 FT 1/SE/ES. The results are summarized in table IV :

TABLE IV - Contamination Fixing Factor.

Number of layers	0	1	2
Total thickness µm	0	25-30	50-60
Activity µCi : Ao	8		
A_1		$4,4 \times 10^{-2}$	
A_2			$9,4 \times 10^{-4}$
Fixing factor : Ao/A_1		184	
Ao/A_2			8600

The results show that one 30 µm layer of epoxy paint gives a fixing factor of approximately 200, and with two layers of total thickness approximately 60 µm, the factor reaches 10^4.

Diffusion Tests.

Considering the aspect of diffusion by the principal radionucleides, waste storage requires details of the quality of the protective coatings to enable the determination of the thickness required. The diffusion studies undertaken concern ^{60}Co and ^{137}Cs.

Diffusion membranes are prepared by applying to a soluble support a determined number of layers of paint to givethicknesses from 70 to 450 um (Blancomme paint 03 FT 1).

The diffusion cell is made up by adhesion of a cylindrical block of epoxy resin to a thin epoxy layer spread on a support. After hardening (polymerisation and baking at 60°C), the support is removed leaving the epoxy diffusion membrane and the cell is ready for diffusion.

The cell is filled with 5 cm^3 of an electrolyte containing the following :

$CoCl_2$ at 10g per litre with 99 µCi of ^{60}Co ;
and CsCl at 10g per litre with 91,5 µCi of ^{137}Cs.

The whole is plunged into demineralised water and closed with an epoxy resin stopper.

Diffusion Control.

Cs^+ and Co^{2+} pass through the membrane into the pure water. The activity of the pure water is measured after 2, 15, 30, 78 and 120 days. The results after 120 days of diffusion are given in table V.

Cell CP1 of thickness 70 µm underwent rapid diffusion. Cells CP4 and CP5, of thickness 100 µm and 200 µm respectively, behaved in a classical manner. Cell CP6 of thickness 450 µm gave results lower than the detection limit.

This signifies that after only 120 days a fraction less than 10^{-5}Ao (initial activity) was eventually able to cross a thickness of 450 µm.

TABLE V - Total released activity ΣA across a diffusion cell containing an initial activity Ao.

Diffusion Cell	CP1	CP4	CP5	CP6
Thickness µm	70	100	200	450
$\Sigma A/Ao$ ^{60}Co : at 60 days	$1{,}7 \times 10^{-2}$			
at 120 days		$3{,}0 \times 10^{-5}$	$5{,}0 \times 10^{-5}$	$4{,}0 \times 10^{-6}$
$\Sigma A/Ao$ ^{137}Cs : at 60 days	$1{,}5 \times 10^{-2}$			
at 120 days		$2{,}2 \times 10^{-4}$	$2{,}3 \times 10^{-4}$	$9{,}0 \times 10^{-6}$

Two types of diffusion are observed which depend on the membrane thickness :

- one of rapid type : CP1 70 µm. The polymer layer is too porous in this case to restrain the ^{60}Co or ^{137}Cs diffusion.
- one of slow type : CP4, CP5, CP6 100-450 µm.

The general thickness of the epoxy resin layers should therefore be greater than 100 µm to ensure the retention of ^{60}Co and ^{137}Cs.

3.2 POLYURETHEN COATINGS.

The epoxy resins however have some disadvantages :

- slow hardening at ambient temperature.
- a thickness limited to the mm.
- a certain fragility.

To limit these disadvantages it seems desirable to research the coatings applicable in a thick layer and resistant to impact.

The idea of protecting the first epoxy resin layer with a second coating containing fibre glass was abandoned. This type of application is carried out on a semi-industrial scale by remote projection. Each application, however, should be followed by a surface conditionning done on a small scale using a roller to obtain a good quality coating. This led to a dispersion of contamination on the moving parts and to contamination of the roller.

The recent progress in elastomeric polyurethen formulation allows the use of very reactive compounds and consequently the rapid application of successive layers. In addition, adjustment to the hardness of these polymers can be made depending on the desired resistance to impact.

This reactivity allows rapid application of successive layers after an interval of one to two minutes between each layer. Rapid attainment of thicknesses greater than a mm is therefore possible.

The two reactants (isocyanate and polyol) without solvent have a flash-point greater than 100°C.

3.2.1 Mechanical Properties of Polyurethen Coatings.

The mechanical properties of elastomeric polyurethen coatings may vary depending on the choice made between a "hard" or "flexible" coating.

The properties of the UREFLEX XPS 801 coating are set out in table VI.

TABLE VI - Mechanical Properties of Polyurethen UREFLEX.

PROPERTY	VALUE	UNIT	NORM
Hardness	83	Shore A	NFT 51 - 109
Tensile Modulus at 100 % elongation	4,25	MPa	DIN 53 504
Tensile Modulus at 300 % elongation	8	MPa	DIN 53 504
Tensile strength	34,7	MPa	DIN 53 504
Elongation on rupture	785	%	DIN 53 504
Loss on abrasion	25	mg/min	DIN 53 516
Persistance of deformation on compression (24h at 70°C)	55	%	DIN 53 517

Adhesion Test.

The study initially anticipated for metal fragments was extended to other types of material. Adhesion for these tests was qualified by the following three criteria :

- peeling ability : the adhesion is considered poor when the coating peels easily by hand.
- good adhesion : the coating adheres but can be torn off by hand over a large surface area.
- very good adhesion : the coating is pratically impossible to peel off by hand. For example, the case of an underlying primer coating (paint, coating) : the primer is torn off without detaching the polyurethen .

The results are set out in table VII.

Resistance to Impact.

Testing resistance to impact is carried out under conditions for the norm NFT 30 039. It involves repeating a dropping test using an 800g weight dropped from an ever increasing height and noting the height at which a defect appears in the coating. The end of the weight is a steel ball of diameter 20 mm.

Comparative tests were carried out on test pieces covered with the Blancomme epoxy paint described earlier and the urethane paint UREFLEX.

TABLE VIII - Resistance to Impact.

SUPPORT	COATING	THICKNESS	HEIGHT OF DROP	DEPTH OF RESULTING IMPRINT	RUPTURE
Steel	epoxy	200 μm	50 cm	/	hairline fractures
Steel	Polyurethen	4 mm	250 cm	~0,2 mm	none
3-ply	Polyurethen	3 mm	250 cm	~3,0 mm	none
Aluminium	Polyurethen	5 mm	250 cm	~0,2 mm	none
Concrete	Polyurethen	3 mm	250 cm	~2,0 mm	none

TABLE VII - Polyurethane Adhesion - Qualitative Tests.

TYPE OF SUPPORT	SURFACE CONDITION	FORM AND NUMBER	DIMENSIONS	NUMBER OF LAYERS	TOTAL THICKNESS	ADHESION
Aluminium	plain	5 plates	200 x 500	5 to 10	2,5 to 5	good
	plain	6 tubes	Ø32 and 63 L 500	4	2	good
	greasy	cover		2	1	peels
Aluminium and primer paint	epoxy paint	2 plates	200 x 500	4 to 8	2 to 4	very good
	epoxy paint	4 plates	Ø32 and 63 L 500	4 to 8	2 to 4	very good
Steel	sheet iron	cover	Ø320	8	4	very good
	plain					
	plain	plate	970 x 420	6	3	good
	cleaned	4 plates	570 x 100	1 to 6	0,5 to 3	very good
	very rusted	2 U-shaped pieces	400 x 100			very good
	rusted	corner	750 x 40 x 40	5	2,5	good
	rusted	2 plates	420 x 160 and 850 x 150	6	3	very good
	rusted and contaminated	blower from dismantling	700 x 500 (51 kg)			
	sanded	hinge	750 x 100	3	1,5	very good
	rusted - greasy	U-shaped	200 x 100	6	3	peels
Polythene	plain	leaf	600 x 400	1 to 10	0,5 to 5	peels
PVC	plain	plate	430 x 200	4	2	good
wood	plain	plate	420 x 260	3	1,5	very good
Fibre cement	plain	plate	700 x 200	3	1,5	very good
Concrete	plain	rough stone	200 x 200 x 200	4	2	very good
Toughened concrete	dusty	plate	400 x 400	4	2	very good

The elastomeric coating resists as anticipated to impact of an 800g weight falling from 2,5 m ; that is, without any cracking of the elastomeric layer, as shown in table VIII.

Contamination Fixing.

The study of the contamination fixing was carried out on three steel plates under conditions described for the epoxy resins. The thickness of each polyurethen layer was approximately 0,5 mm. The values for the fixing factor are set out in table IX :

TABLE IX - Contamination Fixing Factor.

	Ao/A_1		Ao/A_2	
	^{60}Co	^{137}Cs	^{60}Co	^{137}Cs
Plate n° 1	5×10^2	7×10^2	2×10^3	$> 7 \times 10^2$
Plate n° 2	5×10^2	$> 2 \times 10^3$	2×10^3	$> 3 \times 10^3$
Plate n° 3	$> 4 \times 10^2$	$> 3 \times 10^1$	$> 5 \times 10^2$	$> 3 \times 10^1$

It can be seen that fixing of the contamination is effectively obtained for ^{60}Co and ^{137}Cs. The factors are all of the order of 500 for the first layer.

After two layers the fixing factor for ^{60}Co is greater than 1000. The remaining activity for ^{137}Cs is too weak to be measured.

This test confirms the good fixing properties of this type of material. The size of the fixing factors is similar to those of the epoxy resin studied previously (10^4 for two layers of paint).

Containment of contamination.

The tests were carried out according to the method described for the epoxy resins. The results after one month of diffusion are given in table X.

TABLE X - Fraction of activity released after 30 days ($\Sigma A/Ao$).

	CP7		CP8		CP9	
Thickness	1,5		1,9		2,0	
	^{60}Co	^{137}Cs	^{60}Co	^{137}Cs	^{60}Co	^{137}Cs
after 30 days $\Sigma A/Ao$	$<2x10^{-5}$	$<3x10^{-5}$	$<3x10^{-5}$	$<3x10^{-5}$	$<2x10^{-5}$	$<5x10^{-5}$

It can be seen that the protection provided by a thickness of 1,5 to 2 mm is effective at least for this timespan, the activity measured being lower than the detection threshold level.

It is necessary to wait for the first significant activity measurements to calculate a rate of diffusion.

3.3 Comparison with the Blancomme Epoxy Paint (minimum thickness of polyurethen).

The maximum thickness of epoxy resin studied in a diffusion cell was 0,45 mm. It is possible to apply a minimum thickness of polyurethen resin of 0,5 mm.

The previous results have shown that for this same thickness the diffused activity after 30 days ;

- was approximate to the detection limits for epoxy paint ;
- very rapidly crossed the polyurethen paint as a result of the porosity of this 0,5 mm thickness.

It can be concluded that :

- a polyurethane layer of 0,5 mm has properties approximate to those of the epoxy resins as regards fixing ability.
- a single polyurethen layer is however insufficient to play an effective role as a diffusion barrier in the long term.

Optimum Thickness of Polyurethen Resin.

The results of table IX indicate that it is possible to attribute containment properties to the polyurethen resin on condition of a minimum application of three layers 0,5 thick giving a total of 1,5 mm. This acceptable conclusion is reached after 30 days of diffusion. The diffusion study is continuing. The results to come will give qualitative and quantative details of the membrane behaviour.

3.4 Application to Contaminated Waste Produced During Dismantling.

Polyurethen paint was applied to contaminated metallic waste using the coating process involving projection.

The test was carried out on a fragment of a metallic blower cut out of a ventilation shaft resulting from the dismantling of nuclear reactor G1 at Marcoule.

The material had the following characteristics :

- length of 700 mm ;
- width of 500 mm ;
- thickness of 100 mm ;
- weight of 51 kg ;
- form - the "concertina" form was chosen because of its complexity ;
- surface conditions - rough and rusted ;
- activity - β 280 impacts/sec. ;
  γ 2000 impacts/sec. ;
- surface contamination - measured by moist rubbing over 100 cm^2 :
  $1,1 \times 10^{-5}$ µCi < ^{60}Co < $1,3 \times 10^{-5}$ µCi
  $2,6 \times 10^{-5}$ µCi < ^{137}Cs < $1,1 \times 10^{-4}$ µCi.
  The material can therefore be considered to be of low activity.

3.4.1 Projection Method.

According to the description above (cf. projection equipment and utilisation), the polyurethen coating was applied to the material placed on the ground.

After coating one side with a thickness of approximately 3 mm, the part was turned over and painted in the same manner.

A contamination fixing test was carried out before and after coating.

Results.

The photographs included in the annexe show that it was possible to apply the coating to the whole surface, in particular :

- to sharp edges
- to the bottom of the folds of the blower.

This type of projection will therefore allow coating of 10 cm deep hollows through holes of approximate width 1 cm.

The surface activity measurements by moist rubbing gave results lower than the detection threshold levels for ^{60}Co and ^{137}Cs, thereby demonstrating that for actual material this procedure represents an effective means of fixing the contamination.

4. BINDING CONCRETE DUST. WORK UNDERTAKEN BY TAYLOR WOODROW CONSTRUCTION LTD.

4.1 Introduction.

Demolition of concrete structures from nuclear reactors will produce dust activated by neutron irradiation. One method reducing risk of ingestion or inhalation of radioactive dust consists of amalgamating the dust into larger particles which have the following characteristics :

- are of a sufficient size not to be affected by wind ;
- do not tend to disintegrate into dust again ;
- remain in particle form, resisting leaching and other forms of disintegration long enough for the particles to be bound into a stable matrix in a strong protective container.

Sodium silicate meets these conditions and :

- is non flammable ;
- is non toxic ;
- is readily available and inexpensive ;
- undergoes a combination reaction with the dust without the addition of external agents such as heat, pressure or catalysts ;
- allows control and handling with conventional equipment ;
- is resistant to radiation associated with activated dust ;
- is compatible with materials used for ultimate encapsulation and storage, probably concrete or mortar ;
- is fairly resistant to handling conditions, ie: abrasion and water leaching.

Sodium silicate meets most if not all the necessary conditions and was preferred to other binding agents such as the epoxy resins or bitumen.

4.2 Nature and Properties of Sodium Silicate.

The chemical and physical properties of sodium silicate depend on the $SiO_2:Na_2O$ ratio and the solid content of the solution. The chemistry of sodium silicate is not, however, fully understood being dependant upon several complex equilibria. The suitability of a particular solution is then often only determined by a series of trial and error experiments. Sodium silicate solutions contain mixtures of various silicate ions with the ionic type and distribution being dependant upon the $SiO_2:Na_2O$ ratio and concentration. This also affects viscosity and stability.

4.3 Nature of Concrete.

Concrete is essentially a mixture of hydraulic cement, aggregate and water. The cement is formed by "burning" a mixture of chalk and clay. The main chemical constituents of these cements are calcium oxide (CaO : 60-70 %)

silica (SiO_2 : 17-25 %), alumina (Al_2O_3 : 3-8 %) and ferric oxide (Fe_2O_3 : 0,5-6 %). In the hydrated state the main constituents are tricalcium silicate ($(CaO)_3SiO_2$ ~ 47 %), dicalcium silicate ($(CaO)_2SiO_2$ ~ 32 %) and tricalcium aluminate ($(CaO)_3Al_2O_3$ ~ 7 %).

On addition of water the cement hardens by a complex hydration process to form a gel. An important feature of the cement as regards the agglomeration of dust, is that it contains soluble calcium in the form of lime, which is able to react with the sodium silicate to produce a strong binding reaction.

4.4 Behaviour of Concrete Dust with Sodium Silicate.

A range of silicate solutions of different SiO_2:Na_2O ratios was used to determine the most effective concrete dust binding technique.

Three physical characteristics were considered relevant to this end :

- mixing characteristics ;
- setting time ;
- resistance to compression.

In addition, variations in the type of concrete producing the dust were related to the following parameters :

- pH ;
- age or type of concrete ;
- particle size distribution of dust.

Commercially available sodium silicate solutions with a wide range of SiO_2:Na_2O ratios were selected for preliminary tests.

The following table summarises the basic information :

TABLE XI - Sodium Silicate Solutions.

SiO_2:Na_2O ratio by wt.	% Solid Content	Viscosity at 20°C (centipoise)
3.65	33.4	200
3.30	38.1	250 - 500
2.85	43.2	500

It was found that a uniform mixture of the dust with these solutions was not possible because of the very high viscosities. For this reason each solution was diluted to give a viscosity of 25 centipoise and it was found that dust:silicate ratios beween 1:1 and 1,4:1 gave suitable consistencies.

The following conclusions were made :

- an SiO_2:Na_2O ratio of 2,85:1 for the silicate solution with solid content in the range 32-35 % showed the best dust binding characteristics ;
- the effects shown by solutions with an SiO_2:Na_2O ratio lower than 2,85:1 should be studied.

The effect of silicate solutions and solid content on the viscosity variations is shown in Fig.1 with a dust derived from Heysham NPS concrete with a particle size distribution given in Fig.2.

The mixing characteristics of the dust with the sodium silicate solutions were evaluated for SiO_2:Na_2O ratios of 2,00 to 3,65, with the solid content producing a viscosity of 25 centipoise in each solution. These solutions were mixed with various quantities of dust. It was found that an acceptable visual assessment of the mixing character could be made, as change in the characteristics of the mixture led to different results. This is summarised in the following table :

TABLE XII – Mixing Characteristics.

	SiO_2:Na_2O ratio.			
wt. of dust x wt. of solution	2,00	2,85	3,30	3,65
1,4	liquid of low viscosity	viscous liquid	dry	dry
1,2			viscous liquid	viscous liquid
1,8	viscous liquid			

It is difficult to separate the various factors affecting the dust mixing characteristics because of their reciprocal interactions. For example, better wetting behaviour of the lower silica/alkali materials may result from the chemical nature of the solution causing a slower reaction with the conctituents of the concrete dust. For the higher silica/alkali ratios, the reactivity is greater than for solutions of lower ratio, this having been confirmed by measurement of the initial setting times.

The setting time for the dust/solution mixtures depends upon the following factors :

- SiO_2:Na_2O ratio of the silicate solution ;
- solid content of the silicate solution ;
- properties of the dust/silicate solution ;
- chemical nature of the dust ;
- dust particle size ;
- rate of water loss ;

The initial setting times are determined using a Vicat needle according to the method given in BS 4550 part 3. Fig.3. shows the variations of this parameter with respect to the SiO_2:Na_2O ratio and the solid content.

In some cases a "flash set" took place. It is not easy to understand why, as different setting times exist for an intermediate level of solid content. An explanation could be provided by a reaction between the solutions and concrete dust at the dust surface, inhibiting further reaction with the silicate solution.

A similar reaction can be observed on the resistance to compression curves for various SiO_2:Na_2O ratios plotted against solid content. A significant peak can be observed once again for the intermediate values of solid content, as shown in Figs.4 and 5.

Considering the characteristics associated with the concrete such as pH, age and type, it was found that for a high pH value the setting times were low and vice versa. It was therefore postulated that the setting time depends on a reaction between the free lime in the concrete dust and the silicate solution. Age is not a significant factor although the carbonisating effect of the free lime must be considered.

The following tests were carried out on the dust and sodium silicate solutions :

- The penetration of the silicate solutions into dry and damp dust : This problem is associated with that of the dust wetting behaviour and Fig.6. shows that the moisture front moves considerably faster through moistened dust than through dry.
- The resistance to abrasion of hardened mixtures : The variations depend on the pelletising machine used, but for greatest resistance to abrasion there is an optimum solid content of the sodium silicate solutions (formation of minimum quantities of fines). It was observed that secondary impregnation of the pellets with dilute silicate solution increased the resistance to abrasion.
- Dissolving of soluble materials in water : Calcium is not significantly dissolved but a small quantity of silica is taken up by the solution. These last tests are preliminary and should be extended before recommending the procedure for use in the handling of radioactive waste material.

4.5 Engineering the Process.

Primarily, the dust should be separated from the larger particles and then transformed into particles of sufficient size not to become airborne. In addition, the process should "lock" the dust on the surface of these larger particles so that it too cannot become airborne.

It was suggested that the separation of the dust from the larger particles could be effectively carried out using air jets and vacuum exhaustion. Adhering dust not removed by such treatment can be "locked" onto the surface of the larger particles by impregnation (spraying) with a sodium silicate solution, initiating the reactions already described and binding the dust into larger particles.

Several procedures for the production of pellets etc. were studied at makers works. Both liquid and solid phase methods were used. It was found that there were a number of machines capable of correctly producing the solid samples. The resistance to abrasion of the various samples and their pelletising processes indicate no preference for machine type. Extra work should be undertaken in this direction if the process is to open the way for the developpement of a prototype.

4.6 Other Binding Materials.

A number of other materials were considered and the findings may be summarised as follows :

alkyl silicates	- flammable, coatings weak, more costly than sodium silicate.
starch and cellulose ethers	- yield water soluble binders, non-reactive with concrete, biodegradable.
casein	- forms an insoluble film, biodegradable.
urea-formaldehyde resin	- could bind with concrete dust, inconvenient to handle, requires a catalyst, dermatitic.

natural and synthetic rubber emulsions	- used to prevent wind erosion of coal during transport, efficiency uncertain with regard to dust in bulk, would bind dust in larger lumps.
bitumen	- would effectively bind the dust, less easily handled than sodium silicate, carcinogenic, high water resistance, flammable, may require heating.
epoxy and polyurethane resins	- will bind the dust, high resistance to water and radiation, expensive, difficult to handle, require catalysts which may be toxic or dermatitic.
other water soluble silicates	- potassium silicate will react like sodium silicate, is more readily available and less water resistant. The other water soluble silicates are generally more expensive and less readily available.

4.7 Conclusion.

1. Sodium silicate combines effectively with concrete dust in bulk, cheaply and with sufficient strength to produce products in pellet form which are easily transported and stored.

2. The parameters for best binding depend on a number of factors :
- silica/alkali ratio and solid content ;
- viscosity ;
- nature of dust ;
- dust particle size ;
- dust/silica ratio .

3. A number of pelletising methods are suitable without any one having particular advantage over the others.

4. The dust from concrete rubble can be reamoved and collected in a conveyor system using air jets and suction.

5. GENERAL CONCLUSIONS.

The dismantling of nuclear plants leads to the production of large amounts of metallic waste which is uneasily compacted or impossible to pack in the usual containers, along with large quantities of contaminated concrete rubble.

Most of this waste is weakly active. It is economically desirable to be able to ensure protection by locking the contamination directly at the surface and thereby avoiding encapsulation, costly in volume and material, in standardised containers.

The research undertaken with the backing of the E.E.C. has shown that it is possible to attain the two principal aims : fix the contamination and reduce diffusion.

It was demonstrated that on metal waste the epoxy and urethen polymers have good fixing properties with a factor in the order of 10^2 per layer for ^{60}Co and ^{137}Cs. The polyurethen paints which can be the most easily projected were tested with layers of thickness at least 5 mm in minutes per m^2 These thicknesses are necessary when considering storage for a moderate length of time.

Additional tests are necessary to determine the diffusion rates through these materials and storage times. The flexibility or hardness of these materials can be altered by the chemical formulation of the reactants. Coatings with mechanical properties giving rise to resistance to risks incurred during transport or handling operations can thus be produced.

The rusted metallic waste produced on dismantling of a nuclear reactor was covered with a thick layer of polyurethen . The efficiency of the procedure was also tested, particularly for polymer adhesion on a rusted surface and for contamination fixing.

A concrete dust binding technique using sodium silicate can be achieved either by projection onto the concrete surface or by the formation of pellets which can then be surrounded by a suitable matrix. These processes are easily industrialised and would solve the problems of harmful effects due to ingestion of the concrete dust and the radioactive risks associated with contamination.

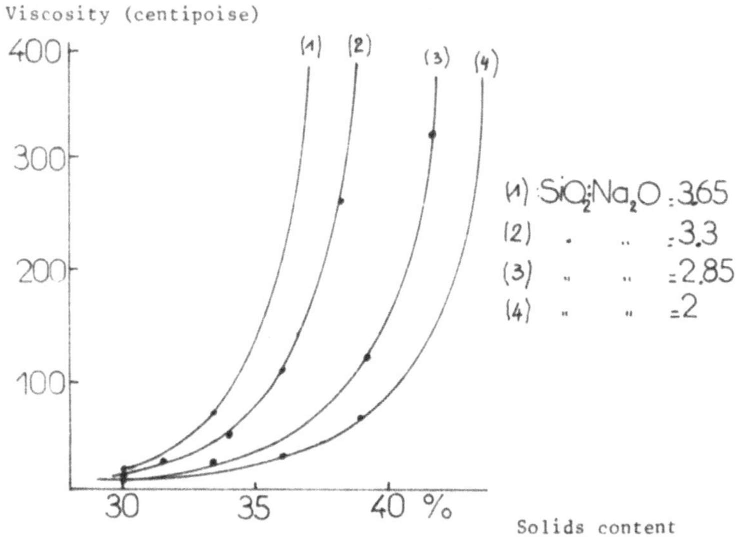

Fig. 1 - Viscosity of sodium silicate solutions

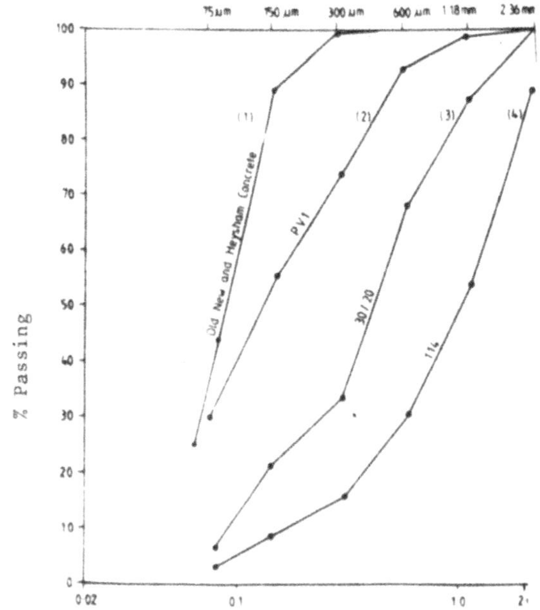

Fig. 2 - Concrete dust grading

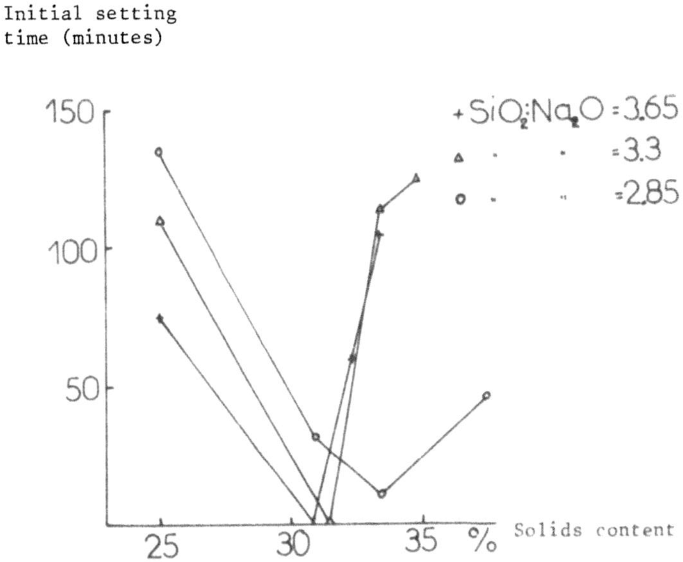

Fig. 3 - Effect of solids content on setting times of treated Heysham dust

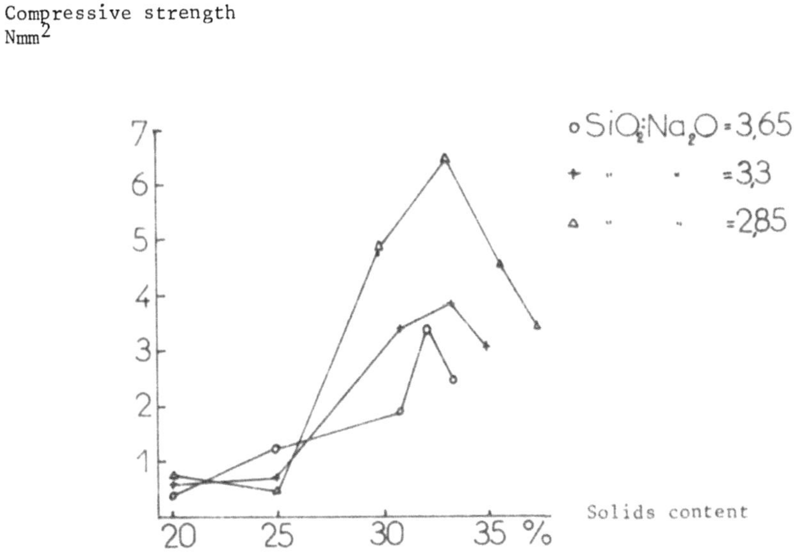

Fig. 4 - Effect of solids content on the compressive strength of treated Heysham dust

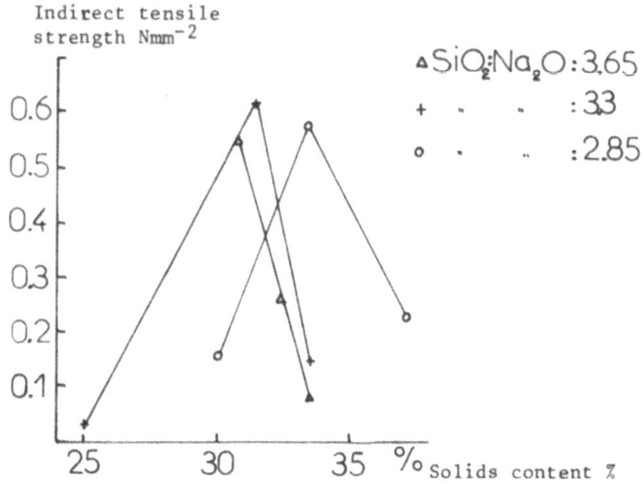

Fig. 5 - Effect of solids content on tensile strength of treated Heysham dust

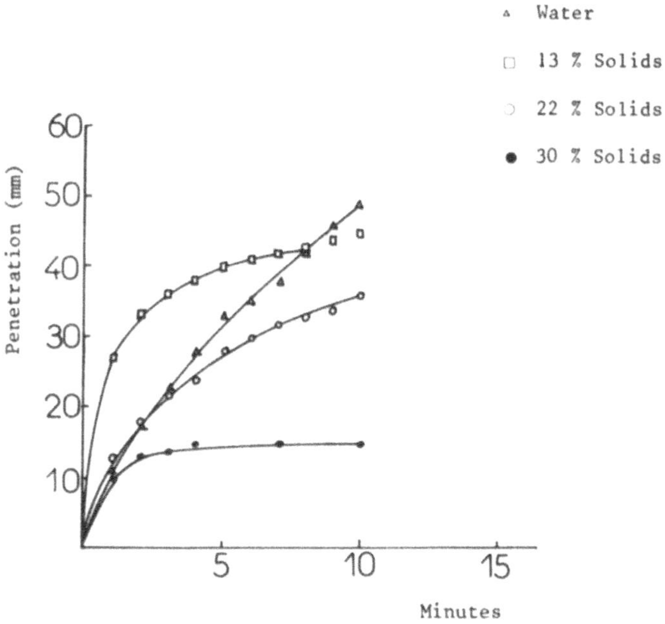

Fig. 6 - Impregnation of PVI dust with sodium silicate with silica/alkali ratio 2.85

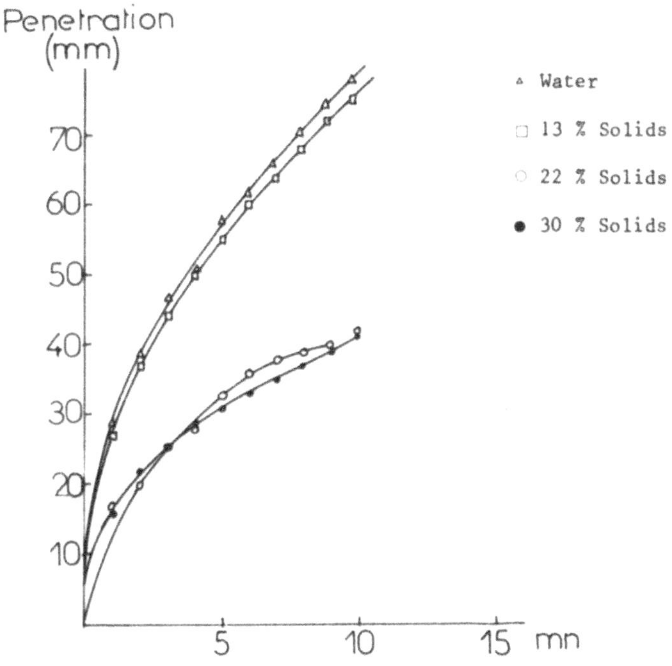

Add to Fig. 6 - 2.85:1 moist dust

TREATMENT OF METAL WASTE BY MELTING

C.R. Gomer, (British Steel Corporation),
M. Sappok, (Siempelkamp),
M. Allibert, (Institut Nationale Polytechnique de Grenoble)
F. Delabbaye (Séri)

Summary

Possibilities in melting are listed under four operation types. Work done by BSC is essentially of the reclaim of metal kind. That by Siempelkamp is largely aimed at waste conditioning under the volume reduction heading and for Séri the aim was waste conditioning by separation of radioactivities. At BSC, steel components removed from several kinds of reactor and designated as waste, have been melted and the metal reclaimed. This was in ordinary furnaces and done without radiological hazards.
At Séri the possibilities for cobalt removal from stainless steel were examined. Electroslag refining was seen as a promising process but no slag was found to take cobalt sufficiently into solution for worthwhile separation. Another study was on retention of cobalt in the metal phase in controlled oxidation of stainless steel. Too much cobalt was included in the oxidation product for worthwhile separation.
At Siempelkamp work is being done on the design and testing of a small furnace system for enclosed melting and ingot casting from radioactive steel waste. The main purpose is to achieve high volume reduction for most efficient storage/disposal of high radioactivity waste. The system enclosure and filtration have shown promise.

1. INTRODUCTION

This paper describes some findings under various kinds of melting practice which might relate to metal wastes arising at reactor decommissioning and applicable in some cases to arisings in service.

The gross amount of metal involved is about 10 kt per reactor at decommissioning and is estimated to average about 100 t per annum in service. Ferrous metals only have been considered, that is chiefly steel of a range of qualities from stainless grades through structural to mild steel.

The methods have been in respect of management procedures mostly at 2nd stage decommissioning (dismantling of plant external to the reactor biological shield and pressure vessel) for highly activated through to very slightly activated steel and to steel with contamination only or contamination combined with very slight activation.

Neutron activation within reactors is amenable to calculation and is predictable. In contrast contaminants can arise erratically in release from faulty fuel and by activation of circulating corrosion products, all being subsequently deposited randomly over metal surfaces, as indeed BSC have found.

2. GENERAL POSSIBILITIES IN MELTING

2.1 Operation Type

2.1.1 Volume Reduction of bulky miscellaneous scrap to achieve

(i) Better fill factors for storage or disposal containers.
(ii) More efficient transportation and vehicle loading, with better radiation self shield characteristics, resulting in a smaller external radiation field for a given radioactivity inventory.

2.1.2 Separation of Radioactivities

(i) Extraction of radioactivities from the liquid metal into a suitable slag phase.
(ii) Retention of radioactivities in the metal remainder following a suitable melt oxidation process.

2.1.3 Fixing of Radioactivities

(i) Incorporation of metal surface radioactivities into the metal mass on solidification.
(ii) Incorporation of metal surface radioactivities into a slag phase.

2.1.4 Reclaim of Metal

(i) By reduction of the concentration of radioactivities to below certain legislation levels by dilution melting so as to free the metal product for unlimited reuse.
(ii) By decontamination of the metal where the slag phase takes up a sufficient quantity of the radioactivities, originally on the metal surface, yielding a metal product as in (i).

2.2 Discussion

These various possibilities are considerably interrelated, but the operations under 2.1.1 and 2.1.2 are essentially waste conditioning procedures. The operations under 2.1.3 and 2.1.4 can be either fixing procedures as under 2.1.3 for high radioactivities, or metal reclaim procedures as under 2.1.4 where only very low radioactivities are involved.

3. PROJECT WORK

3.1 Work at BSC

3.1.1 Preliminary Considerations

At BSC there is a long experience of radiotracer use in steelplant investigations. This valuable work nevertheless involves arisings of radioactive waste including metal waste. Discussion with the UK DoE led to authorised dilution disposal (later changed to registered reclaim) of such metal by means of the largest production steelmaking furnaces. This experience has been augmented by the reclaim at chemical plant dismantling, of a large quantity of vessels and structural steel contaminated with thorium and uranium compounds. Laboratory and full scale melting of this demonstrated the control and radioactivities containment that could be achieved in a modern steelmaking furnace system. These experiences influenced the approach to the present work.

The emphasis has therefore been on the potential for reclaim of metal by melting in production furnaces. Steel and particularly its alloying constituents are high value resources and conservation of these is of increasing importance. It is certain that only very low level activated/ contaminated steel can be considered for the purpose but there are considerable quantities of this, particularly if a certain amount of prior decontamination is done.

3.1.2 The Melt Materials and Programme

Liaison with power station and reactor research establishment staffs have ensured a variety of materials and components taken out of reactor service being available for the various premelt measurements and melting tests.

In the first year of the project mostly small components of a few kg each (e.g. control rod drive gear wheel) were received and gave initial experience of monitoring assessment of these. Also experience was gained in the measurement techniques for the wide range of melt products. These melt products were in each case brought to radioactivity concentration values permitting unrestricted release of the materials.

In the second year the pattern of work was similar and results have largely substantiated the earlier work but the melting was at a much larger scale. An outline of this second year work will therefore be given as a realistic description of the project. Six melts each between 2 and 4 tonnes were made in a BSC 4 tonne electric arc system. Three melts also were made in a small induction furnace, two of these being 500 kg and the other 250 kg. The materials used and the brief purpose of these melts were as follows.

Category I

Melts on activated/contaminated components taken out of nuclear power station service. There were 5 melts in this category each made in the 4 tonne electric arc plant. The purpose of the first two was a general observation of all melt products following the detailed initial monitoring of the components. For the three later melts, by which time it had become apparent that some specific slag control of caesium radioactivities was possible, particular operations and measurements were made to investigate this.

Category II

Melts on metal plate material with known amounts of caesium radioactivity applied as simulated contamination, for the purpose of radioactivity balance in the furnace system for the whole melt procedure. One melt was made in the 4 tonne electric arc plant the other in the small induction furnace.

Category III

Melts with controlled additions of accurately known amounts of radioactivities for external dose quantification from various surface areas and section thicknesses in steel plate. Material from these melts also provided calibration standards for metal samples from the category I nuclear power stations component melts.

3.1.2.1 Melts in Category I - General Furnace Procedure

The subject materials for these melts were all components either removed from service or recovered at decommissioning from nuclear power stations. In each case the melt was made up to between 2 and 3 tonnes in the 4 tonne electric arc furnace with mild steel scrap in finely divided form. The BSC 4 tonne arc furnace system is shown in Fig. 1. The extraction and filter system in particular is shown with those details most relevent to this work. Thus fume sample points at top and bottom of the furnace offtake duct are marked as also are the sites of the high sensitivity radiation detectors used in the later melts and mounted on the duct before the extraction system and also on the gas discharge stack. External to the furnace system were up to 10 air sample collection points (small metered suction pumps drawing air through filter disks with retention ability to below 2 micron particle size). The collection points were variously sited in the immediate furnace workplace.

After the furnace heat exchanger and extractor fan were the permanent fume and dust collection arrangements. These comprise an eight compartment 32 filter bag chamber with a scavenge screw impeller in the chamber base taking fall out dust (dust cake) into a collector hopper through a rotary air seal. Cleaned furnace gas having passed through the filter system is discharged to atmosphere about 30 m above the ground.

From the tilting furnace the metal product was in each case run off into an intermediate ladle and cast from this ladle via a bottom stopper into a mould to form a single ingot. Slag product about 5% of the metal cast weight was removed at various recorded times (as noted in refs. 1&2) the slag being the result of the various steelmaking additions (carbon, lime and ferrosilicon chiefly in these melts) plus refractories erosion and general oxidation products. Slag removal in melt, is done by reverse tilting of the furnace with collection in a separate slag pot. Slag left until the melt ènd is the first material to be cast from the furnace into the metal ladle, it then rises through the metal to be the final material in the ladle after the metal ingot is cast through the ladle bottom nozzle. The slag remainder is tipped from the ladle into a slag pot by upending the ladle after the metal nozzle is stopped. In all cases comprehensive sampling of metal and slag along with the off gas materials and discharge hopper contents was undertaken throughout the operations.

INDIVIDUAL MELT DETAILS AND RESULTS

Melt i - (Row (a), Table 1, ref. 1)

These were CEGB Magnox reactor fuel chute sections (770 kg). The pieces were predominantly activated; Co-60 was identified but there was also some Co-60 contamination. There was no transfer of radioactivity to the slag or any presence in the dust cake.

Only in an 18 minute period during melt down was any level of radioactivity present in the duct fume. The radioactivity concentration in the 3 tonne ingot was uniform.

Melt ii - (Row (b), Table 1, ref. 1)

This was an AERE Winfrith SGHWR steam turbine diaphragm (645 kg). There was no transfer of radioactivity to furnace slag and virtually none carried over to the hopper dust cake. There was an extremely small level of radioactivity in the furnace duct fume but only during melt down. The product ingot had uniform radioactivity concentration.

Melt iii (Table 2, ref. 2)

This was a set of seven AERE Winfrith SGHWR charge face cover plates (126 kg).

Ten of these plates were received in all, having been through phosphoric acid decontamination at Winfrith. After comprehensive monitoring the plates were adjudged to have about equal amounts of Co-60 and Cs-137 contamination, with Cs-134 present at about 10% of either. Actual levels between plates and between faces of the same plate differed widely. An average level was 6.3×10^{-6} µCi cm^{-2} with a maximum of 1.3×10^{-4} µCi cm^{-2}. The seven plates of melt iii were judged to have aggregate activity equal to the three plates of melt iv.

At the time of melt iii, the caesium in slag retention possibilities derived from melts in category II were known, and the furnace procedure was modified accordingly. Thus the slag was made up to ~20% silica - as high as the basic furnace lining would withstand. Also the lime value was reduced to the lowest workable level. To follow caesium retention, slag samples were taken at 10 minute intervals through the melt. It was found that a great deal of radio caesium was retained in slag for 10 minutes. Thereafter most was rapidly lost, the residue declining slowly. Correspondingly, activity was recorded in the duct fume and was high in the hopper dust cake immediately after the melt. The hopper material showed a prolonged radioactivity decline for subsequent non-radioactive casts of the kind also observed for the category II caesium radioactivity balance trial done in this furnace system.

Melt iv - (Table 3, ref. 2)

This was the set of 3 cover plates (54 kg) and was melted in the same manner as melt iii. In order to utilise the initial retention of caesium in slag, the slag was removed as early as possible after meltdown. It was found that about 60% of the caesium present at 8 minutes remained at the unavoidably delayed 16 minutes slag removal time.

Melt v - (Table 4, ref 2)

These were KRB Gundremingen I-BWR control rod drive mechanism tube sections (62 kg). These pieces had 6 to 8 mRh^{-1} surface dose rate falling to 0.7 mRh^{-1} at 1 m. Contamination varied widely in value. It was mainly Cs-137 with some Cs-134 and Co-60.

For the slag, radio caesium amounts were slightly lower than for melt iv, but the pattern for radio caesium was repeated but with a lower fraction of this (35%) fixed in the removed slag.

3.1.2.2 Melts in Category II

Melt i - (line (c), Table 1, ref. 1)

25 µCi of Cs-137 as chloride was dispersed into a steel protective paint and applied to 250 kg of thick steel plate which was added to 2.5 t of charge in the 4 tonne electric arc furnace.

There was no transfer of radioactivity to the steel at any time. There was no retention of radioactivity in the furnace slag. Furnace fume was measurable but at a very low level. The hopper dust cake had a significant radioactivity concentration but represented only 4% of the added radioactivity. Summing the activity in this material over the next 50 non-radioactive casts accounted for the whole 25 µCi of Cs-137. Caesium chloride volatilises below steelmaking temperature, evidently this happened without any incorporation in slag, the caesium then redepositing in the cool parts of extractor filter system with eventual flushing through to the external hopper.

Melt ii - (Table 1, ref 2)

0.1 µCi of Cs-134 as hydroxide was distributed over the surface of 2.5 kg of clean steel plate and melted with 250 kg of non-active scrap in an induction furnace.

The form of application of caesium radioactivity to the plate was a significant step towards realism as regards the in-reactor conditions. Also the slag chemistry of caesium as an alkali metal was assumed, so that an acid slag was used in the lower temperature more quiescent slag, induction furnace. The contaminated plate was added with the steel liquid pool established and with complete slag cover. In this optimised melting procedure it was found that all ceasium radioactivity stayed in the slag phase.

3.1.2.3 Melts in Category III

There is a general lack of data or suitable formulae for calculation of dose rates external to steel with uniform dispersals of radioactivities. This particularly applies to thick sections where various self shield effects have to be taken into account. Since reclaim of steel involving radioactivity dilution is a main prospect for the present work it is important to verify such doseage for benefit versus detriment assessments.

Two melts have been made in which 125 μCi each of Co-60 and Ir-192 have been two stage diluted by induction furnace into 500 kg each of steel ingots. These have been rolled into various shapes and thicknesses of steel plate and have provided the basis for various above surface dose rate measurements.

These accurately known dilutions provided cross reference standards for activity concentration determination in all the product metal of the foregoing melts.

3.1.3 Radiological Safety

A complete schedule of radiological safety has accompanied the work. Material generating any dose rate in excess of 0.75 mRh^{-1} or with removable contamination in excess of 10^{-4} μCi cm^{-2} was handled in controlled conditions by classified radiation workers (Ref. IRR UK 1968 and 69). All persons involved in furnace melting operations wore film badges and radiation dosemeters. All these dosemeters for all workers produced nil readings for the whole of the work. The environmental air monitors running during all melting operations showed no measurable concentration of any radionuclide arising from the metal waste. Measurements in the fume in the furnace duct had a highest value of $6x10^{-9}$ μCi ml^{-1}. This contained fume therefore at no time had radioactivity levels greater than whole time permissible environmental air.

3.1.4 Overall Comments

Amounts of radioactive nuclear power plant components, 1.65 tonne in all, have been reclaimed within UK legislation provisions in the several melts this year.

There have been no radiation doseages or radioactivity contaminations or inhalations to any members of the melting shop personnel carrying out or in any way involved in the work.

Given suitable selection of components and materials it is shown that considerable quantities of steel which would otherwise require disposal can be reclaimed if the concept of dilution to a statutory and accepted de minimis level is operated. It must be emphasised that substantial dilution is best achieved in one stage in a large production furnace and that this will be a steelmaking furnace, most likely electric arc, where operational flexibility towards a defined product can be achieved. Melting costs rise sharply at the small scale and a first small melt to fix radioactivities prior to charge to a large furnace, although giving greater radioactivity concentration certainty, might be uncommercial unless disposal costs also rise sharply or the material is particularly high value.

A point that must be emphasised is that surface monitoring where complex shapes are involved is extremely inaccurate. Thus the two amounts of plates of melts iii and iv in Category I were adjudged equal in Co-60 but the uniform melt products showed them to be in a ratio of 2:1. For Cs-137 the actual ratio was 10:1 where surface monitoring had indicated they were equal.

The work shows:

(1) That Co-60 will almost certainly finish in entirely uniform dilution in the metal.

(2) That Cs radioactivities can be made to stay substantially in slag even in practical steelmaking furnaces.

(3) That significant amounts of low level radioactive metal waste otherwise requiring storage/disposal can be processed for reclaim in ordinary production furnaces with high filtration efficiency fume emission control equipment, without radiological hazards to the furnace operators working normally on the furnace or to any other persons in the vicinity.

3.2 Work at Seri

3.2.1 General Considerations

The aim was assessment, by means of theoretical studies and tests under non-radioactive conditions, of the feasibility of extracting cobalt and possibly nickel from stainless steel. If accomplished it would concentrate the cobalt radioactivity, very strongly generated in reactor service, into a small volume for efficient disposal. At the same time the process might reduce the radioactivity of the main amount of metal to a reclaim level. These objectives evidently depend on the degree of cobalt extraction that can be achieved.

The first process considered was electroslag refining (ESR) offering the possibility of selective extraction of cobalt in a small volume of slag. The literature review and laboratory test work however showed that the extraction potential is not sufficient.

A second process which could possibly bring about separation in the converse way was also considered. It is a controlled oxidation of iron and chromium from the steel melt, at the end of which a large fraction of the nickel and cobalt should remain in a small volume of metal easily cast into an efficient shape for storage/disposal. The tests carried out showed the results not to be as good as hoped for, with too much cobalt taken over into the oxidation product for this to be an entirely satisfactory radioactivity separation procedure.

3.2.2 Separation of Radioactivities by Extraction

3.2.2.1 The ESR Process

This process is for secondary refinement of metals. The metal to be purified must previously be cast to form a melt ingot. A bath of slag is made up in a cooled copper ingot mould being resistance melted by means of an electric current passing from base of the mould to the melt ingot mounted above. The melt ingot material falls in liquid drops through the slag and solidifies on the mould base initiating and then forming the growth ingot.

The refining reactions occur during the three stages of drop formation, transfer and solidification. Modification of the composition of the slag promotes a variety of refining and extraction possibilities.

The main requirements of the slag, that is chemical reactivity, low viscosity and electrical resistivity result in it being chiefly calcium fluoride with additions of lime, magnesium and aluminium. Modifications to bring about desired chemical reactions, are various minor additions of other oxides and metal compounds.

3.2.2.2 The Study and Experimental Work

In this study an element was sought giving with cobalt a compound more stable than with other steel constituents. The compound also having to be soluble in slag. It was found that cobalt phosphide and sulphide were more stable than corresponding compounds of iron, chromium and nickel. The phosphides have the greater reactivity and some experimental work was done on the solubility of these in calcium fluoride and also on the melting of stainless steel with calcium fluoride slag with high phosphide presence to promote substitution reactions with cobalt, thus increasing its partition to slag.

The extraction of cobalt in calcium fluoride slag was calculated to be maximum in the reducing condition of a 1 to 2% lime presence. The maximum partition to slag was calculated however to be very low only 8.10^{-3} in the case of virtually pure cobalt. Melting tests were made using a 0.5% cobalt steel. For this steel the partition coefficient between metal and slag was found to be less than 4.10^{-3} and therefore extremely low, bearing out the thermodynamic calculations.

3.2.2.3 Conclusions on the ESR Process

Iron compounds are generally more stable than those of cobalt and nickel. Exceptionally, cobalt phosphide and sulphide (particularly the phosphide) are more stable than those of iron and nickel. The possibility of using these differences derived from thermodynamic considerations. In practice the ESR extraction of cobalt from stainless steel has been shown from further considerations and tests not to be feasible.

3.2.3 Separation of Radioactivity by Controlled Oxidation

3.2.3.1 The Oxidation Procedure

Unlike ESR there is no established industrial metallurgical process of the present kind. In common with the previous case however, the idea of the particular application of a controlled oxidation procedure to the cobalt problem, was based solely on theoretical considerations which had to be backed up with an experimental study.

The outline procedure for an operating process is now given. It would be a two, (electric induction), furnace operation each having tilt pouring facilities and one having oxygen blowing equipment. A suitable slag to take up the oxidation products of the most oxidisable element, chromium in this case, is melted in one furnace and the subject metal (stainless steel) melted in the oxygen blown furnace. Slag pour into the steel melt furnace and oxygen blow are taken to near complete chromium removal from the melt metal. The chrome slag is decanted into a slag pot. Silicon is fed to the remainder dechromed melt and this is further blown with oxygen to reduce the iron from initial ~70% to ~20% and an overall metal mass reduction of between 5 and 10 times. Final metal and slag are cast into a mould and further slag pot respectively.

3.2.3.2 The Study and Experimental Results

The process was first studied by calculations and tests at the 1 kg melt scale to get more data on chemical exchanges taking place in the processing. From this it was expected that a metal phase reduced to about $\frac{1}{10}$ of the mass of the stainless steel input and containing about 99% of the cobalt and the bulk of the nickel could be achieved. This could be cast, in suitably shielded circumstances in the radioactive case, into an efficient shape for storage/disposal. Also there would be a slag phase, of about $1\frac{1}{2}$ times the input metal mass and perhaps 5 times the volume, consisting of iron and chromium silicates, which could possibly be reused in the production of stainless steel. However, for this slag reuse without radioactivity hazard problems, working from the likely initial activation of any of the steel considered for this treatment and allowing a delay time of 40 years for iron activation (Fe -55) (transferred to the slag) to decay, the cobalt partition to metal needs to be 99.99%.

The early 1 kg melt test results were reworked with respect to changes in slag composition needed to meet fluidity and refractories requirements of larger scale, 20 kg melting. It was found that the 1 kg melt results of only 1% cobalt to slag could not be repeated and that at the 20 kg scale, up to 20% of cobalt transferred to slag.

3.2.3.3 Conclusions on the Controlled Oxidation Process

The initial thermodynamic equilibrium model and partition fractions calculations have had to be discarded. Thus for closely similar cobalt contents of the metal, the slag cobalt content has been found to vary by a factor of 10. Whereas the maximum variation that the calculations would allow over a wide variation in slag composition would be by a factor of 2.

Controlled oxidation has been found to concentrate 80-90% of the cobalt and nickel radioactivities into a metal mass of about $\frac{1}{5}$ that of the initial treated metal.

The main problem with the controlled oxidation possibility when operated at any useful scale is therefore the large fraction, up to 20% of the total cobalt and also up to 7% of the total nickel, which passes to slag and renders this large bulk of slag (i.e. $1\frac{1}{2}$ times the treated metal mass and perhaps 5 times the volume) much too active for it to be utilised or even stored without further conditioning. It is stated again that no cobalt retention in the metal phase however will permit utilisation of the slag until the iron radioactivity has substantially decayed after 30 or 40 years.

3.3 Work at Siempelkamp

3.3.1 General Considerations

Siempelkamp is a foundary for ductile cast iron products with particular specialisation and experience in reactor primary and thermal shielding. They also have experience in manufacture of containers for radioactive waste and particularly a standard container for the dry intermediate storage of spent nuclear fuel elements, pending commissioning of fuel processing facilities, the container being substantially a Siempelkamp design. Siempelkamp are thus in a good position for the design and construction of a plant for melting contaminated metals from nuclear plants. They are doing this in conjunction with NIS a German nuclear engineering consultancy with experience in nuclear plant design, health physics and legislation requirements.

The present project has two aims. One of these has some overlap with the British Steel Corporation work. Thus it is hoped to use the melting plant to recycle metal waste that has a low level of surface contamination, given the appropriate legislation permission. Also it is hoped that a higher concentration level will be accepted for a proposed reuse of metal waste in shield castings for the nuclear power industry, particularly for the standard storage containers for spent fuel awaiting eventual reprocessing.

The other aim of the project is the use of the melting plant to reduce the volume of metal scrap with a high radioactivity contamination or content, by casting the metal in the totally enclosed plant into a convenient block for storage/disposal.

3.3.2 Construction and Operation Sequence of the Plant

The plant comprises a small, about 3 m x 8 m x 5 m (height) completely enclosed melting chamber, inside which the operations are so far as possible remote. The chamber is equiped with an exhaust air filtration system. The metal waste previously sized is loaded into a special container and by means of an installed crane this is placed on top of the ~8 t capacity electric induction furnace. A separate furnace fume extraction system is connected to the in situ container and the furnace interconnecting slide gate is opened, at the same time a bell closure in the base of the container is raised allowing the scrap to fall into the furnace. The furnace slide gate is now closed and melting takes place within the furnace vessel, this being held at reduced pressure. The melting rate is at 1-2 t h^{-1}. For casting, the melting chamber exhaust and filtration system takes over from the furnace system which is inoperative during this procedure. The metal is cast into a metal mould to produce 1.5 t shaped metal blocks. These blocks can be accommodated four to a loading in a highly shielding cast iron transport/storage container with a similarly secure lid.

3.3.3 Initial Melting and Tests in the Plant

The main testing of the system comes under the headings:

(i) General consideration of the filter systems in regards to flow capacity and particularly the degree of removal of dust and gases produced in the chamber. In this the performance of a micro filter at the end of the exhaust air filter duct was most important.

(ii) A study of the behaviour of several radioactivities on melting in the plant. This will comprise determination of fractions in the iron, slag, refractories, ducts and filters.

For the melting work 95 t in all of decontaminated scrap has been obtained from 3 German nuclear power stations. This material retains sufficient radioactivity after decontamination for it to provide radio tracer investigation possibilities for the plant. At present 20 t of this metal has been melted.

3.3.4 Conclusions from Initial Tests

For the several melts made on the decontaminated scrap it has been proven,

(i) That tight enclosure of the melt material within the furnace has been achieved thus yielding the design 1000 mb vacuum.

(ii) That gas leaving the filtering system was free of particulate matter.

(iii) That the majority of particles from the furnace were separated in the preliminary filter.

4. CONCLUDING OBSERVATIONS

Practical work and theoretical studies have been described for three separate research projects. These include work under most of the melting possibilities, headings of para. 2.1.

Comments and conclusions have been given separately at the end of each project description, but some restatement of the work, and concluding observation, will now be made.

At BSC steel components removed from several kinds of nuclear reactor and designated as radioactive waste have been melted and the metal reclaimed within UK (Radioactive Substances Act) provisions, which are based on a sufficient reduction of radioactivity concentration in the product. This has followed a pattern established in earlier steel plant management of radioactive waste including reclaim of substantial tonnages of contaminated process plant. Radiation dosages to persons involved in the present melting have been in all cases below dosimeter thresholds. Radioactivity transfer to the workplace by the melting operations, has been many times (arguably 10^6) less, than could give rise to ICRP based limits of intake.

What has been done shows what is possible in ordinary modern steel plant and the potential for reclaim of valuable metals with corresponding relief on heavily pressed and expensive radioactive waste disposal facilities. It is apparent that skill and judgement must go into selection of material to be treated this way either with or without preliminary decontamination procedures. Taken together these operations could however greatly increase the amounts of low level contaminated steel waste treatable to the present UK required general release (reclaim) level of 10^{-5} $\mu Ci g^{-1}$.

If the operations can be generally accepted in the steelworks then it will be best to treat several tonnes of scrap per melt in large (150 t) electric arc steelmaking furnaces where melting costs are lower and there is a defined metallurgical product. It is realised however that at first a smaller steel plant, designated for the work, and with a work force in full acceptance of the radiological safety factors might have to be used. The costs of the operation here, would be considerably higher, particularly if there was no standard metallurgical product, and further melting required.

The work at Siempelkamp is to construct a melting chamber which is fully enclosed and with a small scale furnace plant operating within. This will ensure by the second line of defence principle, that release of radioactivities to the environment is most improbable. For high radioactivity waste where the purpose is to get maximum volume reduction, for most efficient and least cost use of storage and disposal facilities, this kind of system is essential.

The other proposed use of the system is in non-dilution reclaim, in the more limited sense of reuse of the metal to fabricate components for service in controlled areas within the nuclear power industry, where a higher radioactivity concentration than the UK reclaim level might be permitted. Again the radioactivity inventory, for operations in the system of this kind, would be much higher than for any dilution melting at BSC.

The projects have pointed up in several ways, differences in general approach to release of material in FDR and UK which are based on surface contamination and radioactivity concentration respectively. Difficulties experienced at BSC in predicting the second from the first, for particular materials and melts, gave conviction at BSC of the merit of verifiable concentration, as the more secure criterion. Controlled melting also results in incorporation of radioactivity whether in metal or on slag. It is hoped that some progress towards unification of legislative approach may soon be achieved.

The outlook for melting procedures of the BSC and Siempelkamp kind obviously depend on overall benefit versus radiological detriment agreements. This particularly applies to the reclaim procedures and cannot really be assessed until the amounts of material treatable this way are more clearly determined.

In all cases the cost and acceptance of alternative waste disposal are important factors.

REFERENCES

1. EAEC Contract No. DE-D-002-UK(H) Technical Report No. 3 period January 1 - June 30, 1983.

2. EAEC Contract No. DE-D-002-UK(H) Technical Report No. 4 period July 1 - December 31, 1983.

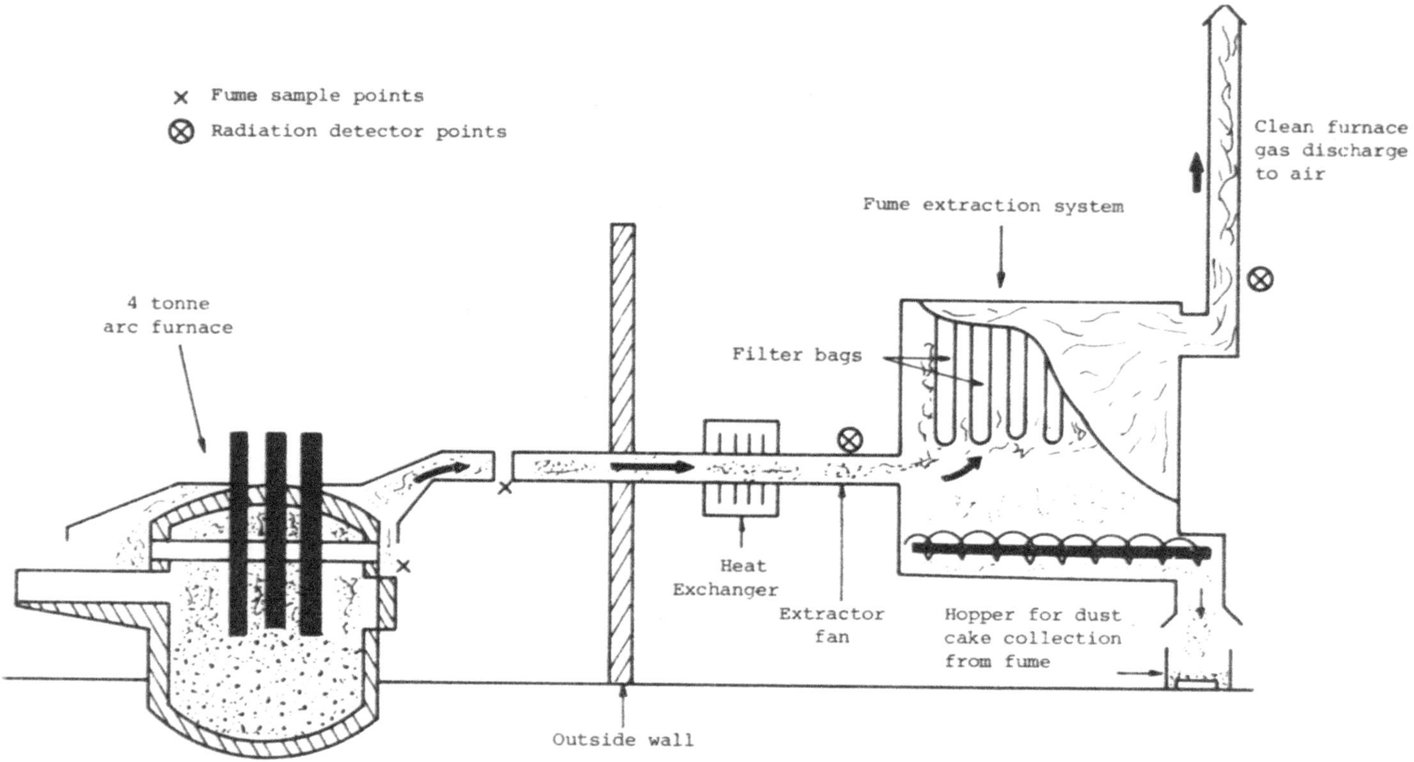

GRANGETOWN PILOT PLANT - 4 TONNE ELECTRIC ARC FURNACE SYSTEM Fig. 1 (R2/231)

MANAGEMENT AND DISPOSAL OF GRAPHITE WASTE

I.F. WHITE and G.M. SMITH
National Radiological Protection Board, UK

L.J. SAUNDERS, C.J. KAYE and T.J. MARTIN
Central Electricity Generating Board, UK

A.N. KNOWLES
United Kingdom Atomic Energy Authority

Summary

A technological and radiological assessment has been made of the management options for irradiated graphite wastes from the decommissioning of Magnox and Advanced Gas-cooled Reactors. The main contribution to the radionuclide inventory of irradiated graphite is from activation of the graphite and its stable impurities. Three different packaging methods for graphite have been described; each could be used for either sea or land disposal, and each is logistically feasible and could be achieved at reasonable cost. Leaching tests have been carried out on irradiated graphite, under a variety of conditions including those of the deep ocean bed; the different conditions had little effect on the observed leach rates of radiologically significant radionuclides. Radiological assessments were made of four generic options for disposal of packaged graphite: on the deep ocean bed, in deep geologic repositories at two different types of site, and by shallow land burial. Incineration of graphite was also considered, though this option presents logistical problems additional to those of direct disposal. With appropriate precautions during the lifetime of the ^{60}Co content of the graphite, any of the options considered could give acceptably low doses to individuals, and all the options would merit further investigation in site-specific contexts.

1. INTRODUCTION

The principal graphite-moderated reactors in EC countries are the UK Magnox reactor and the similar French system, both of which use metallic natural uranium fuel clad in magnesium alloy, and the UK Advanced Gas-cooled Reactor (AGR) fuelled by enriched uranium oxide clad in stainless steel. All these reactors are cooled by CO_2. Irradiated graphite from the cores of these reactors will arise as a radioactive waste at stage 3 of decommissioning.

This study has examined some of the options for the management of graphite decommissioning wastes, considering not only the technological aspects of the various options but also their radiological impacts. Although the study has been generic, related to no specific reactors or disposal locations, it has identified areas where further research is required, and has pointed the way towards the site-specific studies that must precede the actual operations of decommissioning and waste disposal.

The technological aspects of the study were the joint responsibility of the Central Electricity Generating Board (CEGB), the United Kingdom

Atomic Energy Authority (UKAEA) and British Nuclear Fuels Limited (BNFL), the major operators of Magnox reactors and AGRs in the UK. The radiological assessments and project management were the responsibility of the National Radiological Protection Board, the UK organisation responsible for advising on matters of radiation protection.

Full details of the study are given in the Final Report, to be published shortly by the Commission, where all literature references are to be found.

2. ARISINGS AND COMPOSITION OF IRRADIATED GRAPHITE

In order to assess the waste management requirements for graphite, it was necessary to estimate the masses arising and their radionuclide inventories. Magnox and AGR graphites were considered separately.

2.1 Arisings

The entire Magnox reactor programme consists of 26 reactors, 18 owned and operated by the electricity generating Boards and 8 by BNFL. The total mass of graphite arising during decommissioning will be 50,000 t, in the form of blocks, tiles and broken pieces. There is almost a 2:1 size variation between the smallest (earliest) Magnox reactors and the larger, later models. Most of the calculations were carried out for a hypothetical "reference Magnox reactor" of average characteristics, containing 2200 t of graphite. Results for the entire Magnox decommissioning programme were determined by scaling of mass, rating and flux.

Unlike the Magnox programme, the UK AGR programme is not necessarily complete, so only a reference AGR was considered, containing 1600 t of graphite.

2.2 Radionuclide inventories

There are three main sources of radionuclides in irradiated graphite: activation of carbon and impurities, fission of a small natural uranium impurity, and circuit contamination.

Activation is the most important. In order to calculate the inventories due to activation of impurities, reference impurity levels were estimated from the original material specifications and analyses that had been carried out by CEGB. Wide variations in impurity levels had been observed, and the reference levels adopted were those that best represent the probable average over all reactors. There is a systematic difference between impurity levels in AGR and Magnox graphites, as shown in Table 1.

For the inventory calculations it was conservatively assumed that each reactor will operate for 40 years at 70% load factor, and that decommissioning to Stage 3 could begin as soon as 10 years after shutdown. Each reference reactor was divided into zones, and separate inventory calculations carried out for each zone. Results for the reference Magnox reactor, the Magnox decommissioning programme and the reference AGR are shown in Tables 2, 3 and 4. ^{60}Co dominates the radionuclide inventory in the short term, and ^{14}C in the long term. ^{60}Co comes from neutron activation of the cobalt impurity in the graphite, and ^{14}C from activation of the graphite itself. Note that Magnox core is physically bigger than the AGR core, but the higher cobalt levels in AGR graphite and the higher integrated neutron flux lead to larger inventories of both ^{60}Co and ^{14}C in the reference AGR.

Additional radionuclide inventories due to fission of the natural uranium impurity (<0.1 ppm) and due to circuit contamination were also estimated; these proved unlikely to be radiologically significant compared with the activation inventory, and will not be considered in further detail here.

3. TECHNOLOGICAL ASSESSMENT OF MANAGEMENT OPTIONS

Graphite seems well qualified as a solid radioactive waste form in its own right, having good mechanical properties and quite low chemical reactivity and leach rates. Therefore most of the management options considered simply involve suitable packaging followed by direct disposal.

3.1 Packaging options for direct disposal

A variety of packaging options were considered for both ocean-bed disposal and land disposal. All options involve immobilisation of the graphite pieces, and shielding to comply with the IAEA Transport Regulations. Shielding requirements were considered for various times after reactor shutdown.

The packaging options considered were:

- sea-dumping containers to the current UK and NEA norms (capacity 0.4 - 0.6 m^3);
- larger containers (capacity 8-10 m^3) with returnable shielding overpacks;
- 50 t concrete monoliths (capacity 5.7 m^3);
- standard ISO containers (capacity 30 t).

The capacity figures given above make no allowance for packaging efficiency, which should be greater for the larger containers. Apart from the ISO container, all the packages considered could incorporate greater or lesser degrees of shielding. The ISO container is weight-limited and would not be suitable for graphite disposal until several decades after shutdown, when the shielding requirements would have been greatly reduced owing to the decay of ^{60}Co.

Although more information is currently available on the packaging requirements for sea disposal than for land disposal, it was concluded that any of the options considered would probably be suitable for either sea or land disposal (though not necessarily optimum). Cost estimates were made for each of the packaging options considered, and show that any of the options is feasible; but the estimates are not firm enough to permit valid comparisons between the options. It is fairly clear that bigger packages should be more cost-effective, but more detailed work is needed in this area.

3.2 Incineration

Incineration is an alternative to direct disposal of graphite. This option involves discharging the ^{14}C and tritium to atmosphere, and it obviously gives a large volume reduction, leaving all the other radionuclides in the ash to be immobilised and disposed of as a much-reduced volume of solid waste. Incineration is technologically feasible, and a flowsheet has been developed, but there are logistical difficulties and cost penalties. Either a separate incinerator must be built at every reactor site, or the graphite must be packaged and transported to a smaller number of more centralised incinerators. It is worth noting that the packaging requirements for this transport operation would be almost the same as those for transport followed by direct disposal of the graphite.

4. LEACHING OF IRRADIATED GRAPHITE

Graphite is an unusual material, and there was very little information on its resistance to leaching, so a study was undertaken as part of the programme of work. Samples of irradiated graphite from a Magnox reactor were used, and in addition to conventional tests by modified IAEA methods, leach tests were carried out at the temperature and pressure of the deep ocean bed. Table 5 shows the leachants and conditions employed. The various radionuclides are leached at differing rates, but graphite is unlike many other solid radioactive waste forms in that the leach rates of specific nuclides seem relatively insensitive to the chemical and physical leaching conditions. Figure 1 is representative of the results obtained. ^{14}C has the lowest observed leach rate, as might be expected, though the leach rate of tritium is also uncommonly low , emphasizing the unusual chemical nature of graphite.

The leaching tests lasted typically 150 days, and the results require careful interpretation for use in long-term radiological assessments. The observed leach rates of ^{14}C would correspond to an annual release rate of the order of 10^{-5} a^{-1} for particle sizes of about 1 cm. However, in the longer term there would be several possible competing effects tending to alter the release rate. Dissolution would reduce the particle size and hence reduce the release rate, while any crumbling of the graphite would tend to increase the release rate. A further possibility in the case of land disposal is that there would be insufficient groundwater available to leach the graphite at the rate observed in these experiments, in which the availability of leachant was essentially unlimited. Given these long-term uncertainties, it is difficult to justify any particular assumptions about future leach rates, and the approach in the radiological assessments was to examine the sensitivity of the results to a range of assumed leach rates.

5. RADIOLOGICAL ASSESSMENTS

5.1 Introduction

Figure 2 shows the alternative modes of graphite disposal considered in the radiological assessments, and also shows the main radiological pathways to man that were modelled. In all cases, disposal of graphite can potentially lead to the global circulation of ^{14}C.

Doses to individual members of critical groups were calculated, together with collective doses to the world population. The interpretation of these results is discussed in section 5.6.

5.2 Ocean-bed disposal

Disposal of graphite on the deep ocean bed would lead to a slow release of activity into bottom waters by leaching, and then a gradual dispersal throughout the world's oceans. The theoretical possibility was considered of a relatively small volume of contaminated bottom water being somehow transported to the surface, leading to increased individual doses; even with this pessimistic assumption the individual doses would be very low. The main contribution to collective doses would be the global circulation of ^{14}C.

5.3 Deep geologic disposal

Two different types of deep geologic disposal were considered. The first was at an inland site in clay, which could lead to very long transit times for radioactivity transported by moving groundwater. This groundwater was envisaged to emerge into a source of drinking water. Even under the least favourable conditions assumed, individual doses would not be very large, and the transit times for ^{14}C would be so long that much of it would have decayed before it emerged into the biosphere.

The second type of deep geologic disposal considered was at a coastal site in shale. Generically, groundwater transit times would be shorter than in clay, but the groundwater was assumed to emerge from beneath the coastal seabed into the marine environment. Individual doses could be rather higher than for the generic clay site, but would still be quite low.

Collective doses would be largely determined by the extent to which the ^{14}C would have decayed while in transit through the shale.

5.4 Shallow land burial

Shallow burial was assumed to be in an engineered concrete structure incorporating barriers against infiltration of groundwater, and located in a clay formation with low groundwater flows. After the repository had been closed there was assumed to be a period of institutional controls over the use of the site, lasting some decades; at later times no such controls were assumed.

The period of institutional controls would be necessary to prevent accidental excavation, for that would lead to quite high doses to individual building workers from ^{60}Co. Since ^{60}Co has a rather short half-life, it is entirely feasible to prevent the possibility of excavation for the necessary few decades.

For the longer term, the radiologically significant pathways would begin with the leaching of activity from the graphite by groundwater, followed by movement to the soil zone and into crops and drinking water. The individual and collective doses would depend very much on the exact circumstances of land utilization many centuries after disposal. For this generic assessment a range of assumptions were made, the most pessimistic assumption being that farming would take place, and the least pessimistic being that there would be no farming although the groundwater would run off into a stream used by man.

There are a number of uncertainties in the modelling of the behaviour of ^{14}C in shallow land burial. The fact that irradiated graphite consists mainly of ^{12}C invalidates the normal modelling assumption that radionuclides are present in trace chemical quantities. Also it is possible that ^{14}C might be retained by carbonates in the concrete and the soil, leading to doses much lower than those predicted in this study. More investigative work is needed on these subjects, and this must ultimately be carried out for specific shallow burial sites.

5.5 Incineration

The main radiological impact of incineration would arise from the atmospheric discharge of ^{14}C. The greatest individual doses would arise via terrestrial food pathways, but they are quite small; collective doses would be largely due to global circulation of ^{14}C.

5.6 Results

Results for the reference Magnox reactor, the Magnox decommissioning programme and the reference AGR are summarized in Tables, 6, 7 and 8. Ranges of individual and collective doses are given, corresponding to the ranges of assumptions in the modelling.

When comparing the individual and collective doses from the different graphite management options, it should be noted that these doses do not have equal probabilities of being received. For example, the doses due to atmospheric discharges from incineration would have a much higher probability of being received than those from the deep geologic disposal options. In a more comprehensive assessment results would be expressed in terms of risk (ie a combination of probabilities and dose), and specific sites would be considered. From this generic assessment it can only be concluded that none of the options need be ruled out, and that all would merit further investigation.

6. CONCLUSIONS

1. The radionuclide inventories of Magnox and AGR graphites 10 years after the end of reactor life have been calculated. The main contribution is from activation of the carbon itself and the stable impurities.

2. A lesser contribution to the total inventory may arise from fission of natural uranium present as an impurity in the graphite. A measurement on a single sample, and other indirect evidence, suggest that the uranium level is unlikely to exceed 0.1 ppm. Even at this level, the contribution to the total radiological impact of disposal would be relatively minor.

3. Surface contamination of the graphite by circuit crud or by volatile fission products both depend on the design and operating history of the particular reactor. While these contributions to the total radionuclide inventory and the radiological impact of disposal cannot be estimated in a generic study, they will probably be relatively small.

4. Three different packaging methods for graphite have been described. Each is logistically feasible and could be used for either sea or land disposal, though the type at present used for sea-dumping would probably be less suitable for land disposal than the two larger packages.

5. Cost estimates are presented for each of the three packaging methods, and any of the three could be implemented at reasonable cost. The estimates have been made on different bases and cannot be used to make detailed intercomparisons between the packaging methods. However, it appears that larger packages are more cost-effective. Further studies are required in this area.

6. Unirradiated graphite in large blocks retains its mechanical integrity at pressures up to 900 bar. This finding was confirmed during leaching tests on small samples of irradiated graphite at 450 bar and 2.5°C, the conditions on the deep ocean bed at a depth of 4 km. Simulated seawater was used as the leachant, and a parallel test was carried out under ambient laboratory conditions. Other tests under ambient conditions used simulated groundwater and demineralised water. In all cases ^{14}C had the lowest leach rate and ^{133}Ba or ^{134}Cs the highest, but the effects of the different leaching conditions were found to be small.

7. Ocean-bed disposal of graphite is predicted to lead to acceptably low potential individual doses.

8. Two forms of deep geologic disposal of graphite have been considered: at an inland site in clay, and at a coastal site in shale. For a wide range of assumed leach rates and hydrogeologic parameters, either type of generic site could lead to acceptably low potential individual doses.

9. In the case of shallow land burial it would be necessary to restrict building on the site for some decades after disposal, until the ^{60}Co content has appreciably decayed; such restrictions are quite feasible. The long-term radiological impact of graphite disposal by shallow land burial requires further investigation, mainly because of the present difficulties in modelling the behaviour of ^{14}C in soils and shallow groundwater. Ultimately these investigations must be on a site-specific basis.

10. A feasible alternative to direct disposal of packaged graphite is incineration, followed by disposal of a smaller volume of ash. But incineration presents logistical difficulties additional to those of direct disposal: either several incinerators must be built at different sites or the graphite must be transported to a smaller number of centralised incinerators; the corresponding costs could vary widely. The radiological impact of incineration would arise almost entirely from the atmospheric discharge of ^{14}C and is greater than that of many other options, especially in the early term. Incineration does not seem a desirable alternative to direct disposal, unless the reduced volume of ash is judged to override all other considerations.

11. All the graphite management modes considered in this generic study merit further investigation. The choice of the optimum management mode must be made in some specific context, and the optimum will depend on the specific circumstances. When making such a choice (possibly with the aid of cost-benefit analysis) further information would be needed on the time-varying probabilities that predicted individual and collective doses would be received, and on the individual dose rates at which collective doses might accrue.

Table 1

Reference Graphite Impurity Levels

(ppm)

Element	Magnox	AGR
Li	0.05	0.05
Be	0.02	0.02
B	0.1	0.5
N	10	10
Na	1.0	4.0
Mg	0.1	0.4
Al	1.0	4.0
Si	35	35
S	50	60
Cl	2.0	4.0
Ca	35	25
Ti	3	0.7
V	12	0.4
Cr	0.35	0.4
Mn	0.04	0.25
Fe	10	28
Co	0.02	0.70
Ni	1.0	6.0
Zn	0.13	1.0
Sr	0.4	0.4
Mo	0.1	2.5
Ag	0.001	0.001
Cd	0.04	0.07
In	0.05	0.06
Sn	0.05	1.0
Ba	1.5	0.5
Sm	0.04	0.05
Eu	0.004	0.005
Gd	0.005	0.01
Dy	0.008	0.006
W	0.12	0.15
Pb	0.12	0.8
Bi	0.08	0.05

Table 2

Activation Inventory (Bq) of Reference Magnox Reactor

After 40 years reactor operation followed by 10 years decay

^{3}H	$1.2\ 10^{14}$	^{93}Mo	$8.5\ 10^{8}$
^{10}Be	$7.1\ 10^{10}$	^{93m}Nb	$5.5\ 10^{8}$
^{14}C	$8.5\ 10^{13}$	^{94}Nb	$1.0\ 10^{5}$
^{36}Cl	$9.5\ 10^{11}$	^{99}Tc	$1.7\ 10^{8}$
^{41}Ca	$7.3\ 10^{11}$	^{108m}Ag	$2.3\ 10^{10}$
^{54}Mn	$2.7\ 10^{8}$	^{113m}Cd	$1.0\ 10^{10}$
^{55}Fe	$1.5\ 10^{13}$	^{121m}Sn	$4.5\ 10^{10}$
^{59}Ni	$9.3\ 10^{10}$	^{133}Ba	$5.6\ 10^{11}$
^{60}Co	$2.7\ 10^{13}$	^{152}Eu	$2.2\ 10^{11}$
^{63}Ni	$1.3\ 10^{13}$	^{154}Eu	$5.2\ 10^{12}$
^{65}Zn	$2.1\ 10^{8}$	^{155}Eu	$1.6\ 10^{12}$

Total mass of graphite = 2233 t

Table 3

Activation Inventory (Bq) for UK Magnox Programme

(9 twin reactor stations plus 2 quadruple reactor stations)

After 40 years reactor operation followed by 10 years decay

^{3}H	$2.6\ 10^{15}$	^{93}Mo	$1.9\ 10^{10}$
^{10}Be	$1.6\ 10^{12}$	^{93m}Nb	$1.2\ 10^{10}$
^{14}C	$1.9\ 10^{15}$	^{94}Nb	$2.2\ 10^{6}$
^{36}Cl	$2.1\ 10^{13}$	^{99}Tc	$3.9\ 10^{9}$
^{41}Ca	$1.6\ 10^{13}$	^{108m}Ag	$5.3\ 10^{11}$
^{54}Mn	$6.0\ 10^{9}$	^{113m}Cd	$2.2\ 10^{11}$
^{55}Fe	$3.5\ 10^{14}$	^{121m}Sn	$1.0\ 10^{12}$
^{59}Ni	$2.1\ 10^{12}$	^{133}Ba	$1.3\ 10^{13}$
^{60}Co	$5.9\ 10^{14}$	^{152}Eu	$4.9\ 10^{12}$
^{63}Ni	$2.9\ 10^{14}$	^{154}Eu	$1.1\ 10^{14}$
^{65}Zn	$4.8\ 10^{9}$	^{155}Eu	$3.5\ 10^{13}$

Total mass of graphite = 50,000 t

Table 4

Activation Inventory (Bq) of Reference AGR

After 40 years reactor operation followed by 10 years decay

^{3}H	$7.6\ 10^{13}$	^{93}Mo	$5.6\ 10^{10}$
^{10}Be	$2.3\ 10^{11}$	^{93m}Nb	$3.4\ 10^{10}$
^{14}C	$1.9\ 10^{14}$	^{94}Nb	$8.4\ 10^{6}$
^{36}Cl	$2.3\ 10^{12}$	^{99}Tc	$5.6\ 10^{9}$
^{41}Ca	$1.2\ 10^{12}$	^{108m}Ag	$3.6\ 10^{10}$
^{54}Mn	$5.2\ 10^{9}$	^{113m}Cd	$4.4\ 10^{10}$
^{55}Fe	$9.3\ 10^{13}$	^{121m}Sn	$2.8\ 10^{12}$
^{59}Ni	$5.5\ 10^{11}$	^{133}Ba	$3.1\ 10^{11}$
^{60}Co	$9.8\ 10^{14}$	^{152}Eu	$1.0\ 10^{11}$
^{63}Ni	$1.1\ 10^{14}$	^{154}Eu	$2.9\ 10^{12}$
^{65}Zn	$3.7\ 10^{9}$	^{155}Eu	$9.4\ 10^{11}$

Total mass of graphite = 1633 t

Table 5

Test Environments and Leaching Parameters

Disposal Environment	Leachant conditions				Ratio of volume/ surface area (cm)	Ratio of volume of leachant/ surface area of sample (cm)
	Composition	Temperature (^0C)	Pressure (bar)	Volume per sample (ml)		
Land burial site	Simulated groundwater	25 ± 0.5	1.0	250	0.66	1.9
Control test	Demineralised	20 to 23	1.0	300	0.66	2.3
Deep-sea disposal	Sea water (to BS 3900)	2.5 ± 0.7	450	278	0.66	2.1
-	Sea water (to BS 3900)	20 to 23	1.0	125	0.55	1.4

Table 6

Reference Magnox reactor: summary of estimated potential individual doses and collective doses

Ranges correspond to the ranges of assumptions made

Graphite management option	Range of estimated max. potential individual dose, Sv/y	Range of estimated collective dose commitment, man Sv
Ocean-bed disposal	$3.0\ 10^{-10}$ to $3.0\ 10^{-7}$	$5.2\ 10^{3}$ to $8.9\ 10^{3}$
Inland geologic disposal	$3.7\ 10^{-11}$ to $5.7\ 10^{-8}$	$1.4\ 10^{-1}$ to 10
Coastal geologic disposal	$3.7\ 10^{-11}$ to $4.7\ 10^{-6}$	3.1 to $8.7\ 10^{3}$
Shallow land burial	$4.3\ 10^{-7}$ to $1.7\ 10^{-2}$[1,2]	$1.8\ 10^{2}$ to $2.3\ 10^{4}$[1]
Incineration atmospheric discharge	$3.0\ 10^{-5}$	$9.6\ 10^{3}$
Plus disposal of ash to: ocean bed inland geologic coastal geologic shallow land	(negligible in comparison with the atmospheric-dispersion dose from incineration) $4.3\ 10^{-7}$ to $1.7\ 10^{-2}$[1,2]	(negligible in comparison with the atmospheric-dispersion dose from incineration)

[1] These estimates are uncertain and probably pessimistic: see text.

[2] Restriction of building on the site would be required for at least 30 y after disposal (40 y after shutdown), to prevent annual doses in excess of 5 mSv to individual building workers.

Table 7

Magnox decommissioning programme: summary of estimated potential individual doses and collective doses

Ranges correspond to the ranges of assumptions made

Graphite management option	Range of estimated max. potential individual dose, Sv/y	Range of estimated collective dose commitment, man Sv
Ocean-bed disposal	$6.7\ 10^{-9}$ to $6.7\ 10^{-6}$	$1.2\ 10^{5}$ to $1.9\ 10^{5}$
Inland geologic disposal	$8.3\ 10^{-10}$ to $1.3\ 10^{-5}$	3.1 to $2.2\ 10^{2}$
Coastal geologic disposal	$8.3\ 10^{-10}$ to $1.1\ 10^{-4}$	$6.9\ 10^{1}$ to $1.9\ 10^{5}$
Shallow land burial	$4.3\ 10^{-7}$ to $1.7\ 10^{-2}$ [1,2]	$4.0\ 10^{3}$ to $5.1\ 10^{5}$[1]
Incineration atmospheric discharge	$6.7\ 10^{-4}$	$2.1\ 10^{5}$
Plus disposal of ash to: ocean bed inland geologic coastal geologic	(negligible in comparison with the atmospheric-dispersion dose from incineration)	(negligible in comparison with the atmospheric-dispersion dose from incineration)
shallow land	$4.8\ 10^{-6}$ to $1.9\ 10^{-1}$[1,2]	

[1] These estimates are uncertain and probably pessimistic: see text.

[2] Restriction of building on the site would be required for at least 53 y after disposal (63 y after shutdown), to prevent annual doses in excess of 5 mSv to individual building workers.

Table 8

Reference AGR: summary of estimated potential individual doses and collective doses

Ranges correspond to the ranges of assumptions made

Graphite management option	Range of estimated max. potential individual dose, Sv/y	Range of estimated collective dose commitment, man Sv
Ocean-bed disposal	$1.1\ 10^{-7}$ to $1.1\ 10^{-5}$	$1.1\ 10^{4}$ to $1.9\ 10^{4}$
Inland geologic disposal	$1.3\ 10^{-10}$ to $1.4\ 10^{-7}$	$4.7\ 10^{-1}$ to $2.4\ 10^{1}$
Coastal geologic disposal	$8.3\ 10^{-11}$ to $1.0\ 10^{-5}$	6.9 to $1.9\ 10^{4}$
Shallow land burial	$1.0\ 10^{-6}$ to $4.0\ 10^{-2}$ [1,2]	$4.0\ 10^{2}$ to $5.1\ 10^{4}$ [1]
Incineration atmospheric discharge	$6.7\ 10^{-5}$	$2.1\ 10^{4}$
Plus disposal of ash to: ocean bed, inland geologic, coastal geologic	(negligible in comparison with the atmospheric-dispersion dose from incineration)	(negligible in comparison with the atmospheric-dispersion dose from incineration)
shallow land	$1.0\ 10^{-6}$ to $4.0\ 10^{-2}$ [1,2]	

[1] These estimates are uncertain and probably pessimistic: see text.

[2] Restriction of building on the site would be required for at least 56 y after disposal (66 y after shutdown), to prevent annual doses in excess of 5 mSv to individual building workers.

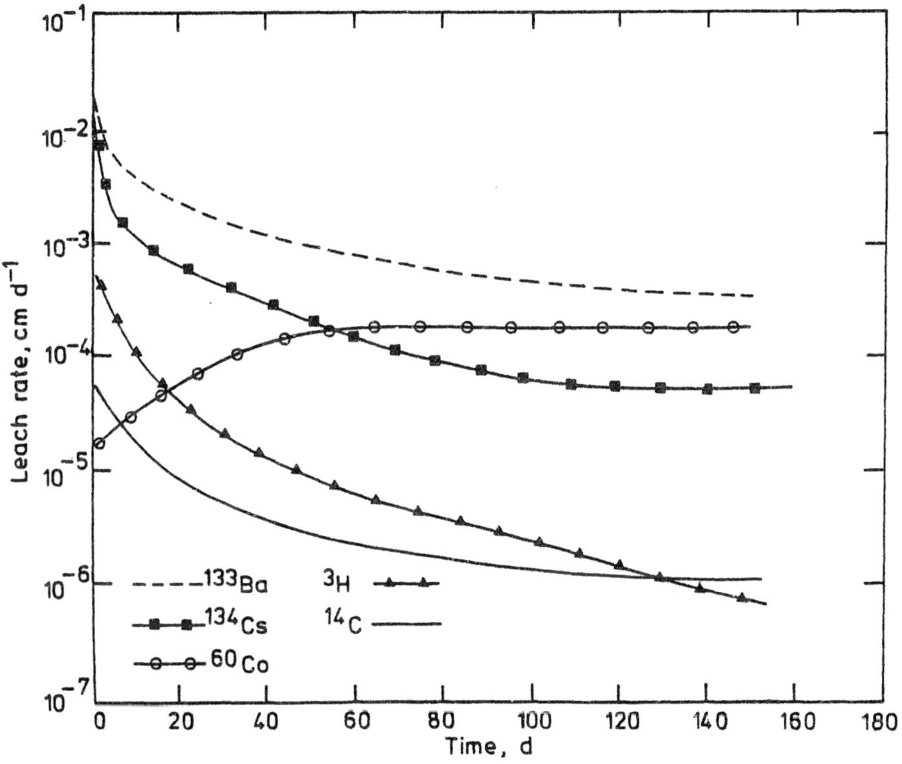

Figure 1 Leach rates for irradiated graphite in demineralised water (1 bar, 20°C)

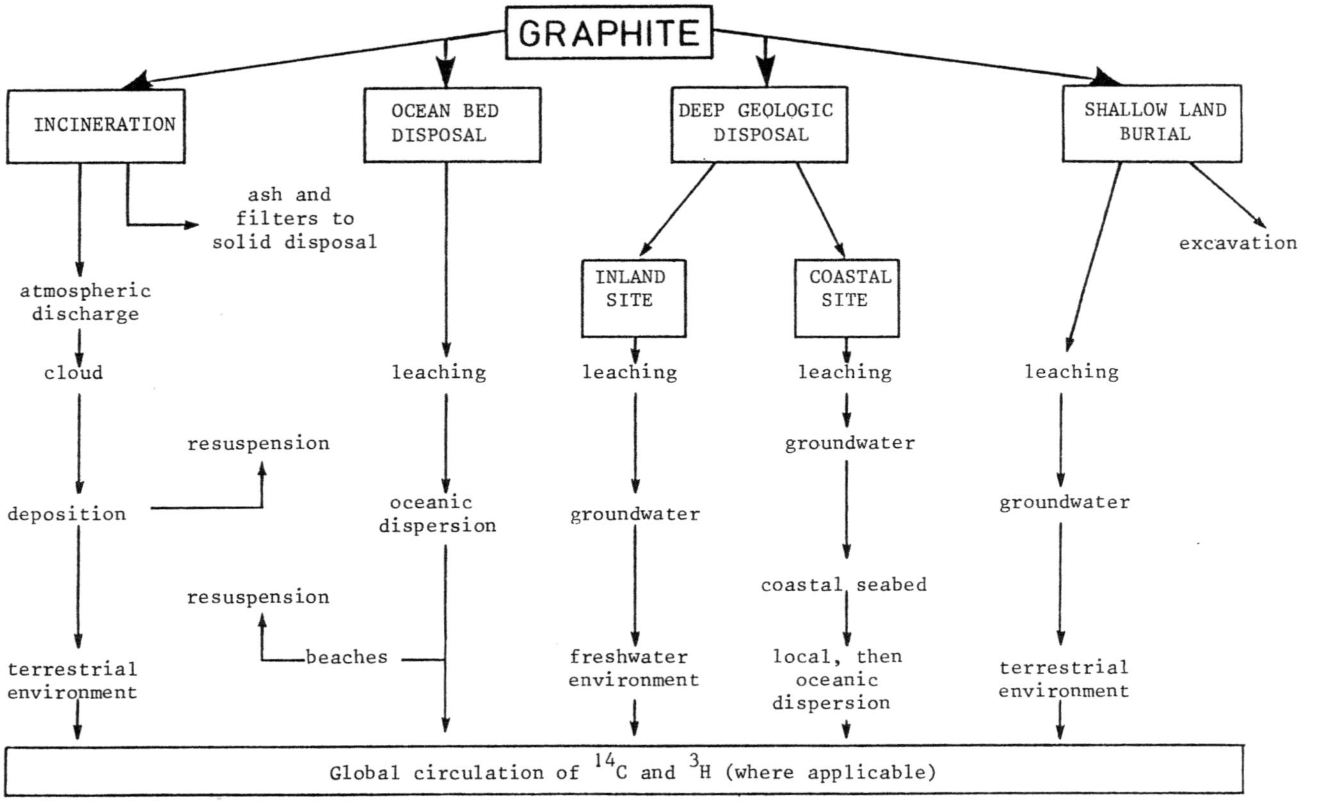

Figure 2 - Options considered for the management of graphite, showing simplified environmental pathways

LARGE TRANSPORT CONTAINERS FOR DECOMMISSIONING WASTE

M S T Price, UKAEA, AEE Winfrith, UK
I Lafontaine, Transnubel SA, Dessel, Belgium
K Pflugrad, Commission of the European Communities, Brussels, Belgium

Summary

A study has been made of the types of large transport container which will be needed for bulky radioactive wastes which result from the decommissioning of nuclear power stations. The work was focussed on the wastes which arise from light water and gas-cooled reactors. Five reference power plants typical of those in the European Community were selected and their associated inventories of neutron activated and contaminated components were assessed. Details of more than 100 existing containers have been compiled and evaluated against a set of generic criteria. The factors affecting transport by road, rail and waterway, including National Legislation, Codes of Practice and International Agreements have been examined for their impact on the design of large containers. Transport costs have been estimated using information representative of Belgium and the United Kingdom. The work has led to the formulation of conceptual designs of Type B and Low Level Solids/Low Specific Activity containers. Such containers were then assumed to be transported from a reactor to a notional repository so that the health impact to the public could be assessed. As a result of the study recommendations are made for the next 5 year programme of the CEC.

1. INTRODUCTION

Radioactive waste resulting from the dismantling of major nuclear reactor components must be transported in larger units than those used at present for other types of waste, in order to reduce the amount of cutting required and consequently the radiation exposure of personnel and the costs of decommissioning.

A system study has been carried out under the first 5 year programme of research on decommissioning of nuclear reactors, which was aimed at defining the types of large transport and/or disposal container needed for bulky radioactive waste resulting from dismantling nuclear power plants.

To carry out the task, two contracts were placed by the Commission of the European Communities (CEC) with Transnubel SA, Dessel, Belgium (TNB) and the United Kingdom Atomic Energy Authority, AEE, Winfrith, UK (UKAEA). TNB was assisted by Belgonucleaire SA, Brussels, Belgium and the UKAEA by PPC Consultant Services Ltd, Crawley, UK (formerly Pollution Prevention (Consultants) Ltd) (PPC).

The work of the UKAEA and PPC was also supported by the UK Department of the Environment as part of its national radioactive waste management research programme. The results will be used in the formulaton of strategy but do not, at this stage, necessarily represent UK Government policy.

The study programme was structured under the following headings:-

1. Reference Nuclear Power Stations
2. Waste Streams
3. Survey of Existing Containers
4. Conceptual Designs
5. Means of Transportation
6. Regulations
7. Transport Costs
8. Radiological Detriment
9. Conclusions and Recommendations

2. REFERENCE NUCLEAR POWER STATIONS

Apart from some experimental heavy water moderated thermal reactors, all the major nuclear power plants in the European Community (EC) belong to one of the following types:-

1. Pressurised Water Reactor (PWR)
2. Boiling Water Reactor (BWR)
3. Gas-Cooled Reactor (GCR)

Accordingly, to focus the work, five reference power plants were selected:-

Pressurised Water Reactors

Type 1 - European PWR of the 3800 MW(Th)/1300 MW(e) class
Type 2 - European PWR of the 2800 MW(Th)/ 950 MW(e) class

Boiling Water Reactors

Type 3 - European BWR of the 3800 MW(Th)/1300 MW(e) class

Gas-Cooled Reactors

Type 4 - UK Magnox of the 840 MW(Th)/205 MW(e) class
Type 5 - UK Advanced Gas-Cooled of the 1500 MW(Th)/620 MW(e) class

In a PWR the reactor core and its pressure vessel are highly radioactive. So are the steam generators and the pipework connecting the reactor to the steam generators. Because the turbine is not directly connected to the primary circuit, it is not usually radioactive, unless there has been tube leakage in the steam generators. The cooling tower and associated piping are normally not radioactive. Much equipment in the auxiliary building, which contains the liquid and gaseous waste treatment systems, the filter exchange vaults and the ventilation systems, is radioactive, as is the spent fuel storage pool and its associated equipment.

The reactor core and its pressure vessel in a BWR are highly radioactive, as is the turbine and the connecting piping. The cooling tower and associated piping are not, since the design pressure of the system is higher, such that any leakage would be from the non-radioactive secondary loop into the primary loop. Much equipment in the radioactive waste and control building, which houses the solid, liquid and gaseous radioactive waste treatment systems, is radioactive, as is the spent fuel pool in the reactor building.

The major components of a GCR are the reactor core and its pressure vessel, the heat exchangers, a steam turbine, an electric generator and a steam condenser system. The reactor core, pressure vessel and the inner part of the biological shield are highly radioactive. The lower power density of GCRs implies a greater mass of activated material compared with light water reactors (LWRs). The use of carbon dioxide as the coolant for GCRs leads to reduced corrosion compared with LWRs. Corrosion of metals in the primary water loops of LWRs produces 'crud' particles which are transported by water through the reactor core where they become radioactive due to neutron capture. The radioactive 'crud' is eventually deposited on both in-core and out-of-core surfaces.

3. WASTE STREAMS FROM THE REFERENCE POWER STATIONS

3.1 General

An accurate estimate of the detailed make-up of the radioactive inventory is important, because it can directly affect the method and timing of the decommissioning. The decommissioning wastes from a nuclear power plant are of two general categories:-

A. Neutron-activated Materials
B. Contaminated Materials

Materials exposed to the neutron flux, eg control rods, the pressure vessel and the inner skin of the biological shield become radioactive. The most significant radioisotopes affecting dismantling, transport and disposal give rise to beta and gamma emissions, chiefly ^{60}Co (half-life 5.27 y), ^{55}Fe (half-life 2.6 y) and ^{63}Ni (half-life 92 y).

^{60}Co, a significant gamma emitter, is of primary concern in planning the decommissioning programme since the dose levels of ^{60}Co radiation will determine the amount of remote operation, the extent of shielding both for workers during dismantling and for the transport containers.

The neutron activation of structural materials is amenable to calculation. Several codes that have the capability of solving the multiple energy group equations are available and have been validated. Large differences between reactors of a similar kind are not to be expected. In contrast, variable and relatively unpredictable sources of radioactivity can arise from phenomena which are the consequence of a variety of mechanisms including release from faulty fuel, corrosion of fuel cans and of steam generators, all of which can be designated contamination.

3.2 Decommissioning Alternatives and their Impact

Once the power plant has reached the end of its useful life, the reactor is shut down, depressurised, defuelled, made safe and kept under surveillance to the satisfaction of the Regulatory Authorities. This can be termed Stage 1 Decommissioning.

Stage 2 involves the dismantling of all active and non-active plant and buildings external to the reactor biological shield or prestressed concrete reactor pressure vessel and the removal of the consequent waste to a suitable storage or disposal site. Continued surveillance of the site is implied.

Stage 3 decommissioning represents the dismantling of all remaining plant and buildings, both active and non-active, with the removal of the generated waste to a suitable storage or disposal site. Following Stage 3, the site may be released for re-use or for more general purposes, subject only to normal planning procedures.

The timescale adopted for the above three stages can vary over wide limits and thereby affect both the radiation exposure of personnel involved in dismantling as well as transport costs. Three general approaches to the question of timescale can be suggested viz:-

(a) immediate dismantling, ie continuous progression through Stages 1, 2 and 3.
(b) 'mothballing' plus deferred dismantling, ie progression through Stages 1 and 2 with a long delay prior to Stage 3.
(c) permanent entombment, ie progression only through Stages 1 and 2.

Immediate dismantling is the option which results in the largest initial commitment of both money and personnel radiation exposure and leads to the greatest transport problems. Safe storage or 'mothballing' may involve a period of continuing care of up to 100 years. Since the activity level due to ^{60}Co will have decayed by approximately 10^{3} in 50 y and approximately 10^{6} in 100 y, this delay will significantly affect the decommissioning strategy. Permanent entombment is unlikely to be attractive because reactors have, in general, not been specifically built on disposal sites and there is likely to be a shortage of sites for new power plants. The ^{94}Nb and ^{59}Ni activities in power reactors which have operated for about 30 y will be above conditional release levels for very long periods as these radionuclides have half-lives of about 20,000 and 80,000 y respectively.

3.3 Inventory Estimation

In studies related to LWRs, neutron activation calculations have been performed using the computer code AKAT (1, 2) only n-gamma and n-beta processes being taken into account, with attention focussed on the production of ^{55}Fe, ^{60}Co and ^{63}Ni. The nuclear reactions considered and the nuclear physics data used for GCRs Reference Type 4 are given in (3). The materials compositions assumed are derived from an earlier study by Lunning (4), modified in some respects by experimental measurements (5). Account was taken of both thermal and epithermal activation and allowance was made for burn-up and complex reactions using STRUMP (6).

The results of the neutron activation calculations are too extensive to be reproduced here but a summary is given in Table I.

Table I

Summary of Neutron Activation Calculations

Reference Reactor Type	Mass of Active Waste (Mg)	Time After Reactor Shutdown (y)	Total Neutron Induced Activity (Bq)
1 (PWR)	1645	1	1.4 E17
		10	3.1 E16
		40	7.8 E15
		100	5.2 E15
3 (BWR)	1425	1	1.0 E18
		10	2.4 E17
		40	4.4 E16
		100	2.8 E16
4 (GCR-Magnox)	20070	10	3.7 E15
	(See Note 3)	100	1.8 E14
5 (GCR-AGR)	51000	10	1.4 E16
	(See Note 3)	100	2.6 E14

Notes
1. In all cases the reactors were assumed to have operated for 25 y.
2. In Tables I and II E represents the exponent. Thus 1.4 E17 signifies 1.4×10^{17}.
3. The masses quoted include concrete which may not necessarily be classed as radioactive waste. For Type 4 (Magnox) the mass of concrete is 14500 Mg and for Type 5 (AGR) it is 41200 Mg.

As has been pointed out earlier, estimation of the contamination inventory is more difficult. Available data has been reviewed and the results are given in Table II.

Table II

Summary of Contamination Inventory Estimates

Reference Reactor Type	Estimated Total Contamination Inventory (Bq)
1 (PWR)	1.1 E13 - 1.1 E14
3 (BWR)	approximately 3 E14
4 (GCR)	approximately 7 E11 (after 10 y decay)
5 (GCR)	2 - 4 E12

4. REVIEW OF EXISTING CONTAINERS

Implicit in the placing of a contract to develop concepts for large transport containers was the assumption that such items do not exist at present. Nevertheless it was considered opportune to carry out a search of existing designs in the United States of Amercia and Western Europe. Further information was obtained from a US Department of Energy Handbook (7).

The containers available in the USA are designed principally for road transport and for solidified drummed wastes typical of operational rather than decommissioning arisings. The designs surveyed in Europe cover a much wider spectrum. They include thin sheet metal boxes, large containers for transporting liquids, flasks designed specifically for long thin items such as control rods and fuel assemblies and even concrete disposal boxes.

General selection criteria were defined covering aspects such as size, shape, mode of handling, materials of construction, container weight and payload, impact limiters, thermal barriers, shielding, fittings and fixtures, design transport category and finning. Details of nearly 100 containers, covering dimensions, materials of construction plus details of possible payloads have been compiled and evaluated against a set of generic criteria.

Those wastes which require to be transported under the Type B category have specific activities in the range 5×10^{11} to 4×10^{14} Bq/Te. Only Type B containers with a shield thickness of at least 8 cm lead equivalent were considered. The general selection criteria were only satisfied by three designs but these gave very small payloads (at maximum 2.2 m^3).

For non-Type B packages the range of specific activities to be covered is between 10^4 and 10^{12} Bq/Te requiring a shielding thickness in the range 0.15 - 12 cm lead equivalent. Nearly 30 package designs met the general criteria. The majority had been designed to comply with the very specific requirements of particular waste streams, eg irradiated fuel, control rods or drummed waste. Few would carry anything approaching an optimum payload.

It was concluded that, whilst a number have useful features, no particular design has been developed fully to take into account

- the range of specific activities likely with decommissioning wastes

- the requirement for packing of large components

- optimisation of shielding thickness and maximisation of payload

- the possibility of developing a range of modules (although it must be noted that the LLS/LSA package proposed by the Windscale AGR Decommissioning Project is approximately half the size of the LLS/LSA package concept developed in Section 8 (8)).

- the desirability of introducing returnable shielded overpacks of good payload efficiency for higher activity waste streams from LWRs.

5. MEANS OF TRANSPORTATION

The wastes generated will be transported to national or regional land disposal sites or stores or to ocean disposal sites and are likely to involve road, rail and/or waterway. For planning purposes the following overall average speeds have been assumed:-

Rail	60 Km/h
Road	20 Km/h
Waterway	1 Km/h

These speeds are based on practical experience of transportation and are overall values allowing for stoppages, etc.

It is not possible at present to relate decommissioning waste arisings to precise storage/disposal facilities and thence to optimise the transport arrangements. Nevertheless because the containers will be large yet need a relatively short transit time the most likely system is a road/rail combination with a short journey from the reactor to a railhead and thence by rail to the repository. Thus the study aimed to define containers with maximum compatibility for transport by road and rail within the largest number of countries in the EC.

The means of transportation have been shown to depend primarily on the following factors

- the dimensions of the load
- its weight
- the type of packaging
- the location of the power station
- the location of the final waste disposal site.

Waterway transportation is particularly interesting for undivided large and/or very heavy components as well as for large and heavy containers. An example of an ultra-large movement of this type was the dismantling and decommissioning of the reactor pressure vessel and shielding tank of the NS Otto Hahn (9). On the other hand, transport by waterway is very slow and the dimensions of locks can be a problem when special vessels are used. There may also be concern since inland waterways can be the source of drinking water. Other comments are made in Section 9 concerning radiological detriment.

Road transport is very convenient, provided that the dimensions and weights are in accordance with national regulations. But to minimise the overall cost of decommissioning, transport packages should be large and this will involve a slow average speed, expensive special truck hire and possible requirements for police escorts. The role of road transport is therefore seen as vital for short distance movements, eg between a reactor site and a trans-shipment area. There is competition between road and rail transport so that the choice of method for longer hauls is a market place decision, strongly influenced by the degree of national subsidy for a particular method.

Rail transport is interesting, providing the following conditions are fulfilled:-

- the weight loading per axle of the railway wagon is within the authorised limit (typically 20 Te/axle in European countries)

- a railhead is available. If not, a short road haul has to be performed.

- the dimensions of the load are within the appropriate railway profile gauge, taking account of the reduction in profile gauge with length.

Rail transport is the fastest of the three methods available, and it gives the possibility of moving components up to 110 - 130 Te over a long distance in a short time.

6. TRANSPORT REGULATIONS

National Legislation, Codes of Practice and International Agreements have been examined for their impact on the design of large containers. The rules are dominated by the International Atomic Energy Agency's Regulations for the Safe Transport of Radioactive Materials (IAEA Safety Series 6), the latest edition of which was issued in 1973. These Regulations are currently subject to expert review and a revised edition is expected to be published in 1985. The implications of this Revision need to be assessed. The IAEA Regulations are so comprehensively integrated into the total structure of International Agreements, National Legislation and Codes of Practice that the only viable approach is to design in conformity with Safety Series 6.

According to the mode of transport to be adopted, there are also non-nuclear limits especially on weight and size.

It can be concluded therefore that in proceeding from the conceptual to the design phase, it will be necessary to ensure conformity with IAEA Safety Series 6 (revised 1985 version), ADR, RID, Draft ADN, IMO and with the National Legislation and Codes of Practice.

7. TRANSPORT COSTS

Transport costs have been estimated for road, rail and waterway transportation using information representative of Belgium and the United Kingdom. The inventory data summarised in Section 3 has been translated into container loads of the Type B and LLS/LSA packages proposed in this report. The estimates of costs given in Table III for decommissioning a PWR Reference Type 1 and a BWR Reference Type 3 are based on the assumption that the waste will be transported 16 Km by road between the reactor and a railhead and thence by rail. The rail distances chosen (160 and 640 Km) are reference distances appropriate to Belgium and the United Kingdom and illustrate the effect of distance. The trains would be operated with, say, 7 wagons carrying full loads of the appropriate Type B or LLS/LSA containers. The costs for both reactors are similar. The major cost-controlling factor is the distance by rail (cost per journey 20,400 and 47,800 ECU for reference distances of 160 and 640 Km respectively).

Table III

Estimated Cost of Transporting Decommisioning Waste from PWR Reference Type 1 and BWR Reference Type 3

Transport Mode	Distance (Km)	Cost per Journey (ECU)	Number of Loads	Number of Journeys	Estimated Cost (rounded to nearest 1000) (ECU)
PWR Reference Type 1					
Road	16	265	106	106	28,000
then					
Rail	160	20,400	106	16	326,000
					TOTAL 354,000
Road	16	265	106	106	28,000
then					
Rail	640	47,800	106	16	765,000
					TOTAL 793,000
BWR Reference Type 1					
Road	16	265	101	101	27,000
then					
Rail	160	20,400	101	15	306,000
					TOTAL 333,000
Road	16	265	101	101	27,000
then					
Rail	640	47,800	101	15	717,000
					TOTAL 744,000

8. CONCEPTUAL DESIGN OF LARGE TRANSPORT CONTAINERS

8.1 Introduction

Decommissioning waste from nuclear power stations will vary widely in size. The container should be as large as possible, commensurate with the provision of adequate shielding whilst not presenting a major transport problem for movement by road or rail to a disposal site.

It is considered that decommissioning wastes fall into only two major transport categories - Type B and Low Level Solids/Low Specific Activity (LLS/LSA). The higher activity waste has to be transported in Type B containers which, because they are relatively expensive, are made returnable. For ease of handling, the waste will be sealed in a thin-walled steel skin before it is placed in the Type B container. The LLS/LSA container will be made of reinforced concrete which will be disposed together with the waste in a shallow burial site.

This part of the study required the development of a total packaging scenario involving prediction of component dimensions and weight as well as activity levels due to activation and contamination after various periods of decay. The components were then notionally sentenced, depending upon their assessed A_2 values*, to be packed in either Type B or LLS/LSA packages.

8.2 Initial Design Concepts

The concepts have been guided by the following features:-

(a) major items of activated or contaminated equipment, steel sections, piping etc will have been reduced in size by dismantling or cutting either locally or remotely, depending on their activity. These will be assayed by non-destructive methods and sorted into either Type B or LLS/LSA container loads.

(b) the LLS/LSA loads will be packed into a steel cage in a remote handling cell. This will then be transferred to a reinforced concrete container whose outer surfaces are protected from contamination. After insertion the voids will be filled with a matrix such as cement grout of suitably low viscosity to permit easy penetration into the interstices. Thus the container will be filled to near the top surface before the addition of reinforcing steel mesh and further grout.

(c) the returnable Type B container requires the load to be separately contained and uncontaminated. The load will therefore undergo a similar process for packing into a sheet metal container which will be sealed after grouting. The inner can, after decontamination and transfer to a clean area will be lowered into the Type B container, the shielded lid will be attached and the package checked for leaktightness. After loading, the Type B container is placed horizontally on a trunnion support frame to enable the impact limiter assemblies to be fitted.

An important constraint is the maximum package weight which can be transported. For rail transport where a 22 Te axle loading is permitted, the maximum package weight is approximately 130 Te. This is the situation for major UK routes and is the long term aim for the European railway system. However at present it is unlikely that a maximum package weight in excess of 110 Te is generally viable for the EC. The UKAEA study used a 130 Te limit because it will eventually be compatible with the European railway system and because changes between a design for a 130 Te and a 110 Te package are minimal. A typical design of a large LLS/LSA container is shown in Figure 1.

For the Type B design two square section configurations were studied initially. In one the internal width was maximised and in the other the internal length was maximised, both designs keeping within the railway size and weight constraints. The subsequent investigation was limited to the "maximum length" configuration using a cast iron basic container and two sizes of lead liner to give a total of three shielding thicknesses.

---

*The method of determining the A_2 values is given in paragraph 404 of IAEA Safety Series 6 (1973 edition).

8.3 The Type B Container

Of the available materials of construction for the outer containers, only cast iron, cast steel and forged steel are worthy of consideration. Although lead is an efficient shielding material, its low strength and relatively low melting point confine its use to inner shielding liners.

Techniques have been developed in the Federal Republic of Germany enabling cast iron to be the most competitive choice. It is therefore proposed that the Type B container should be made in spheroidal graphite cast iron conforming to DIN Specification 1693 grade GGG 40.3.

An alternative is cast steel which has better impact properties and which has been developed in the UK particularly for use on North Sea oil rigs. The material is able to be welded, permitting repair of faulty castings and giving the opportunity to attach features such as lifting trunnions.

The third option is steel forging, which gives optimum strength and impact properties by eliminating all casting defects from the material and improving the metallic grain structure during hot working stages.

Indicative costs suggest that a cast steel vessel would be 20% higher in cost than spheroidal graphite cast iron with a steel forging 30% higher again.

The basic container consists of a square hollow section body with a bolted lid and end, all of 100 mm thick cast iron construction. Four forged steel trunnions are bolted to the body for handling. Figure II shows the outer container together with a typical inner basket. Increasing shield thickness is provided by lead liners. The baskets to suit each cavity size would be of similar construction and all the lead liners would be clad in 10 mm thick steel plate.

A similar approach has been taken in a parallel study of a Type B design, reference TNB 0154 illustrated in Figure III.

The design and manufacture of large containers for decommissioning waste, particularly for those parts of LWRs which might require Type B packaging could occupy several years to bring to fruition.

9. RADIOLOGICAL DETRIMENT

9.1 General Evidence

The radiation doses received by the public due to the transport of radioactive material are very low and mostly ascribable to the transport of radionuclides for medical purposes. A recent national survey by the UK National Radiological Protection Board (NRPB) (10) indicates that only 15% of this dose is due to transport related to the nuclear fuel cycle. The additional effect of transporting decommissioning waste can therefore, on general grounds, be predicted to be small.

The same NRPB survey found that the average dose to occupationally exposed transport workers was 2×10^{-2} man S_v (2 mrem) ie, nearly 100 times lower than their natural radiation dose.

9.2 Radiological Detriment from Normal Transport of Radioactive Waste

An estimate has been made of the average health impact from moving typical loads of decommissioning waste. It was envisaged that the decommissioning waste would be transported principally by rail with only a small fraction of the route by road to connect with the railhead, ie a similar situation to that explored in Section 7. The model studies chose a nominal reactor site and then the health impact to the public was assessed using available transport routes which were representative of the average situation in the UK. Pessimistically each of the reference design concepts of container was assumed to be filled to its limiting radiological capacity (ie, filled to the external radiation limits permissible under the IAEA Transport Regulations). A starting assumption was made that the repository could be due north, east, south or west of the reactor. The direct distance from the reactor to the repository was assumed to be 160 or 640 Km but the actual distance was constrained by the existing transport system. By use of the appropriate UK census data to give the population density, the health impact to the public could be evaluated. The increased risk from the transport of decommissioning waste (see Table IV) is relatively insignificant compared with risks of normal life (eg, driving a motor car). The slower speed of waterway traffic leads to a higher estimated health impact: for the 160 Km northwards case the risk is estimated to be 10 times greater.

Table IV

Health Impact of Waste Transport to the Public per Container Movement (using Road and Rail)

Direct Distance and Route Direction	Collective Risk ($\times 10^{-8}$ Health Effects)*: Collective Risk from Moving the Container – Road Section	Rail Section	Collective Risk from Container Stationary	Total Collective Risk
160 Km				
North	0.2	18	66	84
East	0.5	9.6	55	65
South	4.6	10	85	100
West	1.5	16	29	46
640 Km				
North**	0.5	39	141	180

*Risk Factor = $1.65 \times 10^{-2}\ S_v^{-1}$ (11)

**East, South and West are not physically possible for the UK situation studied.

It should be noted that since the external dose rate assumed in the model calculations is the limiting value for transport it is obviously pessimistic. Wastes will have a spectrum of activity levels and only a relatively small number of container shield thicknesses will be used so that the actual dose rates will normally be significantly less than the limit values assumed.

10. CONCLUSIONS

The following conclusions have been drawn from the work performed under this contract:

- For chosen types of European nuclear power stations (LWR, GCR) the quantity of radioactive solid waste has been estimated to be as follows, 10 years after shut-down:

	LWR	GCR (with prestressed concrete pressure vessel)
Steel	4600 - 8700 Mg	8000 Mg
Concrete	200 - 400 Mg	42000 Mg (but see Table 1 Note 3)
Graphite	-	1500 Mg

The mass ratio of activated to contaminated waste is of about a factor 5 (for LWRs).

- The most appropriate transport mode taking account of radiological detriment, costs, etc) should be a combination of road and railway transport. Waterway permits heavier loads but the average transport speed is slow leading to estimate collective doses which are 5 - 6 times higher.

- An appropriate container system for decommissioning waste should include large disposable LLS/LSA containers of reinforced concrete and Type B (U) returnable containers made of cast iron or cast steel. The limiting design criteria are derived from railway transport constraints.

- With regard to the IAEA Transport Regulations (Safety Series No 6) there is a sensible need of a intermediate transport category. The Type B (U) container would be used for only a small quantity of decommissioning waste (10 - 15% of the total waste).

This system study has formulated the basic conceptual characteristics of large containers suitable for the transport and/or disposal of activated steel, concrete and graphite. Such containers could be commercially produced in the light of current national and international regulations and codes of practice. The health impact to the public is estimated to be relatively insignificant compared with other risks of normal life.

The next step should include the development of such a container system (manufacturing specifications, quality assurance documents, fabrication and testing under real transport and waste disposal conditions (land storage and sea dumping).

11. RECOMMENDATIONS FOR NEXT FIVE YEAR PROGRAMME OF THE CEC

As a result of this study it is recommended that large transport and/or storage containers should be developed. The performance of the matrix/container system under water contacting conditions representative of reference waste repositories should be studied. The control methods for verifying the suitability of the containers for land storage, sea disposal, transport, etc, will need to be considered according to the specific technical requirements for these different uses.

The concepts should be taken to the point at which detailed manufacturing drawings, costs, specifications and quality assurance documents are available to the EC. This task involves optimisation of the shielding thicknesses as well as the preparation of scale models for the performance of regulatory tests. The effects of the changes in the definitive revision of IAEA Safety Series 6 (to be published in 1985) will also need to be taken into account.

In addition studies should be made of the balance of advantage from using ultra large non-standard transport containers and of the relative advantages of detachable and non-detachable shielding for the LLS/LSA type of large container.

12. ACKNOWLEDGEMENTS

The authors would like to thank their colleagues in the four participating organisations, ie Messrs L Dierckx (TNB), P Deramaix (BN), J Demeester (BN), A Emmerton (UKAEA), G R Jamieson (UKAEA), N Berry (UKAEA), B Wright (formerly PPC), A Curran (PPC) and R E Holmes (PPC) all of whom contributed significantly to this study. In addition many organisations gave important assistance particularly ANDRA, CEGB, COGEMA, GNS, HOESCH/ESTEL, NRPB, RWE, SEMO, TRANSNUKLEAR.

13. REFERENCES

1. Stillegung von Kernkraftwerken mit Leichtwasserreaktoren. EUR 5728 d (1977).
2. Bardtenschlager, R et al. Decommissioning of LWR Nuclear Power Plants. Nuclear Engineering and Design, 45, 1, (1978).
3. WOOLLAM, P B and PUGH, I G. The Radioactive Inventory of a Decommissioned Magnox Structure, Part 2. CEGB Report RD/B/R4350 (October 1978).
4. LUNNING, W H. Private Communication (1975).
5. WOOLLAM, P B. CEGB Report RD/B/N 4231 (1978).
6. NAIR, S. CEGB Report RD/B/N 4291 (1978).
7. US Department of Energy. Surplus Facilities Management Programme - Waste Containers for Decommissioning. Report DOE/RLO-SFM-82-6.
8. LAWTON, H. Atom, November 1982, p 228.
9. VIECENZ, H-J, OBST, J, KRIEGER, F and LETTNIN, H K J. Proc 7th International Symposium on Packaging and Transportation of Radioactive Materials, PATRAM '83, Volume II, p1477.
10. SHAW, K B, MAIRS, J H, GELDER, R, HUGHES, J S and HOLYOAK, B. ibid, Volume I, p 705.
11. National Radiological Protection Board (UK), Living with Radiation (1981).

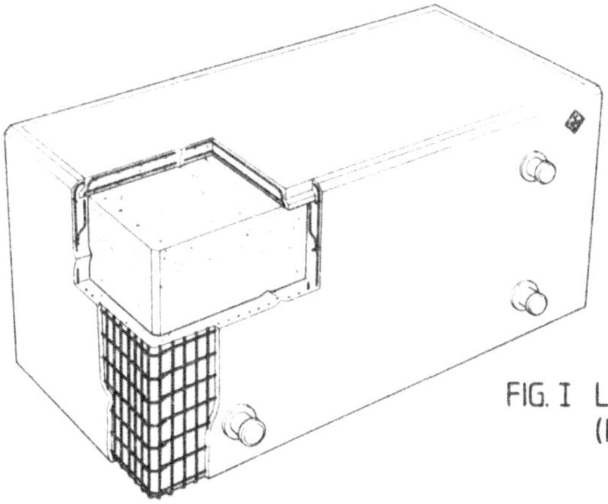

OVERALL LENGTH - 4600mm
OVERALL HEIGHT - 2300mm
OVERALL WIDTH - 2300mm
LADEN WEIGHT - 125 TONNES

FIG. I LARGE LSA CONTAINER (REINFORCED CONCRETE)

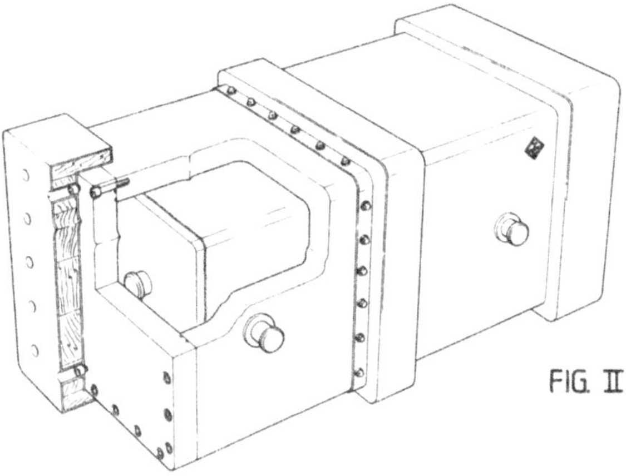

OVERALL LENGTH - 5000 OR 7000mm
OVERALL HEIGHT - 2400mm
OVERALL WIDTH - 2400mm
(INCLUDING IMPACT LIMITERS)
LADEN WEIGHT - 130 TONNES

FIG. II LARGE TYPE B CONTAINER (CAST IRON)

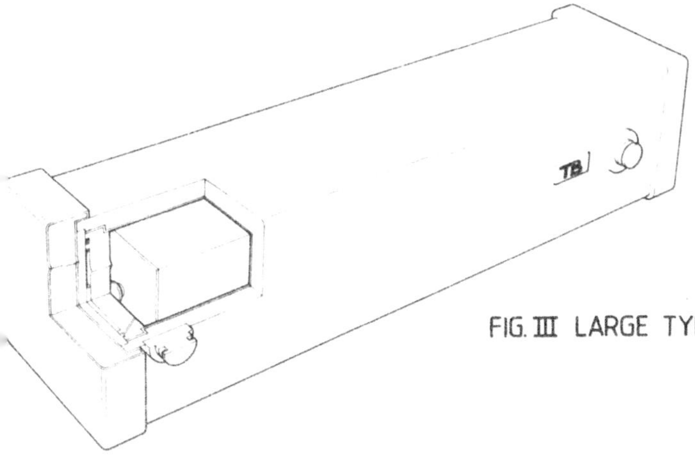

OVERALL LENGTH - 6900mm
OVERALL HEIGHT - 1700mm
OVERALL WIDTH - 1700mm
LADEN WEIGHT - 100 TONNES

FIG. III LARGE TYPE B CONTAINER (CAST IRON)

SESSION IV : SUMMARY OF DISCUSSION

Paper : "Immobilization of contamination on metals and concrete"

The Chairman, after having thanked the authors for the qualified presentations, would like to know what work has been done to engineer the concrete dust immobilization. Mr. Davies, co-author of the paper, replied that a complete system for solidiying the dust industrially has been looked at; this has been mainly a paper exercise but, as part of the process, TWC looked at a number of available machines such as agglomerator, pelletizer and compressor, producing solid pieces easy to handle. None has proved technically preferred. Leachability of the product was checked on samples and the leachate was found to be silicates and not calcium, which would contain the radioactivity. Further investigation on these machines should be made in order to define the most suitable system.

Paper : "Treatment of metal waste by melting"

A U.K. participant asked if BSC intends to set up a commercial operation to recover steels. Mr. Gomer replied affirmatively, providing the industrial relations aspects can be settled. A participant representing an international organization fears that the transfer of radiocesium to the slag could cause more disposal problems. This seems not to be the case at BSC where, up to now, the slag can be freely disposed of.

A U.K. participant asked for the quantity of Co-60 which could be absorbed per year by re-cycling activated material back to the industry; and what is the impact of the longer lived nuclides by this type of re-cycling, and should re-cyling not be restricted to high alloy steels?

Mr. Gomer thinks that up to 100 Ci/year of Co-60 could be absorbed in bulk steels of BSC but the greatest application should be in high alloy steels with perhaps a prior decontamination. The most important aspect of the melting process will be the homogenising of the radioactivity concentrations. In replying to a question of another U.K. participant - what fraction of the radioactivity goes up the stack - Mr. Gomer believes that there are only small amounts, but some is retained in the furnace system and in the off-gas filters with perhaps some preferential transfer to slags.

Paper : "Large transport containers for decommissioning waste"

Upon comments by two participants, Mr. Price explained that no accidents were defined and investigaged but the packages were designed to IAEA standards. The large volumes and weights of the packages described may cause problems for deep geological disposal (particularly in countries where no shallow burial is envisaged. This study concentrated, in fact, on transportation; however, one of the report's recommendations is that the repository constraints should be looked at. It should also be noted that not using the shallow land option would put considerable logistic and economic restraints on disposal.

TECHNICAL PANEL

REQUIREMENTS AND POSSIBILITIES FOR THE DISPOSAL OF DECOMMISSIONING WASTE IN COMMUNITY COUNTRIES

Chairman

S. ORLOWSKI
Head, Nuclear Fuel Cycle Division, Directorate-General for Science, Research and Development and the Joint Research Center, Commission of the European Communities, Brussels.

Secretary

K. PFLUGRAD
CEC, Directorate-General Science, Research and Development

Members

M. CONTI
Responsabile del progetto "Smantellamento degli Impianti Nucleari", Comitato Nazionale per la Ricerca e per lo Sviluppo dell'Energia Nucleare e delle Energie Alternative, Casaccia, Roma.

E. DETILLEUX
Directeur Général de l'ONDRAF / NIRAS.

K. HUBENTHAL
Referat Wiederaufarbeitung und Brennstoffrückführung, Abteilung Kernenergie im Bundesministerium für Forschung und Technologie, Bonn.

A.N. KNOWLES
Central Technical Service, United Kingdom Atomic Energy Authority, Risley Establishment.

J. LEFEVRE
Directeur chargé des Effluents et Déchets Radioactifs, Commissariat à l'Energie Atomique, Paris.

F.J. TURVEY
Irish Nuclear Energy Board, Dublin.

CONTRIBUTIONS BY THE PANEL MEMBERS

Introduction by the Chairman

After presentation of the participants at the Panel, the Chairman invited them to briefly present the situation in their countries related to the requirements and possibilities for the disposal of decommissioning waste.

A.N. KNOWLES

Apart from the decommissioning of the Windscale AGR (we have not heard much about the Windscale AGR at this conference because, of course, it has not featured in the first 5-year Shared Cost Programme) and the Dounreay Fast Reactor, the arisings of decommissioning waste in the UK will be on a small scale until the first of the civil Magnox stations is decommissioned. The Central Electricity Generating Board, which owns eight Magnox stations, has recently reviewed their decommissioning. The oldest of these stations, all of which are operating efficiently and at lower generating cost than comparable coal fired units, are progressively being reviewed against modern safety standards with the objective of extending their life in principle to 30 years. Already Berkeley and Hunterston (owned by the South of Scotland Board) have been so reviewed and similar life extensions seem likely for all the others. Therefore, apart from Windscale AGR and the Dounreay Fast Reactor, major decommissioning waste arisings will probably not occur in the UK until after the year 2000. In fact, it could well be after the year 2000, since it is likely that some delay to allow the decay of radioactivity will occur between Stages II and III, as mentioned by Mr. Gregory on Tuesday.

When Civil Reactor decommissioning does take place, there will be a major surge in the national annual arising of low level and intermediate level waste. It is expected that the annual increase for each of, say, four years would amount to 10% for each of the early stations and more than this for the later stations which are larger. In other words, in the context of a single nation such as the UK, decommissioning of power reactors is essentially a discontinuous operation giving rise to temporary large increases in the annual waste arising.

Long before all this happens, we will have explored the regulatory and disposal situation by carrying out the complete demolition of Windscale AGR. Sixteen thousand tons of waste materials will arise, but of this only 2000 t will be active. The active waste will be about 2/3 low level and 1/3 intermediate level. Most of the low level waste will go to shallow land burial and the intermediate level material will be packed in large reinforced concrete boxes of the type described in the UKAEA/-Transnubel paper presented this morning (in fact, the half module as mentioned by Mr. Price). So far, no definite disposal plans have been made for the 150 or so boxes involved : one feasible option is to store these on site pending the completion of an underground storage facility in the UK.

This action, in connection with the Windscale AGR, is regarded as being important and valuable as it will ensure that all the various considerations covering packaging, storage, transport and disposal are explored in advance of the decommissioning of the civil stations. These civil stations being about ten times larger than the Windscale AGR, will require a major national effort in waste handling, quite apart from the several other important aspects of decommissioning.

F.J. TURVEY

Ireland does not have a nuclear power programme and at first sight would seem to have no reason to be interested in the disposal of decommissioning waste. But Ireland does have an interest and I would like to explain why.

We know already from experience that a problem with the disposal of radioactive waste in one country can result in the exposure of the public nuclides in another - migrating radionuclides do not respect frontiers or other national boundaries. Although such exposures in Ireland are very low, they achieve greater significance because those exposed receive no tangible benefit from nuclear power.

When we look at the massive quantities of intermediate and low level waste to be disposed of from decommissioning reactors in the years to come and when we also look at some of the alternative methods being suggested (I refer here in particular to sea-dumping of solid wastes, burning of graphite and re-cycling) quite naturally Ireland is concerned and we therefore have taken the opportunity presented us of participating in the European Communities' Programme which has, as one of its aims, the safe disposal of decommissioning wastes.

Finally, I should like to add that I do not believe that technical problems will impede the disposal of decommissioning waste, but I do believe that in many countries non-technical problems will greatly delay the progress of decommissioning to stage 3; in fact, I see that the option of progressing to stage 3 without delay is, for many, not a realistic one.

J. LEFEVRE

I shall address three aspects of the question :

- organization and policy;
- technique;
- regulations.

Although it has some special characteristics, the management of waste arising from decommissioning activities does not differ in any essential respect from that of waste arising during operation.

Organization and policy in France :

The studies are conducted jointly within the same budgetary framework and by the same technical units. It should be pointed out that the border-line between "decommissioning" and "waste" was drawn between the dismantling techniques, specific to decommissioning and the decontamination techniques, which give rise to liquid waste during dismantling. Management is also to be viewed from the same standpoint, because the nuclear operating bodies, producers of waste arising during operation, are also responsible for decommissioning operations. Within these great bodies, special structures may be set up such as those at the Commissariat à l'Energie Atomique (CEA), where a specialized unit has been created for the decommissioning of the CEA installations. Finally, ANDRA is responsible for storing the waste arising from decommissioning, as well as storing all other types of waste. The decommissioners thus benefit from a waste management policy for most of their waste.

From a technical standpoint, decommissioning gives rise to the following forms of waste :

- Liquid waste arising from the decontamination operations. The study of decontamination procedures must be conducted with a two-fold objective in mind : the effectiveness of the decontamination and the subsequent treatment of the liquid waste, which will have to be placed in final storage in solid form.

- Solid waste originating from components of decommissioned installations. This, of course, is the main point and the one giving the most characteristic specific aspects. In the case of the high activity waste, the usual techniques will be applied. I may be repetitive, but the three types of waste which are produced in great quantities by decommissioning and in respect of which a special effort must be made are : metals, concrete and graphite. I shall dwell no longer on this topic, since it is the subject of this conference. It is useful to point out, however, that we have to deal with large volumes of low-activity waste and that, in consequence, volume-reduction techniques (melting down the metals, incineration, crushing, etc.) should be developed with priority.

- As regards final storage, special measure will have to be taken, for example the use of crushed low-activity concrete and of metals recovered after melting down as a constituent material of surface storage constructions. In France, the estimates for waste include arisings from dismantling operations up to the year 2000, knowing that for this period waste arising should still be quite small.

Regulatory aspects :

It seems to me that one important point should be stressed : the determination of one or more threshold activity values below which it will be possible either to re-cycle the waste (for example, the metals to iron works) or, by removing all restrictions on the waste, to dispose of it in public facilities. This point will have to be seriously studied both at national and at international level.

E. DETILLEUX

There are no medium-term plans for any dismantling operations concerning power reactors (seven units in operation in 1985-86) or experimental reactors (four units). With regard to reprocessing plants, some dismantling work has been carried out at the Eurochemic plant in Mol. These activities will continue on a fairly large scale over the next few years as part of the plans to resume operations, provided that the relevant decision, which is still under consideration, is adopted in the coming months. If it is decided that reprocessing activities should not be resumed, the installation will be dismantled. In recent years, laboratories and installations used for the production of radium have been dismantled. Belgium has, therefore, acquired a certain amount of experience in the dismantling of industrial installations other than reactors.

After they have been conditioned, the wastes resulting from dismantling operations fall into the same categories as those resulting from normal operation. The following means of disposal have therefore been either used or envisaged :

- Low-activity alpha, beta and gamma wastes and medium-activity non-alpha wastes. Until 1982, such wastes were dumped in the North Atlantic under the supervision of the OECD Nuclear Energy Agency (NEA). This type of disposal has been suspended for the time being. In view of the difficulties involved in dumping at sea, investigations are now being of the possibilities of shallow burial in this country.

- High-activity and alpha wastes. In order to solve the problem of disposing of waste of this type resulting from the reprocessing of irradiated fuels, an extensive programme is being conducted in collaboration with the European Community and has now been in progress for over ten years.

It involves evaluating the potential of clay beds, in particular the one underlying the Mol site. If necessary, disposal of the wastes produced in dismantling operations could also be considered under this heading.

- Wastes only slightly contaminated and non-active wastes. No waste produced during any disassembly or dismantling operation (including the work on the radium installations) has ever been placed in a conventional industrial repository. The problem has arisen everywhere and the only solution adopted so far has been to store the material on the site of a nuclear installation provision being made for radiological monitoring. The matter will be dealt with under the heading of the general waste management policy, due to be drawn up by the public body responsible for waste management (ONDRAF-NIRAS), which has recently been given the task.

As I have just said, the general policy as regards radioactive wastes is currently subject to revaluation and will necessarily cover the question of the primary and secondary wastes generated by dismantling work. Waste decontamination and conditioning techniques are being examined in detail and consolidated by the operators concerned, together with the authorities responsible for the relevant regulations and monitoring procedures and ONDRAF, the national body responsible for the final disposal of wastes.

K. HUBENTHAL

In the Federal Republic of Germany the three stages of decommissioning, final shut-down, entombment and complete decommissioning have to be licensed according to the German Atomic Act. The safety criteria, established by the Minister of Interior in 1977, requires that a decommissioning concept should be produced before getting the license for decommissioning operations, even before starting up a power plant. The logical consequence is that the appropriate waste management mode has to be decided on a case by case basis within the license. There are only two possibilities in the Federal Republic of Germany for the disposal of waste : the Asse salt mine, closed since 1978, and the Konrad iron mine. Actually, we are planning to use the Konrad iron mine for the disposal of radioactive waste under the responsibility of the Federal Government. Waste which does not go into the final repository, if it is below certain limits, could either be disposed unrestrictively or, under certain conditions, might be recycled. A condition for the disposal of waste in the Konrad mine is the use of containers with dimensions of meters 3 x 1.6 x 1.6 (much lower than the containers presented this morning in Session IV) and a total weight of 20 tons. The amount of radioactive waste coming from decommissioning will be rather limited, because the number of reactors to be decommissioned in this century is very low. Otherwise, in a very rough estimate we found that, by decommissioning one reactor we will create about the same amount of waste we have from all the other sources in one year.

Work practically done in Germany on decommissioning is the following: in the early seventies, the Alkem production facilities for mixed oxyde in Karlsruhe have been decommissioned, radioactive waste has been brought to the Asse mine in 200 or 400 liter drums, with or without concrete shielding, according to the content. The nuclear ship Otto Hahn has been decommissioned; this has been mentioned here before. The nuclear part of the hot steam reactor HDR at Grosswelzheim has been practically taken out and the waste brought to the Asse mine; the concrete structure was used for safety investigations.

For the decommissioning of NPPs to stage 3, dividing the waste into three categories is envisaged :

- waste which is considered to be non-radioactive;
- metal scrap which up to certain activity limits will be re-cycled through melting;
- and radioactive waste to be disposed in the Konrad mine.

The number of nuclear power plants to be decommissioned in this decade in Germany is quite small : KKLs, Lingen, KRB Gundremmingen and MZFR Karlsruhe.

I think that the problem is not urgent, as far as time is concerned. The public acceptance is also to be solved. I also think that the lack of final repository in Germany and the uncertainty of successful licensing procedure, are more hindering to go to stage 3 of decommissioning, than the lack of technology to do that.

M. CONTI

There is no great urgency concerning the disposal of decommissioning wastes in Italy today. Most of the installations being decommissioned are small zero-power research reactors, which have generally been in operation for only relatively short periods.

One or two other research reactors may be dismantled at a later stage, but these will be fairly small, having a capacity of not more than 5 MWth. Since, unlike many other European countries, Italy does not have any medium-sized reactors (e.g. 40-50 MWth); the reactors to be considered for decommissioning are the 160 MWe BWR at Garigliano, the Magnox reactor at Latina (160 MWe) and the Trino Vercellese PWR (270 MWe), the two latter being near the end of their useful lives. They have been in operation for almost 20 years. In the short term, therefore, Italy's decommissioning wastes will clearly amount to only a few dozen or a few hundred tonnes of low-and very low-activity solid material, although the amount will increase to thousands of tonnes when the above-mentioned power reactors are dismantled. Nevertheless, the problem of medium- and low-activity wastes in beginning to outgrow the experimental phase and gradually assume industrial proportions.

Above all, there is the need to cope with substantial quantities of wastes produced during normal operation which cannot be stored for long periods on the sites of nuclear installations. Italy can cope with the management of low-activity wastes because it has not only the technology required for treatment and conditioning, but also temporary storage sites and an industrial-level operational structure (the firm NUCLECO). The problem of acquiring one or more sites for permanent storage remains, however, unresolved.

The availability of one or more solutions to this problem would complete the management cycle and would provide basic references for establishing technical specifications for conditioned waste products.

The ENEA has conducted studies to identify suitable sites for shallow geological burial. The results indicate a range of appropriate solutions : shallow burial (trenches) and disposal in underground cavities like chalk, clay and salt formations. The survey covered all parts of Italy and was based on strict hydrological, lithological and geological parameters, which can be summarized as follows :

- the material has to be fairly loose and highly permeable;
- the site must have a deep ground-water table in order to prevent any prolonged contact between water and waste, even during occasional rainy periods or when there is any build-up of underground water;

- there has to be a reasonable possibility of ion exchange with any radionuclides that become mobile.

Thus the situation in Italy can be described as follows :

- the quantity of decommissioning waste is relatively small and does not, neither for the time being and nor in the near future, constitute a problem;
- in the long term, the problem will become much more serious because three first-generation power plants of about the same age will probably be dismantled;
- the technology required for the management of low-activity waste has been acquired;
- the ENEA has studied suitable sites for final disposal;
- the ENEA is stepping up its activities as regards the qualification of conditioned products and of shallow geological formations;
- government bodies will have to take appropriate steps to allow the sites to be purchased.

Conclusion by the Chairman

The Chairman summarized the presentations as follows :
The question of decommissioning waste does not represent an urgent management problem because the volume of such a waste would be limited to a small percentage of the total nuclear waste produced up until the next century.

It must be noted that all EC countries are concerned with the problem because all of them have to protect health and environment. This common concern, the level of which depends of the national nuclear commitments, is the basis of the present R&D Community action. It should be also pointed out that all types of nuclear installations are being operated industrially within the European Community and will therefore contribute to generate various types of decommissioning waste, including long lived waste (alpha, beta, gamma). However, most of the decommissioning waste will be low level solid waste (steel, concrete, graphite) and therefore, as said by Mr. Lefevre, the definition of free release limits for decommissioning waste would be an important step to reduce drastically its quantity.

Dismantling and decommissioning waste treatment and conditioning need the development of specific techniques, in particular within the CEC R&D programme on decommissioning. This programme has to be considered in relation with the CEC radioprotection programme and the CEC radioactive waste management and storage programme, as far as radiological questions, waste management strategies and disposal options are concerned.

The research programme on radioactive waste management especially constitutes a frame and support for the decommissioning programme, particularly concerning the development of final disposal repositories.

In conclusion, the Chairman stressed that all solutions which could be found by common effort (as already said here before by Mr. Hübenthal) should be better accepted either by the authorities or by the public. He then invited the floor, as well as the panel participants to raise questions.

TECHNICAL PANEL ON REQUIREMENTS AND POSSIBILITIES FOR THE DISPOSAL OF DECOMMISSIONING WASTE IN THE COMMUNITY COUNTRIES

SUMMARY OF DISCUSSION

A U.K. participant asked if the CEC should not press to keep the Sea-Dumping option open. The Chairman replies that from the Commission's point of view, decisions on management of radioactive waste have to be taken on a scientific basis. Consequently, despite a recent European Parliament's resolution to banish Sea-Dumping, the Commission waits for the results of the reviews being carried out by the responsible international bodies, like the OECD/NEA's coordinated research and environmental surveillance programme relevant to dumping of radioactive waste at the current site, and the IAEA current re-assessment of the permissible radioactivity to be disposed of, before taking position.

An FRG participant asks if a second shallow land burial facility will be developed in France, anticipating loading to full capacity of the "Centre de la Manche". Mr. Lefèvre answers that, as the facility is loaded only to half of its capacity, there is enough time to arrange for a new site. Decisions of the French Government on the development of new repositories are expected soon.

A French participant underlines that looking for solutions for disposal of decommissioning waste is urgent. At the moment, a political decision on a strategy has been taken, the time needed to solve the technical problems could easily take ten years. He thinks that even if the contrary has been said before, during the panel, it is not too early to start solving these problems now. Mr. Knowles supports this point and adds that a real decommissioning exericse is therefore carried out in the U.K., so that technical, political and regulatory problems can be solved.

A participant representing an international organization expresses his opinion that decommissioning waste must be placed in the same context as other industrial wastes; even non-nuclear waste could benefit from radioactive waste management. The Chairman adds that in the European Community countries about 40 MTe of highly toxic chemical waste has been produced in 1983, a quantity which has to be compared to about 1000 m^3 of high level radioactive waste to be expected for the year 2000.

A U.K. participant states that threshold values and "de minimis" values are important but asks what form these limits should take, e.g., a general limit or specific limits for specific isotopes and materials. Mr. Lefèvre means that the threshold value could not be a single value, but has to be modulated for differing isotopes and materials. Mr. Turvey stresses that the dose to individuals and populations counts and that the limits should be assessed on a case-by-case basis. It is important to get a repository and then tailor the waste to go there.

On a question by an FRG participant, asking why repository requirements for final disposal were not a subject of the Conference, the Chairman points out that this field is covered by the second CEC research and development programme on radioactive waste management and storage, including experimental investigation of repositories for final disposal; the results will be presented at the second radioactive waste management Community Conference to be held in April 1985, in Luxemburg. Three experimental underground facilities (in Belgium, Germany and France) for final disposal are foreseen to be tested in the framework of the third five-year programme on radioactive waste management of the CEC. This is the reason for not discussing repositories and general waste problems at this Conference.

SESSION V

DESIGN FEATURES FACILITATING DECOMMISSIONING OF NUCLEAR POWER PLANTS

Chairman : F. MOTTE, CEN/SCK

Secretary : K.H. SCHALLER, CEC, Directorate-General Science, Research and Development

Introduction by the Session Chairman

Concepts aimed at minimizing the activation and contamination of concrete

Reduction of exposure to radiation by substitution of Co alloys and reduction of the concentrations of Co, Nb and other companion elements in steel

Review of plant design features facilitating decommissioning

Summary of discussion

Concluding Panel : Decommissioning of nuclear installations - an outlook

- Contributions of the Panel Members
- Summary of discussion

INTRODUCTION BY THE SESSION CHAIRMAN

F. MOTTE
CEN/SCK, Brussels, Belgium

The designers of nuclear power plants are interested in designing installations which perform reliably and well, and thus have low downtimes, which offer all possible guarantees of safety for their personnel and the environment, and which provide a sufficiently attractive return on capital outlay.

Any suggestion that other requirements should be borne in mind in addition to these primary considerations produces a reaction which is initially rather negative. This attitude was fairly widespread when the subject of this session was first raised : it would not be easy to persuade designers to consider ways of easing the tasks of those who would be called upon to dismantle their nuclear power plants 50 to 100 years hence.

The sceptics' usual line of argument is that not only will dismantling methods depend on a social, political and economic context which might be very different from today's, but there will be enormous developments in technology, and the provisions we adopt a century in advance to facilitate the dismantling of an installation are quite likely to prove to some extent inappropriate.

While I would not suggest that they are completely wrong, it should be borne in mind that it is becoming necessary in several countries to allow for dismantling in the designs of nuclear power plants and that, in accordance with an approach which, although slightly different, is nevertheless indicative, many countries are introducing legal guarantees covering the owner's solvency with regard to dismantling operations.

On a more fundamental level, it soon emerged from the preliminary studies on design requirements to facilitate decommissioning that the general objectives pursued - namely to minimize the time and cost of dismantling and the radiation dose and volume of waste produced - are such that concrete provisions are introduced which make it easier to maintain and repair the installations in a power station and which therefore contribute directly towards improving its availability, and thus its profitability.

It should be added, for the sake of completeness and objectivity, that it is clear that constant improvements in the design of nuclear power plants aimed at better technical performance also contribute, in some cases very directly, towards facilitating dismantling. To take an example already referred to, there have been steady improvements in the sealing of fuel claddings and steam generators in PWR's; as they greatly limit the transport of radioactive products within the power station, such improvements reduce contamination, thereby making dismantling easier.

Before concluding these brief general remarks on the subject of this session, I would like to make the point that the nuclear industry must be the only sector of industry which is concerned with dismantling its installations before they are even built. This displays a high degree of social awareness, and it is to be hoped that a similar attitude will be adopted by other sectors, in particular the chemical industry, since it seems to me that politicians and the media focus their attention all to rarely on waste management in that sector.

The progress achieved to date under ten research contracts signed by the Commission with ten research firms or bodies will be outlined in the three reports to be presented during this fifth session.

The first, entitled **"Concepts aimed at minimizing the activation and contamination of concrete"** - concrete providing the reactor's biological shield - summarizes the results of four research contracts signed with the following :

- Commissariat à l'Energie Atomique, CEN Saclay-France;
- Dyckerhoff und Widmann Aktiengesellschaft, Entwicklungsabteilung, Munich, Federal Republic of Germany;
- Electrowatt Engineering Services Limited, United Kingdom;
- Agip Nucleare, Italy.

Summary

These contractors have been investigating the activation and contamination of the biological shield with a view both to minimizing personnel exposure during nuclear power plant operations, dismantling and waste disposal following demolition, and to reducing the volume of material to be treated as radioactive waste following demolition.

Reducing activation calls for greater care in the choice of materials, and the design of the biological shield could perhaps be modified by the addition of an initial "sacrificial" wall of a neutron-attenuating and absorbing material other than concrete. Reducing contamination calls for research on the features required for protective linings for concrete structures which have to remain effective throughout the working life of a power plant and can then be easily disposed of.

The report will be presented by Mr Bittner of Dyckerhoff and Widmann.

The second report, entitled **"Reduction of radiation exposure by substituting cobalt alloys and by controlling cobalt, niobium and other trace elements in steel"**, presents the results of work carried out under two research contracts with :

- Kraftwerk Union Aktiengesellschaft (KWU), Erlangen, Federal Republic of Germany;
- United Kingdom Atomic Energy Authority, Atomic Energy Research Establishment Harwell, Metallurgy Division.

Summary

The aim of this research was to replace the cobalt-based alloys used in primary circuit components which need to be particularly hard-wearing (e.g. Stellite) by alloys free of cobalt but with similar properties as regards resistance to friction, mechanical wear and corrosion. A large number of such alloys - potential replacements, at least - appear to exist.

In connection with the stainless steel type 304, which is extensively used as a structural material in the primary circuit of a PWR nuclear power station, in particular in the internal parts of the reactor vessel, the authors note that, while Co-60 constitutes the main source of the radiation in the first few decades following the closure of the power station, other isotopes present in the steel as trace elements, such as niobium and silver, may become predominant sources after about 100 years.

They therefore advocate the use of a stainless steel produced using materials containing very little cobalt and trace elements. The additional cost of producing such a steel and the greater capital outlay involved would be readily offset, if only by the shorter duration, and consequent lower cost, of the safe storage period of the power station before final dismantling.

This report will be presented by Mr Wieling of KWU, Erlangen.

The third report, which is of a more general nature, has the same title as Session V, namely the **"Review of plant design features facilitating decommissioning"**. It outlines the progress achieved to date on four contracts concluded with :

- National Nuclear Corporation Limited; United Kingdom;
- Taylor Woodrow Construction Limited, United Kingdom;
- NIRA, Genoa, and the ENEA National Research Centre, Rome;
- the Technische Hochschule Darmstadt, Federal Republic of Germany.

Summary

The subjects to be dealt with bring us back to the central theme of the discussions. In the first part, the authors discuss the importance of defining dismantling strategies when nuclear installations are designed and outline the relationship between nuclear installation design and dismantling strategy. In the second section, which refers to advanced gas-cooled reactors, the authors demonstrate, on the basis of a number of concrete examples, that it is important that designers be given the task of recognizing those features of a proposed design which might be modified to facilitate dismantling, without significantly complicating construction problems, without increasing operational hazards and without affecting the economics of the project to an unacceptable degree.

In the third section the authors discuss the very concrete question of a computer-assisted documentation system providing access to any information necessary for decommissioning preparations.

This report is to be presented by Mr Regan of the National Nuclear Corporation.

CONCEPTS AIMED AT MINIMIZING THE ACTIVATION AND CONTAMINATION OF CONCRETE

A. Bittner / D. Jungwirth
Dyckerhoff & Widmann AG, München, Germany

M. Bernard / L. Gerland
Centre d'Etudes Nucleaires de Saclay, France

G. Brambilla
Agip Nucleare, Medicina, Bologna, Italy

J. Fitzpatrick
Electrowatt Eng. Services, Horsham, United Kingdom

Summary

The following two concepts have been investigated as part of the Commission of the European Communities Research and Development Programme for Decommissioning:

- minimizing the activation of the biological shield,
- removable coatings to protect concrete surfaces from contamination.

The first concept was investigated independently by Dyckerhoff & Widmann AG, Germany and Electrowatt Engineering Services (UK) Limited, United Kingdom. Agip Nucleare, Italy and Commissariat à l'Energie Atomique, France independently investigated the second concept. The object of all investigations was to determine if certain measures could be included in the design of the plant such that a reduction in waste arising and personnel exposure at decommissioning could be achieved. To study the feasibility of the first concept, designs for light water reactor plants (both BWR and PWR) were examined. Neutron flux levels were calculated using the ANISN program and subsequent computations for activation were performed using ORIGEN 2. A range of possible neutron absorbing materials was considered and their suitability for the concept determined. During research for the characterization of coatings for the protection of concrete structures, mainly materials generally available on the open market have been tested. These tests were in accordance with given standards, guidelines and specifications for contamination, decontamination, adhesion, elasticity, flexibility, mechanical and chemical resistance, easy removal after the action of active media and artificial aging.

1. INTRODUCTION

1.1 Investigations into activation

The object of the investigations carried out by Dyckerhoff & Widmann AG (DYWIDAG) and Electrowatt Engineering Services (UK) Ltd. (EWE) was to minimize the activation of the material subjected to a neutron flux in a 1300 MWe PWR. The investigations performed by EWE also included a 900 MWe BWR.

Reduced activation of the materials surrounding a reactor would facilitate access to the reactor during the operating phase, dismantling operations during the decommissioning phase, and disposal of the demolished material.

The investigations carried out by DYWIDAG primarily concentrated on the biological shield area of the Kraftwerk Union (KWU) PWR as constructed to date in Germany. Particular attention was paid to the fact that the material of interest (concrete) came from different nuclear power plants (NPPs) with the mixtures originating from various German regions and conseqeuntly reflecting the water used from different rivers.

In addition, materials other than concrete and available on the market were examined and selected for alternative concepts. Determination of the chemical elements was conducted at different laboratories using various methods including activation analysis partly extending into the ppb-range.

The work performed by EWE was an investigation into the concept of replacing the concrete in the inner biological shield with a material which has more favourable neutron attenuation and activation properties. This work has been reported in full (4).

Numerous candidate replacement materials were identified and these were compared by performing neutron attenuation and activation calculations. Flux profiles in the structures surrounding the core were determined using ANISN and activity levels in the biological shield were computed using ORIGEN 2.

1.2 Investigations into Contamination

Agip Nucleare, Italy and Commissariate à L'Energie Atomique (CEA) of Centre d'Etudes Nucléaires de Saclay, France, have carried out investigations to characterize paints, coatings, etc. for the protection of concrete surfaces against contamination. Products available on the market were investigated.

The amount of radioactive waste arising from the dismantling of contaminated NPP structures could be reduced significantly if the concrete surfaces were effectively protected against contamination. Protection against contamination during the operation of the plant could be achieved by means of temporary coatings, easily removed prior to dismantling operations.

All investigations were conducted in accordance with existing standards. The investigations revealed that the paints designated for decontamination can be divided into two groups. One group consists of paints which absorb very few activated particles, the other group consists of those paints absorbing activated material. The latter group may require an additional intermediate layer.

Special testing methods were applied for measuring tensile strength, stripping and contamination.

2. DESIGN CONCEPTS TO MINIMIZE THE ACTIVATION OF THE BIOLOGICAL SHIELD

2.1 Aim of Investigations

The aim of the study was to identify and evaluate alternatives in biological shield design that would pose fewer radiological problems at decommissioning than existing designs. Two possibilities exist for achieving this aim, namely:

- by improved control of the trace elements in steel and concrete, and
- by modified design which facilitates the dismantling operations without endangering structural integrity.

At present the biological shield in a nuclear power plant is constructed of concrete. Two main parameters lead to the activation of the concrete. These are the flux of neutrons escaping from the reactor and the presence of isotopes in the concrete with a potential to transmute to long-lived radionuclides as a consequence of a nuclear reaction. In order to reduce the activation in the concrete, one or both of these parameters must be reduced in value.

Other aspects concerning the reduction of the neutron flux in the concrete require the insertion of some neutron attenuating material in the space available between the reactor and the concrete.

Such material must be capable of being inserted in the space available, withstanding the environment, and of reducing the activation of elements in any concrete which is beyond that material to below the datum value separating radioactive wastes from non-radioactive wastes. It must not give rise to an even greater problem at the dismantling phase (for example due to activation within the attenuating material).

2.2 Construction and Design

The initial part of the work studied current trends in biological shield designs in light water reactors, both PWR and BWR. As a result of this work a PWR design and a BWR design were selected for analysis.

The two designs selected were:

- General Electric BWR/6 (900 MWe) design, (Leibstadt),
- KWU four loop (1300 MWe) design, (Unterweser, Grohnde, Busher).

Both of these power station designs have a biological shield comprising two walls. The inner wall is non load bearing and can be considered to be sacrificial.

These designs in particular were selected as a basis to investigate the concept since by simply replacing the inner wall concrete with the neutron attenuating material no structural analysis is required other than demonstrating the stability of this wall during normal operation and accident situations.

The gap between the reactor pressure vessel lagging and the biological shield is restrictive and is used for such things as ventilation (to maintain the pressure vessel and the concrete at acceptable temperatures) and for in-service inspection of the pressure vessel itself. Any proposed alteration to the gap width would require an assessment of the effects on the ventilation.

In this study it has been assumed that the external dimensions of the biological and supporting shield remain unchanged.

For the selected designs the concept involves using a suitable neutron-attenuating or absorbing material other than concrete for the construction of the inner wall.

The following load cases were considered in the design:

- load cases during reactor operation: temperature difference ($50^{o}C$); maximum conceivable accident (MCA); loss of coolant accident (LOCA); leak of the reactor pressure vessel; break of a reactor coolant pipe; quasi-static and dynamic jet forces; water pressure ($78\ kN/m^2$);
- special load cases (external incidents): vertical earthquake intensity: $\pm$ $2.4\ m/s^2$; horizontal earthquake intensity: $\pm$ $2.0\ m/s^2$;
- tilt and sliding resistance.

In addition to the investigation into the arrangement of candidate materials, the design of the structure was examined for easy removal of the biological shield during decommissioning.

A precast structure with prestressing steel on the outside has been examined. For the production of precast elements the contact construction method ("Kontaktbauweise") patented by DYWIDAG has been suggested. In contact construction 3 mm wide steel plates are inserted at the joints in order to separate the precast concrete elements during production. These plates are removed after the concrete has set. The locating surfaces of the joining concrete surfaces remain. The precast elements which may also be produced of various material layers can vary in dimension so that they comply with various requirements.

2.3 Materials

At first the concretes used for biological shields in different German NPPs and for Leibstadt were examined for their chemical composition. Separately from this examination, the individual components such as sand, gravel, cement etc. were analysed. Various laboratory tests and methods were involved in these analyses. Among the methods used were X-ray fluorescent analysis (BGR Hannover), ICP-method (Garching and Saclay) and neutron activation analysis (BAM Berlin, Imperial College London).

Although the main chemical composition of the materials can be classified fairly easily, it is very difficult to measure trace elements occuring in parts per million (ppm, 1E-6) or impurities in parts per billion (ppb, 1E-9) and having a considerable influence on the activation of the concrete. The determination of trace element impurities does not only apply to concrete but to the other materials proposed as substitutes. This means that alternative materials to concrete have yet to be examined thoroughly to determine the impurity levels.

EWE in particular have examined a multitude of different materials for their good shielding and low activation properties, but found the data on impurity levels sadly lacking.

Some of the materials examined for DYWIDAG and EWE are shown in the tables I, II, III.

2.4 Calculations

2.4.1 Analysis of attenuation performance

Attenuation calculations were carried out using the computer codes ANISN (6) and XSDRN (11) in conjunction with the data libraries BUGLE 80 (9) and CASK (10).

The computed fluxes for neutrons were reduced to three groups, namely, fast, epithermal and thermal. These three groups were defined according to the following energy bands:

by EWE (BUGLE 80)		by DYWIDAG (CASK)	
fast neutron flux:			
17,3 MeV	i=19	15,0 MeV	i=12
Ø (E) dE = S Φi;		Ø (E) dE = S Φi;	
0,82 MeV	i=1	0,550 MeV	i=1
epithermal neutron flux:			
0,82 MeV	i=43	0,550 MeV	i=20
Ø (E) dE = S Φi;		Ø (E) dE = S Φi;	
1,885 eV	i=20	1,12 eV	i=13
thermal neutron flux:			
1,855 eV	i=47	1,120 eV	i=22
Ø (E) dE = S Φi;		Ø (E) dE = S Φi;	
1E-5eV	i=44	1E-4eV	i=21

Available comparative values (5) show that the values established by other institutes are in the range of the above flux rates.

For each reactor type, BWR and PWR, a normalized source was used to compute the neutron flux shape. This was then scaled to give realistic flux levels.

The total flux at the inner boundary of the BWR (EWE) was 4.4 E9 n/cm2 s. The total flux at the inner boundary of the PWR was 5.0 E9 n/cm2 s (EWE); 1.9 E9 n/cm2 s (DYWIDAG).

The attenuation performance of various candidate materials were measured. Outputs and plots are available for both individual (fast, epithermal, thermal) and the total of n- and γ-fluxes, and local dose rate.

2.4.2 Analysis of activation of material

Activation calculations were carried out using the isotope generation and depletion code ORIGEN 2 (8). Elemental composition of different materials, including known impurities were used.

The ORIGEN 2 libraries include the elements stated by the CEC (Ag, Ca, C, Cl, Co, Cs, Eu, Fe, Nb, Ni, Sm) and the radionuclides resulting from these elements.

Reactor operating periods considered were 1, 5, 10, 20 and 30 years and the decay periods of interest were 1, 5, 10, 30, 60 and 100 years.

2.5 Results

Many chemical elements of concretes and their individual components constructed in German NPPs were determined. Individual values see table II. These results are available for the river basins: Elbe, Isar, Main, Rhein, Weser.

The chemical elements of the reinforcing bars and for stainless steel as it is applied for thermal insulations between reactor pressure vessel and biological shield were also determined; see table III. The chemical elements of other materials are also available, e.g. radiation protection concrete (Haematite, Serpentine), ceramics, steel, glass, boron carbide (B4C).

EWE in particular examined a large number of hydrogenous materials enriched with boron. These investigations range from borated water, borated wood to borated silicon plastics.

The computer results show with about 5 % water has a better neutron-attenuation effect through the present hydrogen than radiation protection concrete with only 3 % water, see fig. 3.

2.6 Comparison of Standard Alternative Results

Since no material has both good thermalising and absorption properties a design consisting of two layers was examined whereby the inner material would be a good thermalising medium and the outer material a good absorbing medium. Calculations were carried out for this two tier system for the PWR conceptual design.

EWE chose a Ricorad-layer for the inner tier and a neutron absorption layer enriched with boron behind this tier. Additionally it appears that, due to its composition the Ricorad-layer is only activated very slightly. If a concrete layer is placed behind an absorption tier this concrete layer will have little activation.

Fig. 3 shows the ratio of neutron flux to activation in the case of B4C-layer placed before or behind the concrete layer.

The activation values are proportional to the neutron flux. The smaller the neutron flux the less the activated material. A major reduction of activated material can be seen from the values of the 100 mm thick B4C-layer placed before the concrete layer, see Fig. 7. Low activation values were also achieved if a Ricorad-layer had been applied. However, data on impurity levels was not very extensive. After certain years from shut down of the reactor the activated materials are also smaller than in the case of other materials, see Fig. 6, 7.

2.7 Discussion

The investigations conducted have shown that selective materials and their specific arrangement can improve the attenuation of the neutron flux and, thus, lead to less activation of the biological shield.

The following designs stated in the order of increasing priority are possible:

- normal concrete with high water content and mainly silicate constitutes, few impurities, such as Co, Cs, Eu, etc.,
- normal concrete, as above, however only with little mild reinforcement, possibly with prestressing reinforcement on the outside,
- multilayer designs with thermalising and absorbing components,
- installation of artificially produced materials which have both thermalising and absorbing components.

Irrespective of the material composition, the investigations showed that various design arrangements may also facilitate dismantling of the biological shield. These possibilities include:

- installation of prefabricated elements, perhaps, also application of prefabricated multilayer elements,
- dimensioning of prefabricated elements in accordance with specified requirements to conditioning, transport and ultimate storage,
- use of easily removable tensioning tools such as prestressed concrete, etc.

Due to the fact that the biological shield should be designed to take up accident loads, it is still felt by EWE and DYWIDAG that part of the biological shield needs to be built of concrete. A boron carbide layer should be arranged before the concrete layer; this arrangement allows the quantity of activated material to be reduced even more than in the case of conventional designs.

3. RESEARCH AND CHARACTERIZATION OF COATINGS FOR PROTECTION OF CONCRETE SURFACES AGAINST CONTAMINATION

3.1 Scope and Investigation

The aim of the research work carried out was to find protective coatings for concrete surfaces with the following requirements:

- to assure a lasting and effective protection against contamination,
- to be easily removable, if possible, with remote control techniques in order to save on decontamination costs, both concerning the doses absorbed by the operators and the total amount of waste arising.

3.2 Materials

For several years it has been possible to find special paints on the market, which have an elasticity and flexibility such as to assure sufficient adhesion to the support in situ as well as easy removal, if necessary, with mechanical means.

These paints, which are defined as "plastic paints" or "stripping paints" are currently used in industry as a temporary protection for ferrous metals in general machinery and equipment to be transported.

In the nuclear field they could have a triple function: as stripping decontaminants, as a precautionary measure for the protection of surfaces susceptible of being contaminated and as contamination fixing agents by isolating the contamination itself.

Research on the use of removable coverings in nuclear industry has found out that they are mainly used as decontaminants both of carbon steel and concrete surfaces.

In order to achieve the present aims a protective coating should have the following features:

- safe and easy application,
- good behavior to ageing, especially in the presence of ionizing radiations,
- penetration resistance and chemical inertia with respect to contaminants,
- easily decontaminable and resistant to the corrosive action of decontaminant agents,
- easily removable before dismantling.

In order to set a testing programme for the specific characterization of removable coatings different types of stripping paints, found on the market in France and in Italy, have been examined.

3.3 Tests

The selected paints have undergone several tests on their chemical and mechanical resistance, according to the Standards and Regulations concerning protective coatings used in laboratories and in nuclear plants:

- water resistance,
- resistance to chemical agents (acids, bases, organic solvents),
- impact resistance,
- abrasion resistance.

The resistance to tensile stress of paint films stripped from the support and the stripping capacity of the paint from concrete and similar materials such as "eternit" (asbestos cement) supports have been determined by means of equipment for tensile test, allowing the recording of a load-elongation diagram (tensile stress) and a load-time diagram (stripping capacity).

For the purpose of estimating the behavior of paints in time, the tensile stress and peeling tests have been repeated on samples artificially aged by exposure to UV-rays and sun lamps corresponding to a period of approx. 2 years of outdoor exposure.

For the purpose of a nuclear characterization, the decontaminability of paints has been measured at the laboratories of CEA in Saclay and of ENEA in Rome, according to the chosen Standards or Regulations, see (12), (13).

To determine the activity of α and β rays proportional counters filled with gas are used, see Ref. (1, 12). The device consists of preamplifier, amplifier and discriminator with scales.

For determining the contamination the samples were, depending on the application of different standards, submerged in or coated with a nitrate solution containing uranium (i.e. 4,3 % U-235 for 24 h) and then allowed to react. Thus the initial value Ao has been established.

After rinsing of the sample with running water (i.e. 1 min) and subsequent drying over a period of 24 h the value A1 has been established. To determine the absorbed activity A2 the samples are then wiped by means of a decontaminating agent, rinsed again with running water and dried for 2 hours.

The percentage of the contamination absorbed has been calculated with the subsequent formula:

$Sc\ (\%) = A1 \times 100/Ao$ (12)

According to French certificates of approval this value shall not exceed Sc = 15 % for class A if the coating is to be applied in NPPs. The percentage of decontamination has been calculated by the formula:

$p\ (\%) = (Ao - A2) \times 100/Ao$ (12)

The decontamination shall not be less than p = 95 % for class 1. Specific data are stated in the applied standard e.g. Norm Francaise NFT 30901.

A3 (residual activation) indicates the activity of the sample after one week. The remaining residual activity compared to the activity after decontamination (i.e. after 2 hours) is determined by the subsequent formula:

R (%) = (A3 - A2) x 100/A2 (12)

The protective power of the coating against contamination, has been checked by recording the surface activity before and also after stripping the paint film.

3.4 Results

In general the results show that in the case of suitable paints against contamination the values of the activity filtered through have been very low, in any case smaller than 1% of the deposited activity.

Table IV gives an example of measured values. Details about examined products are specified in the respective final reports (1, 2).

3.5 Effects

The tests showed that various types of paint absorb activated material and do not release it again. Thus, you may proceed on the assumption that depending on the requirements paints

- which are good against contamination and those,
- which are used for decontamination purposes

are applied.

In the case of the most products a prime coat is applied. The prime coat is then followed by another coating which may be removed again by chemical or mechanical means without damaging the actual surface.

3.6 Discussion

The results of the tests lead to the following conclusions:

- it is preferable to apply removable coatings on concrete surfaces already previously painted, for instance with a normal epoxy cycle, instead of raw surfaces ("multilayer" technique), in order to always get good conditions with respect to the stripping capacity of the coating,
- the stripping paints should be applied in a thick layer, not less than 150 - 160 μm, that is with the application of two following coats,
- the use of paints without solvent, not inflammable, is recommendable for obvious safety reasons. Anyhow paints of this type, containing water, should be suitably developed in order to improve their chemical resistance characteristics as they are emulsions and not solutions,
- the life of removable coatings actually is only affected by their exposure to anomalous mechanical stresses (shocks, continuous wearing, etc.) without which they can be conveniently used in nuclear plants.

4. CONCLUSION

This work is an initial investigation into concepts for minimizing activation of the concrete in the biological shield and for reducing the amount of radioactive waste arising from the dismantling of activated and contaminated structures.

In the first case this can be achieved by reducing the neutron-flux by means of well-aimed application of selective materials and by the arrangement of tier designs of attenuating and absorbing materials for neutrons.

In the second case the concrete surfaces were to be protected effectively against contamination. This can be achieved by means of temporary coatings easily removable before the dismantling of structures.

From the information available on the shielding materials and the analysis carried out, it can be concluded that the outer shield wall (i.e. supporting shield) can be reduced to non-radioactive waste and that the activation in the inner shield wall is less than for concrete.

At present, there is no clearly defined limit between radioactive and non-radioactive material and therefore a borderline case may well be regarded as radioactive.

Although CEA, Saclay alone tested more than 20 different paints and coatings, only a few could be classified as decontaminants after the first tests. It was only on consultation with the manufacturers and further tests on French, Italian and other products from Germany and the Netherlands that a large number of useful test numerical values could be determined, which then indicated good suitability of various products.

The various contractors examined some aspects more intensively than others.

To exactly understand the results obtained in detail it is recommendable to also know individual input values. Since a demonstration of the latter would be too extensive here it seems to be necessary to refer to the 4 final reports of the individual contractors (1, 2, 3, 4).

5. ACKNOWLEDGEMENT

The authors would like to express their gratitude to the Commission of the European Communities for the financial support of the work. In particular we express our thanks to Mr. Schaller, Scientific Secretary and official responsible for Project 7, for his continuous technical assistance and co-ordinating activities. We also feel obliged to all institutions for the provision of construction materials, chemical analytical values, decontaminants and to all persons who by word and deed contributed to make the work a success.

6. REFERENCES

1. AGIP Nucleare and CNEN, Italy; EC-Contract: DE-G-005 I; Removable Coatings for the Protection of Concrete against Contamination
2. Commissariat a l'Energie Atomique, Saclay, France; EC-Contract: DE-G-006 F; Characterisation and Improvement of Coatings Protecting Concrete against Contamination
3. Dyckerhoff & Widmann AG, München, Germany; EC-Contract: DE-G-009 D: Biological Shield Design with Dose Reducing Effect for Decommissioning
4. Verrall S., Fitzpatrick J. "Design Concepts to minimize the Activation of the Biological Shields of Light Water Reactors", EUR 8804, August 1983.
5. Battelle Pacific Northwest, (USA); Technology, Safety and Cost of Decommissioning a Reference Pressurised Water Reactor Power Station,

NUREC June 78

6. Engle Jr., W.W., ANISN; A one dimensional discrete ordinates transport code, RSIC Computer Code Collection, Oak Ridge National Laboratory, March 1979
7. Oak, H.D. et al. Technology, Safety and Costs of Decommissioning a Reference Boiling Water Reactor Power Station, NUREG/CR-0672, June 1980
8. Croff, A.G., ORIGEN 2; A revised and updated version of the Oak Ridge isotope generation and depletion code, ORNL 5621, July 1980
9. DLC-75 BUGLE-80: Coupled 47 neutron, 20 gamma-ray, P3, Cross section library for LWR shielding calculations, RSIC Data Library Collection, Oak Ridge National Laboratory, June 1980
10. DLC-23E CASC: 40 Group Coupled Neutron and Gamma Ray Cross-Section Data RSIC Data Library Collection Oak Ridge National Laboratory, March 1975
11. Control Data, CYBERNET Services, XSDRNPM, User Information Manual, USA, 1981
12. Norme NF T 30 901 Essai de decontamination
    Norme NF T 30 049 Essai de vieillissement artifiel
13. Norma Italiana, UNI 7497, UNI 7800, UNI 7801, UNI 7802, UNI 7803, UNI 7804 - (Determination of the ionizing radiation resistance)

7. APPENDIX

Tables and Figures

Element	CONCRETE (grams/kg)	AQUA-GEL (grams/kg)	TYPE-207 (grams/kg)	TYPE-277 (grams/kg)	RICORADD (grams/kg)
H	8.0	103.503	70.7	33.7	121.0
B	-	9.994	10.0	15.9	20.0
C	1.0	137.304	287.2	-	844.0
N	-	17.33	-	-	-
O	495.0	731.475	404.2	586.2	-
Na	6.3	0.086	0.8	5.9	-
Mg	-	-	0.9	5.0	-
Al	-	-	206.1	239.1	0.1
Si	277.0	-	1.3	21.3	0.3
S	-	0.205	0.1	1.9	-
Cl	-	-	-	-	0.2
K	7.1	0.102	0.2	-	-
Ca	195.0	-	17.7	88.3	0.01
Sc	0.0032	-	-	-	0.01
Cr	0.036	-	-	-	-
Fe	10.3	0.003	0.1	2.7	0.05
Co	0.01	-	-	-	-
Ni	0.008	-	-	-	-
Cu	0.006	-	-	-	0.001
Rb	0.031	-	-	-	-
Sr	-	-	0.7	-	-
Cd	0.0002	-	-	-	-
Sn	0.01	-	-	-	-
Cs	0.0017	-	-	-	-
Ba	0.15	-	-	-	-
La	0.011	-	-	-	-
Sm	0.002	-	-	-	-
Eu	0.00011	-	-	-	-
Tm	0.004	-	-	-	-
Yb	0.001	-	-	-	-
Hf	0.004	-	-	-	-
W	-	-	-	-	0.01
Pb	0.011	-	-	-	-
Bi	0.003	-	-	-	-
Total	999.992	1000.000	1000.002	1000.000	985.671

TABLE I, (4, EWE) Elemental composition of candidate materials including known impurities for activation analysis

From German NPPs

Concrete (Weight %)

Element	KBR 1 Elbe	KKG 2 Main	KWG 3 Weser	KKI-2 4 Isar	KKP-2 5 Rhein	KKU 6 Weser	Rad.Prot. 7 Concrete
Si	33,2	30,4	29,06	29,28	32,01	35,31	2,43
Ti	0,1	0,168	0,096	0,054	0,084	0.09	0,054
Al2	2,56	2,47	2,34	1,12	2,69	2,41	0.66
Fe2	0,85	3,56	0,88	0,56	0,99	0,93	58,20
Mn	0.023	0,031	0,039	0,015	0,023	0.015	0,054
Mg	0,20	0,50	0,38	1,41	0,34	0,21	0,07
Ca	8,52	8,65	14,05	12,56	9,28	6,22	3,55
Na2	0,74	0,39	0,35	0,12	0,57	0,47	0
K2	1,13	1,00	1,14	0,415	1,28	1,39	0,133
P2	0,031	0,048	0,031	0,022	0,031	0,061	0,031
S	0,22	0,16	0,15	0,19	0,23	0,16	0,09

TABLE II, (3, DYWIDAG); Elemental composition of candidate materials without impurites

Steel Samples (Weight %)

Element	Stainless Steel	Carbon steel for reinforced structures 1	2	3	4	5	6
C	0,049	0,170	0,176	0,187	0,388	0,706	0,713
Si	0,99	0,23	0,02	0,48	0,49	0,79	0,67
Mn	1,52	0,75	0,78	1,13	1,43	1,53	1,37
P	0,041	0,017	0,020	0,015	0,022	0,023	0,020
S	0,034	0,019	0,020	0,026	0,023	0,029	0,021
Cr	15,98	0,01	0,01	0,10	0,01	0,01	0,01
Mo	0,21	0	0	0,02	0	0,01	0
Ni	8,82	0,07	0,07	0,16	0,07	0,09	0,07
Cu	0,51	0,05	0,04	0,29	0,04	0,04	0,05
Sn	0,027	0,005	0,005	0,022	0,005	0,009	0,005
Al	0,093	0,005	0,005	0,005	0,005	0,006	0,005
V	0,11	0	0	0,06	0	0	0,06

TABLE III, (3, DYWIDAG); Elemental composition without impurities and main components like Fe

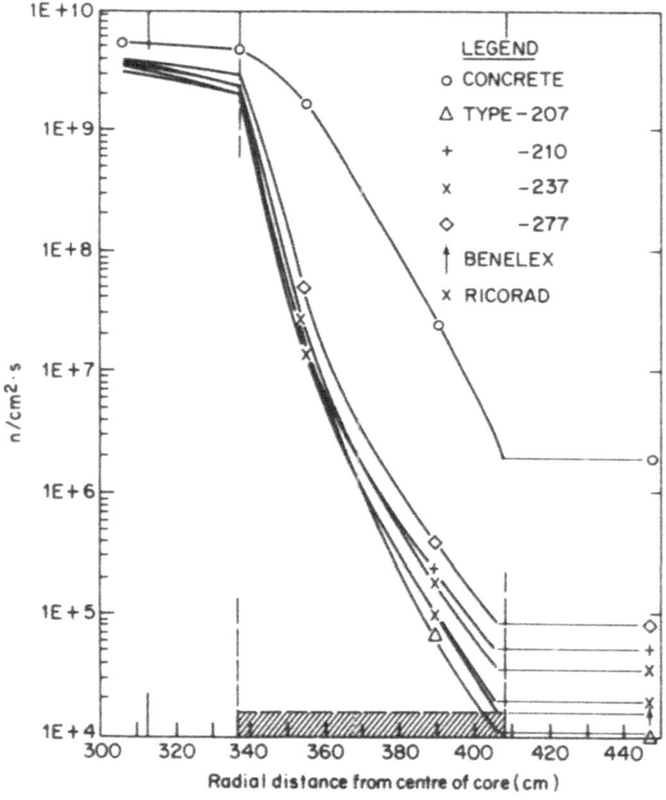

FIG. 1 (4, EWE) - Flux shapes in inner wall of PWR plant

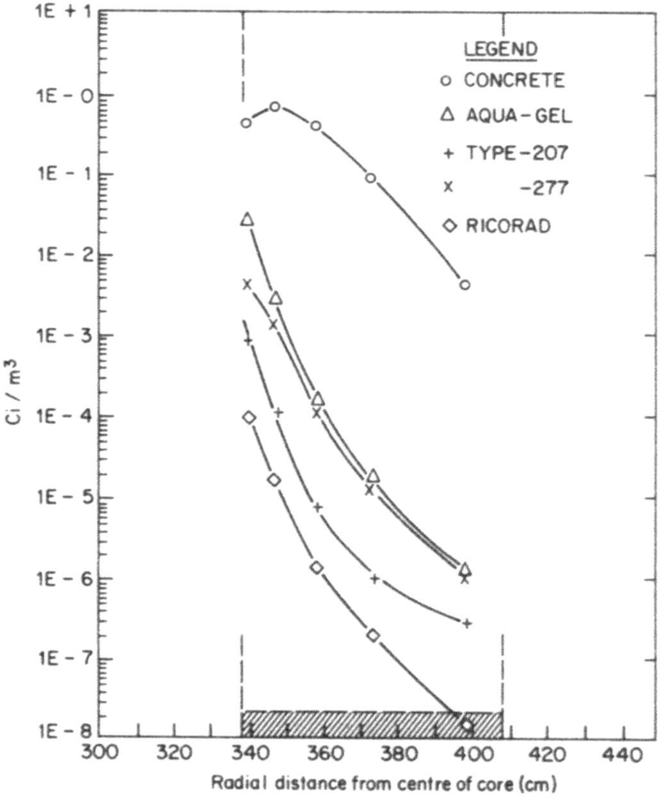

FIG. 2 (4, EWE) - Activation in inner wall of PWR plant

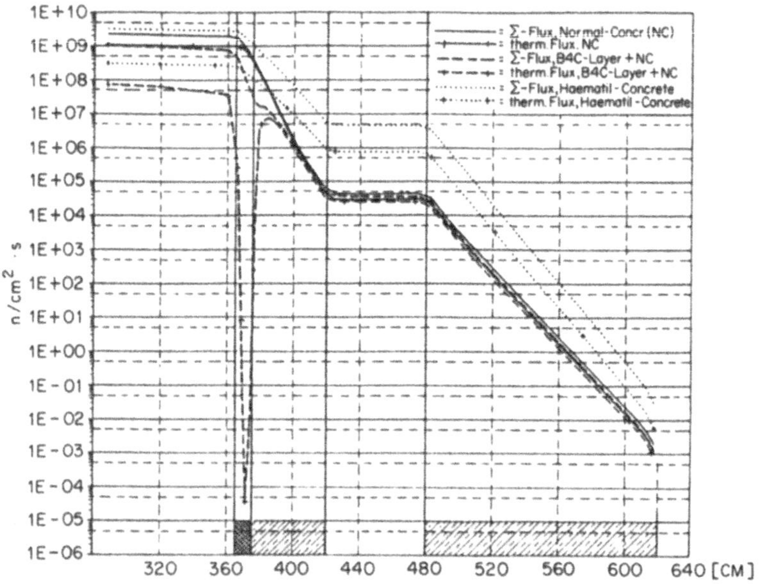

FIG. 3 (3,DYWIDAG) - Neutron flux shapes of different materials

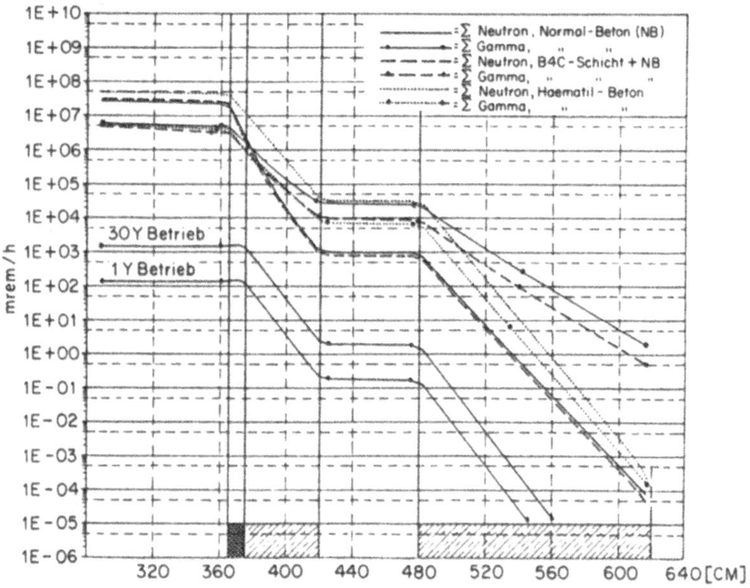

FIG. 4 (3, DYWIDAG) - Dose rate curves of different materials

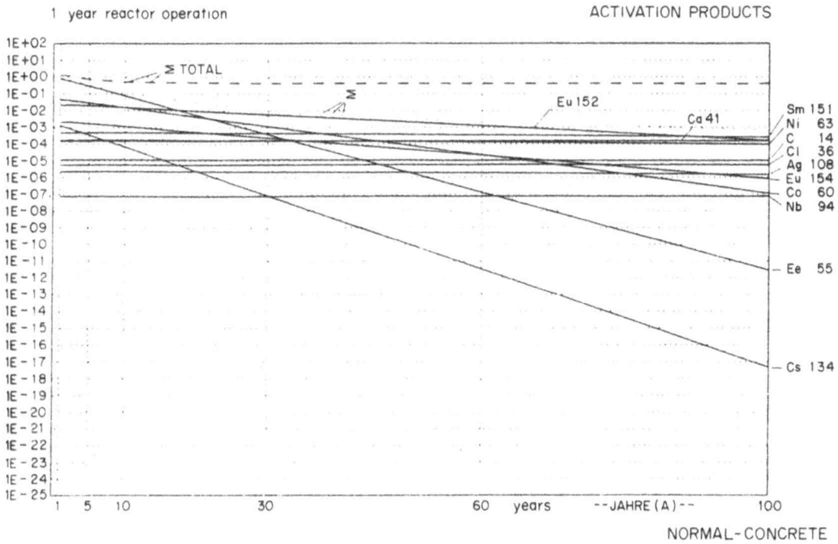

FIG. 5 (3, DYWIDAG) - Radioactivity outside edge in relation to decay periods

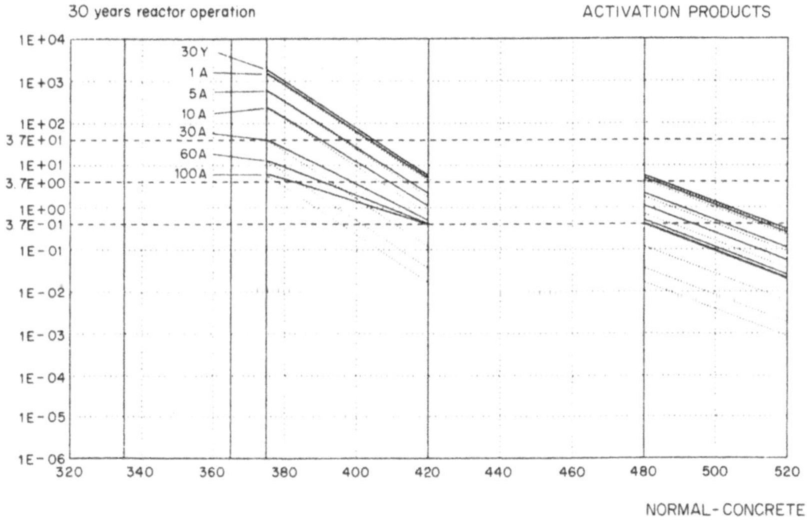

FIG. 6 (3, DYWIDAG) - Radioactivity in relation to decay periods

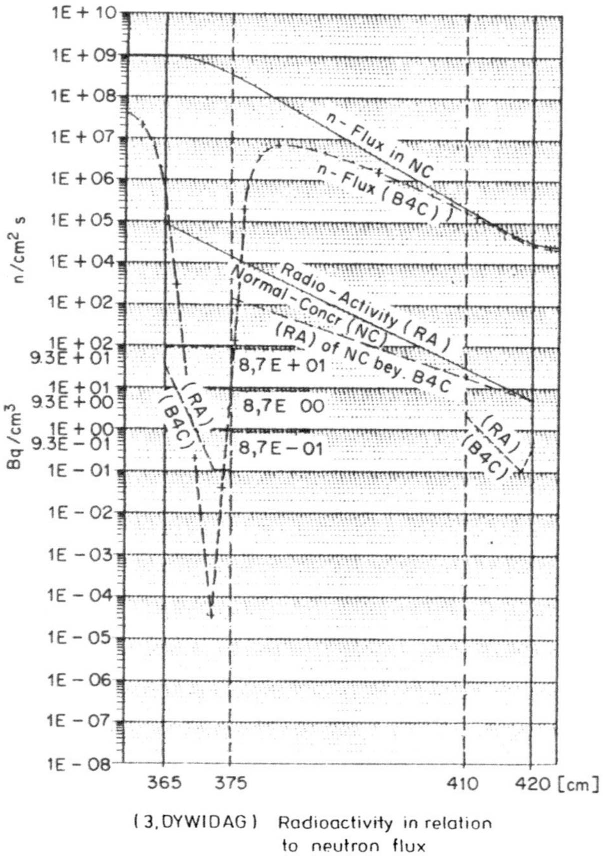

FIG. 7 (3, DYWIDAG) - Radioactivity in relation to neutron flux

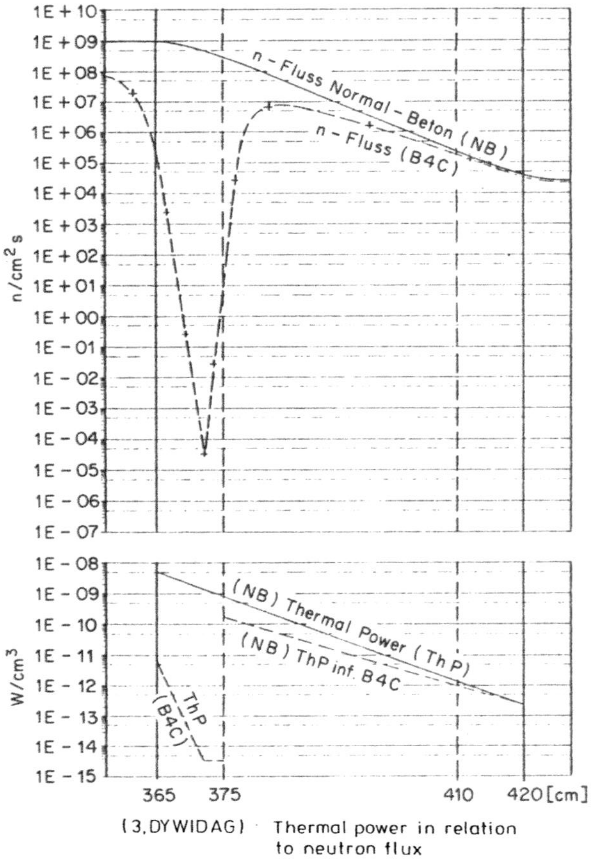

FIG. 8 (3, DYWIDAG) - Thermal power in relation to neutron flux

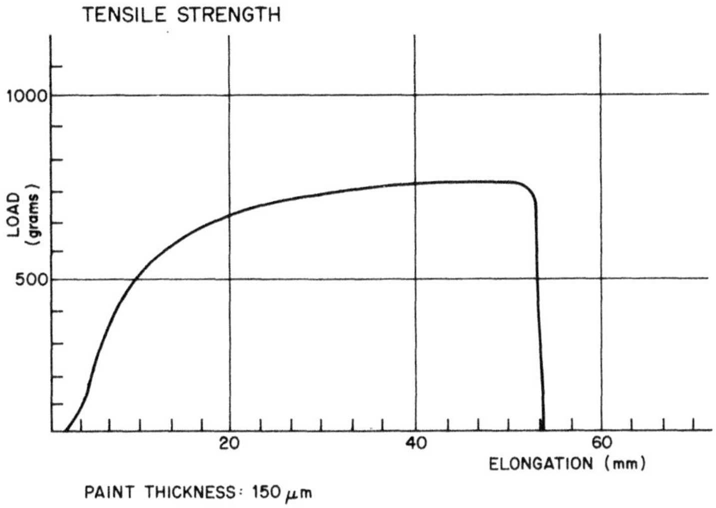

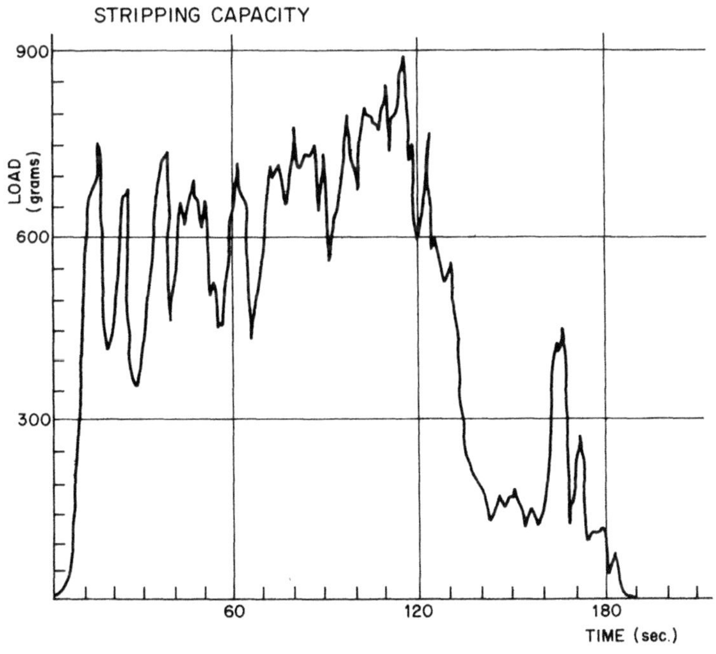

FIG. 9 (1, AGIP NUCLEARE) - Test diagrams

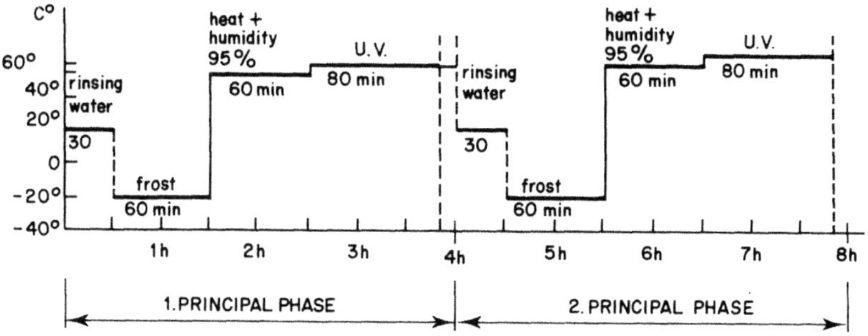

FIG. 10 (2, CEA) - Artificial weathering test of paints and related prod.

Product No		Ao	Al	Sc	A2	P	A3	R
		deposed activity	fixed activity	suscepti. bility	residual activity	decon-tamin.	remain.	residual activity
		D = 1E-9Ci/dm²	D	%	D	%	D	%
1	Pf	1392	898	64	744	47	742	-1
2	Pf	390	390	100	315	19	275	-13
3	Pf	618	459	74	431	30	425	-2
4	Pf	991	900	91	735	26	2	-99
	Pu	1289	855	66	135	90	2	-99
5	Pf	1275	102	8	70	95	69	-1
	Pu	885	311	35	112	87	85	-1
6	Pf	1121	99	9	80	93	83	-4
	Pu	1610	770	48	222	86	1	-99
7	Pf	1246	85	7	2	99	2	0
8[x]	Pf	1285	118	9	108	92	108	0

x after artificial weathering
Pf = Solutions with fission products
Pu = Solutions with plutonium products

TABLE IV (4, CEA) - Decontamination measuring values

REDUCTION OF EXPOSURE TO RADIATION BY SUBSTITUION OF Co ALLOYS AND REDUCTION OF THE CONCENTRATIONS OF Co, Nb AND OTHER COMPANION ELEMENTS IN STEEL

N. WIELING and P.J. HOFMANN
KWU Erlangen

R.M. BOOTHBY and T.M. WILLIAMS
AERE Harwell

Summary

Activation of the alloys used in primary circuits of nuclear reactors increases γ-radiation. This hampers repairs and dismantling operations. The level of radiation can be reduced in two ways :

1. substitution of Co-free alloys for those containing Co;
2. reduction of the levels of Co and other companion elements responsible for long-lived γ-radiation from steel.

The following studies were carried out with European Community support.

At KWU, on the basis of earlier KWU studies confirming the resistance of a number of Co-free alloys to friction, abrasion and corrosion, a study was made of the resistance of such construction materials to erosion, corrosion and hydro-abrasive wear.

At AERE Harwell a study was made of the sources of γ-radiation from steel. the results showed that the most important source of such radiation over the first 100 years following a final reactor shut-down is Co, but that Nb and Ag subsequently become more important. Investigation of possible ways of reducing the concentrations of these elements led to the conclusion that purer raw materials will have to be used.

1. INTRODUCTION

The γ-radiation emitted by the radioactive isotopes of certain metals hampers the inspection and dismantling of nuclear power stations.

The main isotopes at fault are those of Co, Ag, Nb and Ni. The materials in the primary circuit containing these materials are :

- construction materials, i.e. high alloy Ni- and Fe-based materials;
- hard materials used to strengthen surfaces exposed to high tribological stresses.

There are, therefore, two possible ways of reducing post-shut-down radiation levels in nuclear power stations :

1. substitution of Co-free alloys (or alloys with a low Co-content) for alloys with a high Co content;
2. reduction of the Co, Ag, Nb and Ni-content of construction materials.

The focus of contract DE-G-003-I with KWU is on point 1 above and that of contract DE-G-004-UK with AERE Harwell is on point 2.

2. KWU-STUDY

2.1 Studies of erosion corrosion

2.1.1 Materials studied

The study covered Co-free friction, abrasion- and corrosion-resistant construction materials, together with Co-free alternative materials used by arms manufacturers and other materials considered likely, by virtue of their mechanical properties (hardness) and/or structure, to provide useful solutions, when the results were analysed, to the basic problems under consideration. The name and composition of each type of material tested is given in Table 1.

2.1.2 Test installations / Test conditions

The installations in which the tests were carried out permitted the simultaneous testing of 8 samples at pressures up to 15 bars, temperatures up to 150°C and flow rates up to 65 m/s. Water conditions (temperature, O_2-content, pH, conductivity) could be continuously monitored.

The following test conditions were established :

a) flow rates
65 m/s, 44 m/s, 2 m/s

b) medium

composition	:	dist. H_2O + 750 ppm B (in the form of H_2BO_3) + 2 ppm Li (in the form of LiOH) + 10 ppm Hydrazine N_2-Degassing
O_2-content	:	< 0.01 ppm
CI-content	:	< 0.5 ppm
Temp.	:	150°C
Duration	:	700 h (Inspection after 300 h).

2.1.3 Analysis of the results

The following test values were measured :

- changes in weight
- changes in surface profile
- REM-values
- metallographical abrasion
- comparative values for abrasion and hardness.

2.1.4 Results

Changes in weight

Weighing of random samples after the second test cycle confirmed the results of the first cycle showed the weight changes of the samples to be of a lesser order of magnitude than the accuracy of measurement of approximately 1 mg.

Surface roughness

From the many data collected, values were selected for the roughness of the test samples exposed to the highest flow velocities (65 m/s). Table 1 shows the R_z-, R_a and R_t values for the samples at the start of the test against those recorded after 300 and 700 h.

Any comparison of the measured values must take the following factors into account :

- the values are individual measurements characterized by the relatively marked scatter that is commonly encountered in the measurement of surface roughness;
- when the test materials were exposed to erosion corrosion, oxide layers were formed and had to be removed by suitable products (e.g. "Endox", "AP2") before any measurement of roughness could be carried out.

Significant changes were measured only at PH 17-4, at which value a smoothing effect was found.

Scanning electron microscope

In this case, the starting values for the samples subjected to a flow rate of 65 m/s were compared with those measured at the end of the tests (without or with deoxidation).

The results may be summarized as follows :

- Colmonoy 4
  Table 2 shows that after the 2nd cycle a thick oxide layer had developed. This was only partly removable with a deoxidation treatment that does not affect the underlying material. The original surface was significantly smoothed by exposure to erosion corrosion. Selective corrosion was also found;
- 17-4 PH
  Table 2 shows that steel 17-4 PH was significantly smoothed. There was no significant selective attack:
- other materials
  Little or no change was recorded.

Metallographical sections

These tests were intended to provide reliable evidence of whether and (if so) to what extent selective corrosion is caused by the flowing medium. Evidence of singificant attack was found only in the case of Colmonoy 4.

Comparative values for abrasion and hardness

No connection was found between abrasion and hardness.

2.1.5 Conclusions

The results warrant the following conclusions :

- the test materials all showed high resistance to erosion corrosion;
- on the basis of the small differences observed in their resistance to erosion corrosion, under the conditions described above, the test materials can be placed in the following order :
  Pantanax 25
  Stellite 5 (Reference-material)
  Antinit DUR 300
  Nitronic 60
  Tribaloy T 700
  LC-1C
  17-4 PH
  Colmonoy 4
- There is no connection between the hardness of a material and its resistance to erosion corrosion, which clearly depends on the chemical properties of the material rather than its mechanical characteristics.

2.2 Hydroabrasive wear tests

The medium flowing through the reactor components is not entirely free of particulate matter and can in fact carry measurable quantities of corrosion products (mainly Fe_3O_4).

The resulting hydroabrasive process is not very closely comparable with that of erosion corrosion because mechanical removal of material occurs in association with anodic dissolution of the metal.

The supplementary test programme gave the following initial results (cf. fig. 3):

- abrasion of all the test materials by magnetite particles at v = 44 m/s
- inhibition of the formation of the thick oxide layer observed in the tests of Colmonoy 4
- acceleration of the (selective) corrosion of Colmonoy 4.

3. TESTS CARRIED OUT AT AERE HARWELL

3.1 Neutron activation in steel reactor components

Certain trace elements in steel are the source of long-lived γ-radiation in neutron-activated reactor components. This hampers the decommissioning of nuclear installations. Although the decay of ^{60}Co is the main source of γ-radiation at the time of final shut-down and for several subsequent decades, a study of thermal reactor installations described in the literature (1) has shown that the relative contribution to γ-radiation made by long-lived ^{94}Nb (2) and ^{108m}Ag (3) -isotopes becomes more important in the longer term and that in components with a relatively low neutron flux the Eu-isotope must also be taken into account. In fine steels with nickel as the main alloying element, the decay of ^{59}Ni also contributes to the production of such radiation.

To ascertain the relative significance of this gamma-emitting isotope, calculations were made of activation (1) for reactor components of special steel type 304 (69,2 Fe, 17,5 Cr, 10% Ni by weight as base metal, with assumed trace element contents of 0.1% Co by weight (i.e. 1000 ppm), 100 ppm Mb, 1 ppm Ag and 0.4 ppm Eu). The calculations were based on total neutron fluxes of 5.6 x 10^{13} and 8.2 x 10^{10} $cm^{-2}s^{-1}$, which are typical values for certain components of an 1175 MW(e) PWR (4), and a reactor operating time of 30 years under full load. The values calculated for specific activity (Curie/g) and the estimated dose rates (rem/h) at the time of reactor shut-down are shown in table 1. Fig. 4 shows the post-shut-down dose rates over time for components with high and low neutron flux. The results show that the dose rates attributable to 100 ppm Nb and 1 ppm Ag remain comparable over the longer timescale and that when the decay period excedes about 100 years these elements make a higher contribution to the dose rates than 1000 ppm Co. The high burn-up cross-section of the γ-radiation emitting Eu isotopes makes them relatively harmless under conditions of high neutron flux but renders them all the more harmful in components with low neutron flux. After 30 years irradiation with neutron flux of 8.2 x $10^{10}cm^{-2}s^{-1}$ and a subsequent decay period of a hundred years, the ^{152}Eu resulting from the presence of 0.4 ppm Eu in the reactor steel would be the most dangerous source of such radiation, although its absolute dose rate (10^{-2} rem/h) would be low (fig.4b).

TABLE 1

Calculated values for specific activity and dose rates subsequent to reactor shut-down, after 30 years' irradiation, for neutron fluxes of 5.6 x 10^{13} and 8.2 x 10^{10} $cm^{-1}s^{-1}$

Isotope	Half-life	$5.6x10^{13}$	$n.cm^{-2}s^{-1}$	$8.2x10^{10}$	$n.cm^{-2}s^{-1}$
	in years	Curie/g	rem/h	Curie/g	rem/h
^{55}Fe	2.7	$2.14x10^{-1}$	$6.84x10^{-2}$	$5.86x10^{-4}$	$1.89x10^{-4}$
^{60}Co	5.27	$1.09x10^{-1}$	$2.51x10^{-5}$	$3.54x10^{-4}$	$8.13x10^{2}$
^{59}Ni	$8x10^{4}$	$1.09x10^{-4}$	$5.27x10^{-2}$	$4.54x10^{-7}$	$2.19x10^{-4}$
^{94}Nb	$2x10^{4}$	$3.74x10^{-7}$	$5.42x10^{-1}$	$7.64x10^{-10}$	$1.11x10^{-3}$
^{108}Ag	127	$6.79x10^{-7}$	$9.99x10^{-1}$	$9.95x10^{-10}$	$1.46x10^{-3}$
^{152}Eu	13.3	$2.09x10^{-12}$	$2.16x10^{-6}$	$6.23x10^{-7}$	$6.44x10^{-1}$
^{154}Eu	8.6	$9.62x10^{-8}$	$1.11x10^{-1}$	$8.59x10^{-8}$	$9.84x10^{-2}$

3.2 Trace element concentrations in reactor steels

In practice, the concentrations of trace elements in steels can vary considerably. In certain earlier calculations of radioactivity, cobalt contents of up to 0.2% for steel type 304 were used; but it is now safe to assume specific cobalt contents of max. 0.025% (250 ppm) for the reactor steels used in nuclear power stations. The Nb content of steel type 304 is known to vary between 10 and 200 ppm (5); little information is yet available with regard to the Ag and Eu contents (prior to the present study there were no data for silver and only 1 value for Europium (0.4 ppm) (2) for steel 304. For this reason, supplementary studies were carried out at Harwell (1), in which neutron activation processes were used for the determination of Co, Ag and Eu contents and inductive coupled plasma-emission spectrometry for Nb. Analysis of melted samples of steel 304 (suitable for nuclear installations) from four different meltings gave average Co values in the range of 140 - 350 ppm and average Nb values < 25 ppm (but 85 ppm in one sample). The silver content of three of the samples analysed was < 0.1 ppm (but that of the fourth sample was 0.8 ppm). The highest silver/Nb values were found in different samples. It may be concluded, on the basis of the activation calculations presented above, that either of these two elements could be the most prolific source of gamma radiation in reactor components after a decay period of, for example, 100 years. The Eu content of all four melted samples was very low, at < 4 x 10^{-4} ppm, and would not constitute a major radiation hazard whatever the corresponding neutron flux.

3.3 Minimization of trace elements at the steel production stage

Relatively high trace element concentrations in steel are mainly traceable to the incorporation of scrap material and, to a lesser extent, to the standard use of ferro-alloys in the steel production process. The most important step in the production of steel is the removal of carbon from the molten raw material by controlled oxidation. The probability that an element will be removed from the molten metal with the slag will thus depend on the relative stability of its oxide (6). Iron and elements with less stable oxides, including cobalt and silver, are not oxidized in the production process and therefore remain in the molten metal. Somewhat more stable oxide builders, such as chromium and niobium, are partly transferred to the slag in the carbon removal process; but in order to minimize the expensive loss of chromium in the production of stainless steel these elements are subsequently recovered from the slag by means of a strong reducing agent (e.g. iron silicate). (It must furthermore be noted in this context that the reduction of niobium oxide, which is present in small quantities in the rutil-(titanium oxide) coatings employed in the electrode welding process is a potential source of local and relatively high niobium concentrations in steel. Niobium values of around 200 ppm attributable to this source have been found in stainless steel welding materials) (7). Like all rare earth metals, europium has a very marked affinity for oxygen and should be completely and finally removed with the slag whenever it is present in the charge materials. It should not be forgotten that rare earth metals mixtures mainly comprising cerium and lanthanum, but possibly also small quantities of europeum, are deliberately added at the last (non-oxidizing) stage of steel production in order to strengthen the steel by changing the form of the incorporated elements. The discrepancy between the results of the analyses carried out at Harwell (1) and the much higher europium values cited in earlier reports (2) is possibly attributable to this practice, which obviously ought not to be adopted in the production of steels for nuclear reactors.

Hence the presence of cobalt, niobium and silver, in so fas as they are present in the raw materials used for steel production, in the steel end product. One would hardly expect to find high silver contents in practice, because the element is not very soluble in molten steel. Neither the cobalt content nor the niobium content can be reduced by repeated vacuum melting, as both elements are less fluid than iron. The remaining silver content could theoretically be reduced in this way; but the procedure would not be a viable proposition in view of the relatively small amount of Ag remaining in the steel. Hence the need for rigorous selection of raw materials so as to minimize the presence of residual amounts of potential sources of long-lived γ-radiation in reactor steels.

Stainless steels are normally made from a combination of scrap (mild or stainless steel, nickel-based alloys, etc.) plus smelted alloys (ferro-alloys) with Cr, Ni, etc. Scrap, whether from within the steelworks or of external origin, is a relatively cheap source of iron, chromium and nickel, but may also contain large quantities of cobalt and niobium. The typical cobalt content of stainless steels not earmarked for nuclear reactors is 0.2%, with specified highest levels of 0.5%; nickel-based alloys may have a cobalt content of several percent. Up to 1% niobium is added to a few stainless steels to improve their resistance to intercrystalline corrosion and creep. Bought-in scrap may contain silver in the form of silver filler (for brazing). It follows that stainless steel scrap or scrap with a high alloy content should never be used for the production of steel for use in nuclear installations.

Mild steel scrap, used in steel production to provide the necessary iron content, normally contains about 0.01% (i.e. 100 ppm) of Co. This cobalt concentration is acceptable in certain nuclear applications, but purer raw materials for iron are available. One analysis (1) of a sample of commercial sponge iron produced by the direct reduction of iron ore, gave a cobalt content of 45 $\pm$ 5 ppm and low values for niobium (6 ppm) and silver (0.04 ppm); the europium content of the sponge iron was relatively high at 0.27 ppm, but this value would probably be reduced in the course of the steel production process.

The cobalt content of nickel iron is too high for reactor steels (it is normally 1/30 of the nickel content), but high-purity nickel is generally quite acceptable. For example, an analysis of crystalline nickel of 99.7% purity showed impurity levels of only 7 ppm Co, 1 ppm Nb, 0.15 ppm Ag and a negligible value for Eu (1). Standard type ferro-chrome alloys contain up to 800 ppm Co; it is therefore possible to produce a ferro-chrome alloy enriched to 80% with a maximum Co content of 200 ppm that is quite suitable for reactor engineering purposes. In the case of stainless steel components exposed to a high neutron flux, the use of high-purity chromium (producer analysis : 0.3 ppm Co, $<$ 0.01 ppm Nb, $<$ 0.02 ppm Ag) is fully justified in spite of the relatively high cost. It can be assumed that the trace element levels achieved by using high-purity chromium, nickel and sponge iron for the production of 18%Cr/10%Ni-based steel could be $<$ 50 ppm Co, about 5 ppm Nb and $<$ 0.05 ppm Ag. The long-term exposure attributable to Nb and Ni in stainless steel construction components would then be lower than the values for ^{59}Ni. Reduction of the 1000 ppm cobalt content assumed in the activation calculations to 60 ppm would in fact extend the decay period for a given ^{60}Co dose rate by 4.3 half-lives for 23 years.

3.4 Conclusions

1. The decay of ^{60}Co is the main contributor to the γ-dose rate from steel components for several decades after reactor shut-down; but the long-lived ^{94}Nb, ^{108m}Ag and ^{59}Ni isotopes have a greater effect in the longer run. In components exposed to a low neutron flux, the gamma-emitting europium isotopes are quite notable sources of such radiation but would not make a high contribution to the dose rate in absolute terms.
2. Analysis of stainless steel type 304, curently in use, showed cobalt values of between 140 and 350 ppm, niobium contents up to 85 ppm and silver values in various melts and warrants, the conclusion that either of these two elements could in practice make the highest contribution to the γ-dose rate from reactor components after a decay period of, for example, 100 years.
3. Cobalt, niobium and silver are not oxidized in the steel production process and in so far as they are present in the raw materials these elements will also be present in the steel end-product. the sources of trace elements in steel are mainly scrap and to a lesser extent smelted alloys such as ferro-chrome. Use of purer raw materials for steel production would substantially reduce the quantities of potential γ-emitting sources. Use of high-purity chromium, nickel and sponge iron for the production of steel 304 should ensure values of 50 ppm Co, about 5 ppm Nb and 0.05 ppm Ag.

LITERATURE

1. R.M. BOOTHBY and T.M. WILLIAMS, European Applied Research Reports, 5 (2), 121, (1983).
2. J.J. STEPHENS and R.O. POHL, Nucl. Eng. Design, 47, 125 (1978).
3. P.B. WOOLLAM and I.G. PUGH, in "Decommissioning of Nuclear Facilities" (IAEA Vienna, 1979) p. 353.
4. R.I. SMITH, G.J. KONZEK and W.E. KENNEDY, NUREG/CT-0130, (Battelle Pacific Northwest Laboratory, 1978).
5. V.K. SIKKA, A.J. MOORHEAD and C.R. BRINKMAN, ASTM STP 679, (1979), p. 69
6. G.N. GOLLER, ASTM STP 418 (1967) p. 37.
7. A.W. MARSHALL and J.CM. FARRAR, "The effect of residual, impurity and micro-alloying elements on weldability and weld properties", (Welding Institute Conf., London, 1983), paper 24.

Name of the material	Composition (%)													Vickers hardness HV 20
	C	Si	Mn	Cr	Mo	Ni	N	W	V	Fe	Co	Cu	B	
Welded layers														
Pantanax 25/0 Mo-G	2.5	0.4	0.9	25	3.1	-	-	-	0.4	bal.	-	-	-	470...480
Tribaloy T 700	-	3	-	15	32	bal.	-	-	-	-	-	-	-	450...470
Antinit DUR 300	0.11	5	6.5	21	-	7.8	-	-	-	bal.	-	-	-	300...340
Colmonoy 4	0.3	3.0	-	8.0	-	bal.	-	-	-	2.0	-	-	2.0	450
Detonation gun coating														
LC-1C	Cr_3C_2+20%NiCr8020													
Structure materials														
1.4542 (≙ 17.4 PH)	0.006	0.36	0.46	17	-	3.9	-	-	-	bal.	-	3.6	-	350
Nitronic 60	<0.1	4	8	17	-	8	0.13	-	-	bal.	-	-	-	250...270
Reference material														
Stellite 6	1.1	1.5	1	30	1.5	3	-	4.5	-	3	bal.	-	-	430...460

FIG. 1a - Composition of the test materials

Condition of test surface	Roughness (μ m)	Direction of measurement	Pant.25	T 700	DUR 300	Colm. 4	LC-1C	17-4 PH	Nitr.60	Stell.6
Pre-test	R_z	longitudinal(*	1.91	2.32	2.46	4.33	1.55	4.53	2.26	3.23
	R_a	longitudinal(*	0.28	0.31	0.42	0.54	0.20	0.57	0.33	0.48
	R_t	longitudinal(*	2.22	3.44	3.28	6.26	2.05	6.49	3.39	4.40
	R_z	lateral(*)	3.49	3.78	5.61	6.96	2.24	5.85	5.27	8.78
	R_a	lateral(*)	0.37	0.45	0.64	1.11	0.28	0.63	0.57	1.32
	R_t	lateral(*)	4.79	4.86	7.55	8.66	3.04	9.55	8.77	12.89
After one cycle	R_z	longitudinal	1.86	1.21	1.85	3.06	1.77	3.00	3.02	2.37
	R_a	longitudinal	0.27	0.19	0.31	0.53	0.23	0.56	0.44	0.30
	R_t	longitudinal	2.81	2.17	2.93	4.61	2.15	5.40	4.25	3.98
	R_z	lateral	3.60	3.80	5.00	6.65	2.43	5.57	6.13	9.70
	R_a	lateral	0.42	0.42	0.64	1.07	0.31	0.63	0.71	1.73
	R_t	lateral	4.47	4.64	7.86	9.91	2.93	7.84	7.98	11.96
After two cycles	R_z	longitudinal	1.92	2.32	2.03	2.75	1.63	3.25	1.43	2.91
	R_a	longitudinal	0.24	0.39	0.33	0.42	0.22	0.53	0.21	0.60
	R_t	longitudinal	2.62	2.87	2.66	3.59	2.10	4.18	2.40	4.07
	R_z	lateral	3.49	3.85	5.08	7.16	2.53	4.85	6.14	10.00
	R_a	lateral	0.37	0.42	0.61	1.05	0.28	0.57	0.64	1.57
	R_t	lateral	4.32	5.05	7.75	9.06	4.36	6.03	8.46	13.41

(*) in relation to the direction of flow

FIG. 1b – Change of surface roughness caused by erosion corrosion (65 m/s)

APPEARANCE OF THE COLMONOY 4 SURFACE

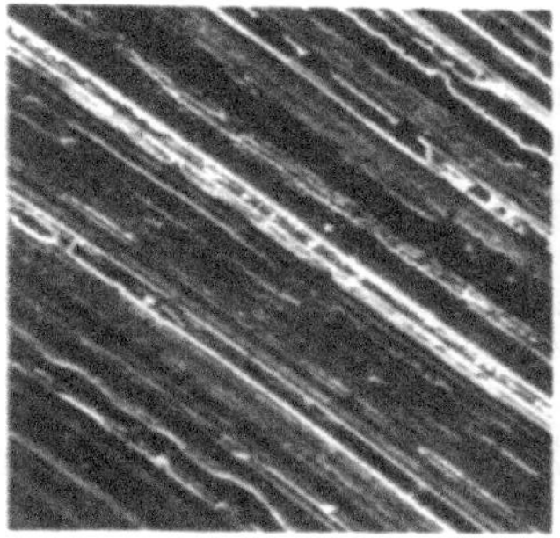

a) Pre test

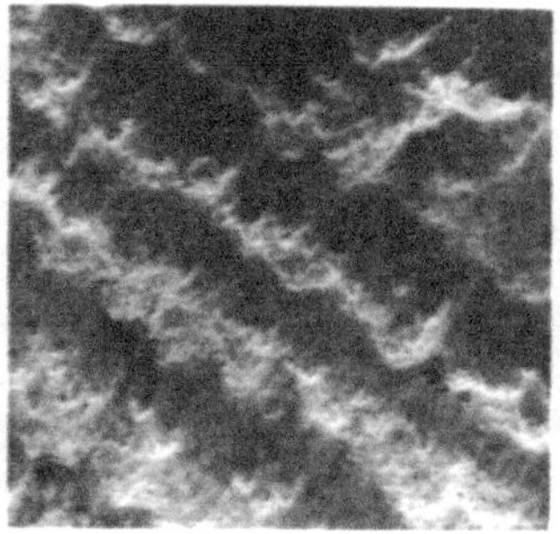

b) After two cycles

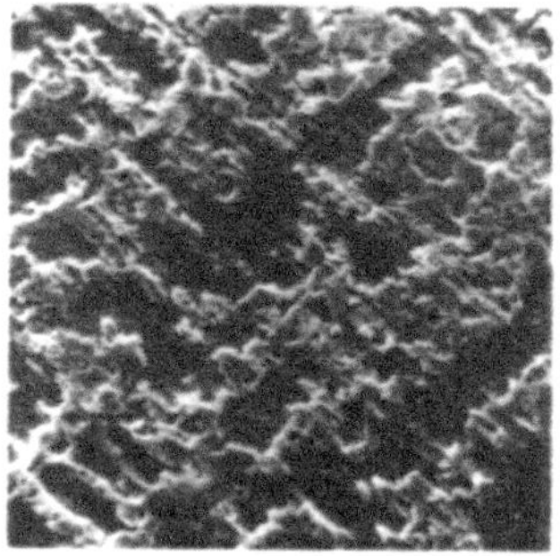

c) After deoxidation

APPEARANCE OF THE 17-4 PH SURFACE

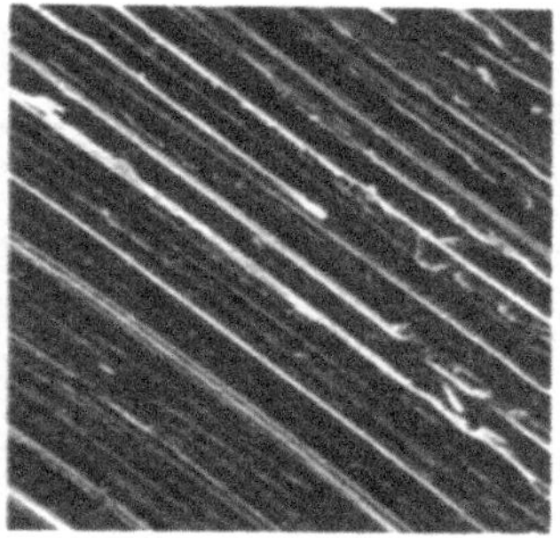

a) Pre-test

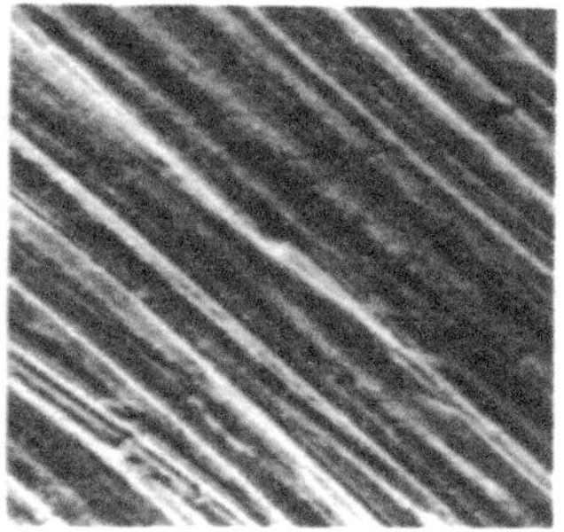

b) After two cycles

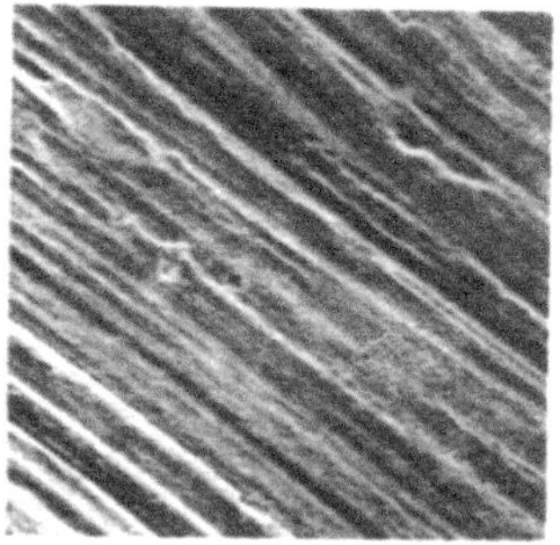

c) After deoxidation

FIGURE 2

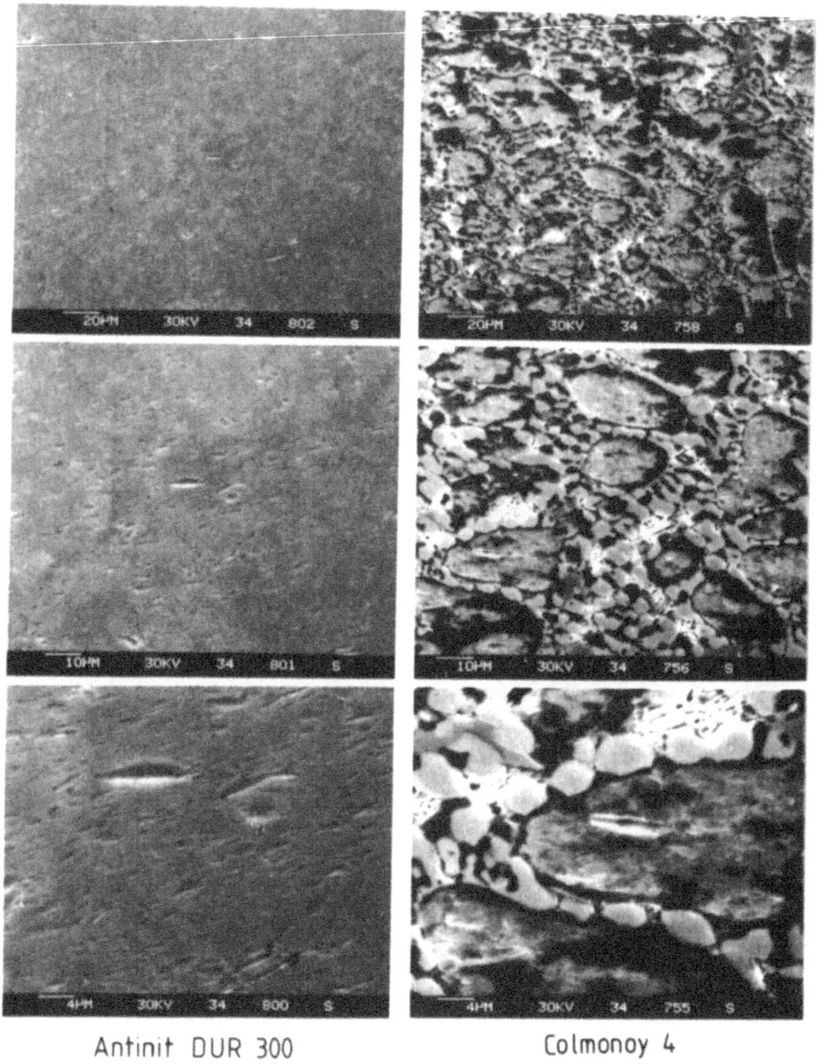

Hydroabrasive wear tests

Surfaces after approx. 100 h

FIGURE 3

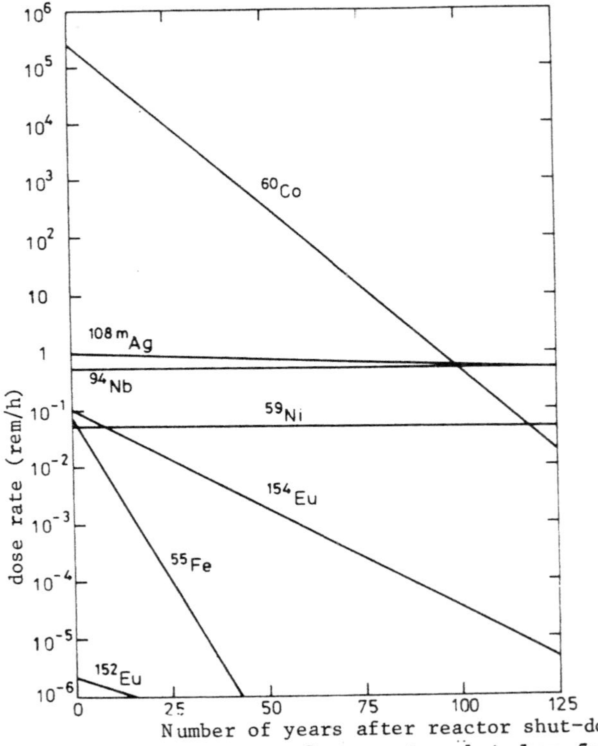

Fig. 4.a - Dose rates after reactor shut-down from components of stainless steel type 304 exposed for 30 years to a neutron flux of 5.6 x $10^{10} cm^{-2}s^{-1}$.
The assumed trace element values are 1000 ppm Co, 100 ppm Nb, 1 ppm Ag and 0.4 ppm Eu.

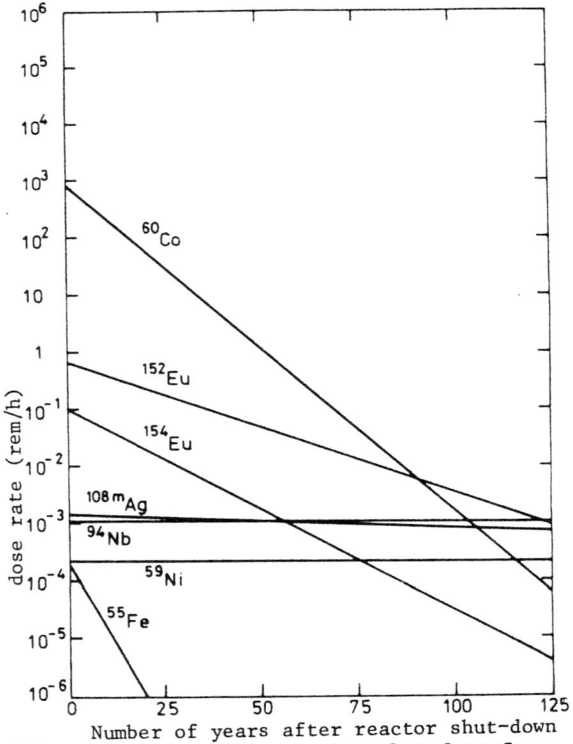

Fig. 4.b - Dose rates after reactor shut-down from components of stainless steel type 304 exposed for 30 years to a neutron flux of 8.2 x $10^{10} cm^{-2}s^{-1}$.
The assumed trace element values are 1000 ppm Co, 100 ppm Nb, 1 ppm Ag and 0.4 ppm Eu.

REVIEW OF PLANT DESIGN FEATURES FACILITATING DECOMMISSIONING

J D REGAN (NNC); A A PATON (TW);
P S AVANZINI (NIRA); H SCHWARZ (TH DARMSTADT)

Summary

At the design stage it is necessary to define the decommissioning strategy. This offers the opportunity to influence the design and hopefully deduce the optimum dismantlingsequence, in terms of cost, dose rate and time.

The decommissioning activity will be some 50 years to 150 years in the future, though the strategy must be based on present known or developing technology. In the event, the developments in remote handling techniques must mean the assumptions in this paper are pessimistic.

Throughout the paper, delayed decommissioning is assumed, though a major accident or pressure for early re-use of the site, could require earlier decommissioning. This aspect is not considered in the paper, nevertheless, the conclusions would not be invalidated by such a requirement.

A schedule of design features which facilitate decommissioning and details of some specific items installed to aid decommissioning of Advanced Gas Cooled Reactors is presented.

The problems of long term storage of decommissioning documents and the selection of material for storage, can be influenced at the design stage. Some suggestions regarding storage and eventual recovery of information to aid decommissioning is included.

1. INTRODUCTION

At the design stage of the nuclear power station it has become necessary, in many countries, to submit a decommissioning proposal. This provides the designer with the opportunity to examine the design approach, and influence the project in a manner which will ease the task of those, who, in 50 to 100 years time, will have to decommission the nuclear power station.

The rapid advance of technology over the past 50 years has brought about the development of the nuclear power station. It is not unreasonable to anticipate that, in the next 50 years, remote handling techniques will be developed to the point where dismantling of the same nuclear power stations will not be hazardous to the dismantling team or the general public. Nevertheless it would be wise to show that the station could be decommissioned using current dismantling/demolition techniques. The description of the dismantling/demolition techniques proposed should form part of the permanent record. In the event, such guidance may be of value despite improvements in remote handling.

1.1 Strategy to influence design to facilitate decommissioning

The overall strategy for decommissioning a nuclear power station will depend on social, economic and technical demands. Whether a delayed decommissioning programme is adopted or whether an early decommissioning scheme is preferred can be accommodated within the existing technology, and will probably be decided by social and/or economic factors.

The technical decommissioning strategy, the procedures adopted and the mode of demolition will depend on, the plant design, the operational history, the radiological risk, and the available dismantling/demolition techniques. The means of establishing the optimum dismantling sequence has been examined with the object of defining the several possible "dismantling trees", and then relating each route to the cost, in economic and radiological terms. When the optimum sequence is identified, the critical items are analysed and any improvements thus identified, fed back into the design process. The logic sequence and the application of this method is shown in Fig 1.

To verify the validity of this method, it has been employed to test the preliminary design of the fuel building of a reference power plant, developed as part of the Italian nuclear power programme.

The great number of possible sequences suggest the need to assist the decommissioning engineer with computer techniques. The computerisation of these techniques is in the process of development.

1.2 Design features which facilitate decommissioning

Nuclear power stations along with other engineering plants are designed to meet the operational requirements, be reliable, and safe to operate. They must also be capable of being maintained without undue risk to the maintenance staff.

Design considerations to facilitate decommissioning will sometimes conflict with these requirements and relevant design provisions may not be automatically accommodated in view of the many constraints (not least the financial ones) involved in commercial design and manufacturing work. Nevertheless, the aim should be to provide a design, which as well as meeting operational and maintenance requirements is also consistent with the plan for decommissioning. Additional features may often be incorporated in the initial design specifically to facilitate decommissioning but more often it is a case of recognising the aspects of the design which have an effect on decommissioning and optimising these aspects. The overall aims in optimising the design to facilitate decommissioning are to reduce decommissioning costs, minimise decommissioning time, minimise the production of radioactive waste and minimise radiation dose to both decommissioning staff and members of the general public.

The Advanced Gas Cooled Reactors (AGRs) which formed the basis of this aspect of the study are of a twin reactor design, each reactor capable of producing 660MW(e). The reactor core structure, boilers and gas circulators are housed in a single cavity Prestressed Concrete Pressure Vessel (PCPV) Fig 2. The PCPV is a vertical cylinder with helical multi-layer prestressing tendons in the walls. The vessel inner surface has a steel liner insulated and cooled to 50-70°C to limit the concrete temperature. The insulation consists of layers of ceramic fibre held in contact with the liner by a system of steel foils and cover plates each of which is retained in position by primary and secondary retention features.

The reactor core, is a sixteen sided stack of graphite, connected at the periphery to a steel restraint tank which is itself situated within a steel envelope called the gas baffle which provides for the re-entrant cooling of the graphite structure. The graphite structure incorporates integral graphite and steel shields to reduce radiation levels so that personnel may enter the pressure vessel (during reactor shutdown) for inspection and to facilitate decommissioning. The upper and lower shields consist of graphite and steel bricks, and the side shielding includes steel rods located in the two outer rings of graphite bricks. Additional side shielding is provided by steel plates fastened to the restraint tank. The core graphite structure rests on a diagrid which is supported by the skirt forming the lower end of the gas baffle cylinder.

Twelve boiler units are positioned in the boiler annulus between the gas baffle and the pressure vessel, and after passing through a boiler the coolant gas discharges into the circulator annulus which forms the entry plenum for the gas circulators.

The main design aspects of the AGRs which are utilised to facilitate decommissioning are listed below. The principles, in some situations, apply to other reactor designs.

(i) The station layout is designed to facilitate radiation and contamination control. The main principles of this control are to minimise radiation dose by appropriate segregation and shielding of radiation areas and to minimise the spread of contamination by appropriate containment, zoning, ventilation and a system of changerooms and sub-changerooms local to the sources of contamination. Decommissioning is therefore facilitated by both the layout (which can be utilised as a basis for a similar control system during commissioning) and the segregation and limited spread of contamination produced during reactor operation. By reducing activation and contamination more material will be available for re-use from the dismantled station. (It may also be noted that the AGR has several other inherent features which assist in the minimising of the amount of plant which becomes significantly contaminated during reactor operation, e.g. relatively low contamination of coolant circuit and a non-active turbine).

(ii) Shielding is provided around the reactor core inside the PCPV, These shields (known as internal shields) are primarily included to reduce the radiation levels inside the PCPV during reactor shutdown periods to allow personnel access via permanent man-access penetrations for inspection (as a back-up/supplement to the remotely operated in-service inspection equipment). However, this shielding has the dual advantage for decommissioning of reducing induced radiation levels outside this shielding to very low levels (mrem/h) during decommissioning, (thus enabling extensive personnel access to be made for manual dismantling or setting up remotely operated dismantling equipment) and reducing the neutron induced activation of the structures outside the shielding, which greatly reduces the radiological problems associated with dismantling and disposal of these structures. (The activation of the internal shields themselves does not present a major radiological problem for decommissioning, as the construction is based on the use of graphite blocks (which do not activate significantly), which are simply keyed together and supplemented by steel inserts and plates: all of which can be removed and transported without difficulty.

(iii) There is tight control on the materials used inside the PCPV, with particular attention paid to the levels of impurities which could produce significant amounts of long lived neutron induced radioactivity. It is not feasible or economically viable to reduce the levels of relevant impurities to zero or insignificant values, since some trace elements are only present in very low concentrations and the economics of avoiding them can be excessively expensive if possible at all. As a result of a cost benefit analysis the maximum levels of cobalt specified in all stainless and mild steels used inside the PCPV have been restricted to 1500 ppm and 300 ppm respectively. (The average actual values of cobalt level in steels used have in fact been shown by sample analysis to be significantly lower than these). This material control, in a similar way to the provision of internal shields, facilitates decommissioning by both reducing radiation levels for access inside the PCPV and reducing the activation of the components themselves.

(iv) In high radiation areas (where direct personnel access may not be possible during reactor operation) consideration is given in the initial design to the provision of such features as will facilitate future operations (e.g. mounting features for remote handling equipment and boxed-up openings in the cast concrete structures filled with more easily removable material such as concrete blockwork). One specific example of this is the provision in the High Activity Debris Vault (HADV) of support beams for use during decommissioning in the removal of the high activity waste. These support beams extend part way through the concrete shield wall of the vault, where they are accessible by removing concrete blockwork during decommissioning of the vault (see Fig 3).

(v) Where change rooms, access ways, etc which are required for decommissioning only, can be identified, floors and walls should be installed which permit easy modification, e.g. precast concrete slabs rather than "in situ" cast structures. This has been applied adjacent to the HADV openings.

(iv) In all potential contamination zones the surfaces are designed to facilitate decontamination. Although the surface finishes (chlorinated rubber paint, epoxy paint etc) will deteriorate during station life (and will need maintenance/replacement), they will have served their purpose in enabling a significant build-up of contamination on surfaces to be avoided and the basic surfaces will present a reasonable base for decontamination during the early years of the decommissioning activities. It is not considered necessary to improve on standard surfaces specifically to aid decommissioning, since life-time performance is subject to many variables and equipment is already available for cleaning seriously contaminated surfaces (particularly) when there are no onerous requirements on the standard of the post-decontamination surface finish during decommissioning).

(vii) Plant access routes are provided in the layout and construction of the station either as permanent operational routes or temporary construction routes. In most cases the decommissioning plan is essentially a reversal of the construction plan and use can be made of construction routes (sealed for reactor operation) for the removal of large mechanical items prior to the dismantling/breaking-up of the main building structure (which for contamination areas can serve as a useful containment for contamination control, although apart from the PCPV and HADV basic structural integrity is not relied upon for long term decommissioning plans).

(viii) Where the need for the use of remote demolition tools is identified, consideration must be given to ensuring adequate space is available for placing the tools in the decommissioning situation. To leave space for this purpose only would be uneconomic, therefore the possibility of removing adjacent plant items before the remote operation is commenced, should be planned where possible.

(ix) At any stage of the decommissioning activity it may be considered necessary to re-assess the radioactive inventory. To allow this to be achieved, it is recommended that a materials bank be established. This should consist of samples of all materials that will become activated during the station life. These samples would be retained with the decommissioning records.

(x) If a delayed decommissioning route is chosen, it will be necessary to leave the most active components for a long period, to allow such gamma emitting nuclides as ^{60}Co to decay. Although the structures which remain during the Stage 3 period would be sealed, it may be necessary to provide adequate surveillance. This cost could be reduced to an absolute minimum by ensuring the structure left to Stage 3 decommissioning, can be made secure against intruders. For example, choose an outer boundary of the structure to be left, which is part of a massive concrete component, with the minimum of penetrations. Such a boundary would also keep in a safe condition over long periods. These features can be recognised at the design stage.

It will be seen from Fig 2 that the major demolition problem will be the removal of the PCPV which is a dominant feature of the current generation of gas cooled reactors and has been the subject of a separate civil engineering study.

The vessels consist of about 50,000 tonnes of concrete and 3000 te of pre-stressing steel tendons. The vessel is activated by the epithermal and thermal neutron fluence to a depth of less than 500 mm, from the inner face. The first layer of prestressing tendons is centred at 500 mm from the inner face, therefore this is the only prestressing steel which could become mildly radioactive. The remaining 39 layers of prestressing tendons will not become radioactive.

The weight of concrete within the activated zone (that is the first 500 mm from the inner face) is 2700 tonnes, the remainder of the PCPV concrete would not become radioactive. It is clear from this, that if the inner 500 mm of concrete can be removed the remainder could be treated as non-active, and probably re-used in discrete quantities as the mean activity is less than one becquerel per gram.

One line of investigation was the possibility of a plane of weakness between the activated and non-active concrete. This may present some problems as far as anchoring the steel liner but such a development appears to be credible.

It is considered feasible to eliminate reinforcement and prestressing tendons from the activated zone. Whether or not this option would be adopted in future designs is a matter to be considered at the design stage, taking into account possible effects on design, plant layout, construction and overall cost. Several possible methods of forcefully breaking up the activated region of the PCPV have been considered and some of these appear feasible.

Traditional explosive techniques would seem to be a very effective way of breaking up homogenous concrete and a considerable amount of work has been done investigating this aspect under the CEC programme on decommissioning. This technique does, of course, have the disadvantage of noise, vibration, dust, flying debris and the like, depending on the size of the charges used. Where necessary, these disadvantages could be avoided by using, instead of explosives, chemical demolition agents, such as "BRISTAR" or mechanical devices such as hydraulic bursters.

Use of any of the above three techniques would be most effective if, during construction of the PCPV, preformed holes were provided in the activated zone to accommodate them. Preformed holes, however, give rise to two possible problems. Firstly, in current PCPV construction, it is not permissible to leave, in the structure, any void that could form a shine path for neutrons from the core. Preformed holes, therefore, would need to be plugged with a material capable of maintaining the shielding property of the concrete, able to survive immediately behind the liner during reactor operation, yet able to be easily removed when required. Lead or graphite are two potential filler materials. Secondly, a large number of preformed holes immediately behind the liner may raise doubts on liner anchorage integrity and liner support. Further investigation of this idea could best be carried out against known requirements in a detailed design phase.

The removal of the steel liner from the inner face of the PCPV presents a major problem to the decommissioning engineer. The design assessment has led to two possible avenues of investigation, they are as follows. First, further reduce activation, by additional shielding around the core. This may not be cost effective as the specific activity of the liner 10 years after shutdown, at the end of the reactors operational life, is only 7.8 Bq/g. The other proposal involves the use of a free standing liner. The current liner designs are bonded to the concrete by a multitude of closely spaced stud or hook anchors. These attachments will make the removal of the liner difficult and it is considered that it could be advantageous to partially isolate the liner from the concrete of the PCPV by only a minimal use of liner anchorages. This approach is possible, but presents problems regarding stresses set up in the liner. If the liner were not robust enough, such a design change could result in leaks which would be difficult to detect or repair. It would pose many new problems both from a design and a construction point of view.

The primary aim of the designer is to recognise those features of the proposed design which can be modified to assist decommissioning without significantly increasing operating risks of constructional problems. Changes to proven designs may only offer a marginal decommissioning advantage, but result in a disproportionate increase of other risks.

1.3 A documentation system to assist decommissioning

To support the specific plant design features which facilitate decommissioning, it is necessary to decide which documents and drawings will be necessary to assist the decommissioning engineer to implement the proposals.

The documents and records that will be required at the decommissioning phase will consist of a selection from the original design and safety documents, drawings, operational records, information on any changes that may have been made to the plant, and information relating to accidents, spillages and unusual events. In addition to the written records, material samples will also be required.

For nuclear power stations built between 1960 and about 1980, that is those stations that will be among the first to be decommissioned, the stock of data which is essential to the process of decommissioning will comprise of 100,000 to 300,000 sheets of paper and about 30,000 engineering drawings. The stations that were designed after this date will probably have a more extensive data bank.

The very large numbers of documents will need to be stored for a long period. The selection of the material, in sufficient detail for the decommissioning engineer, will require careful planning. The method of data retrieval will also be an important part of the overall effectiveness of the decommissioning proposals.

As part of the decommissioning study, a documentation system for assisting in the decommissioning of nuclear power stations has been investigated. The system, known as INFOSBAU/INFODOK was originally devoted to project data retrieval in building organisations, and was developed in Germany by the "Bundesminister fur Forschung und Technologie".

The systems constitutes an adaptable information retrieval system based on a powerful data base management software.

INFOSBAU/INFODOK makes accessible the information which is stored in documents of arbitrary form. It consists of:

- A documentation language to describe the information contents of any documents for the purpose of filing them as well as for the purpose of searching for them:

- Computer programs to define the special semantics of that language in every application, to store and to manage the information about the contents of documents and to retrieve the actual address under which a document is to be found;
- Methods to formulate one's own language according to the professional application environment; to employ that language when describing the information contents of documents and when formulating queries to search for information sources, and to organise the storing and the retrieval procedures.

Four application programs are needed:

- DOKDEF to define and to maintain the vocabulary of the documentation language;
- DOKEIN to feed the document descriptions to a database file;
- DOKAUS to assist the user when searching for informations by means of a guided dialog procedure and to control the retrieval process in the data base.
- DOKAUS/STDFR to support and formulate standardised queries to extract parts of data stock from the file or to retrieve statistical data.

These programs utilise the user interface of a data base management software.

If there are large amounts of documents to be made accessible it will be useful to concentrate them on microfilm. This will speed up the access to the information and at the same time reduce the expenses for a duplication of the whole stock as well as for copies of specific documents.

The whole system including computer assistance and microfilm equipment is illustrated by Fig 4.

Most of the existing documentation and registration systems are unsuitable to retrieve the information relevant to decommissioning, much of which is hidden in the documents originating from the plant's specification, design, construction and operating records.

To develop specifications of a system which is suitable for the purpose of decommissioning, co-operation with a company running nuclear power stations and preparing to decommission its first reactor has been established. From discussions with people engaged in decommissioning and documentation processes, from initial studies of information material and from an analysis of its information content the following results were derived:

The total amount of data generated during the phases of a power station's lifetime may be divided into the following parts:

(1) Project Data (including design data and material specifications)
(2) Data documenting the operating process
(3) Data documenting the shutdown phase
(4) Data documenting the decommissioning process

Discussions with organisations running nuclear power plants already being engaged in decommissioning, led to modifications of the documentation language, of the documentation unit's record description and of the application programs of INFODOK. Due to these modifications the documentation system has been decomposed into a model of layers separating infological from technological layers.

Two video display forms are proposed to support the documentation process in practical application.

Requirements to the software system are functionally defined within a model of layers (Fig 5), and the interfaces between the system's main components the application programs of the system INFODOK, the data base management system, and the operating system as well as the interfaces between the units of function in the interior of a layer are described.

An automatic microfilm storage and retrieval system integrated into the whole system and coupled via an interface to the computer is specified. To organize the computer aided documentation process a basic sequence of operations is proposed.

It is intended to install the INFODOK hard and software system as a pilot documentation system for a project of dismantling and decommissioning of a Nuclear Power Plant.

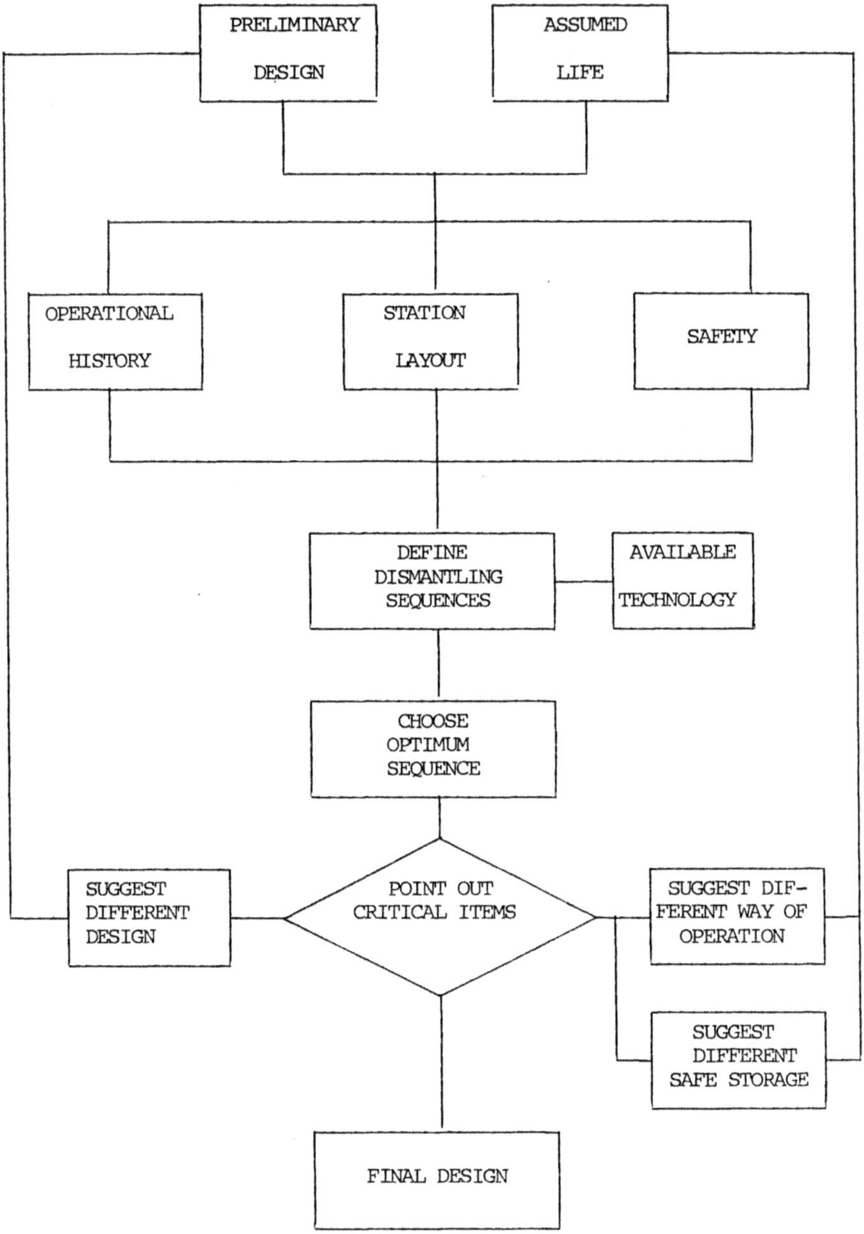

FIG. 1 - STRATEGY LOGIC SEQUENCE

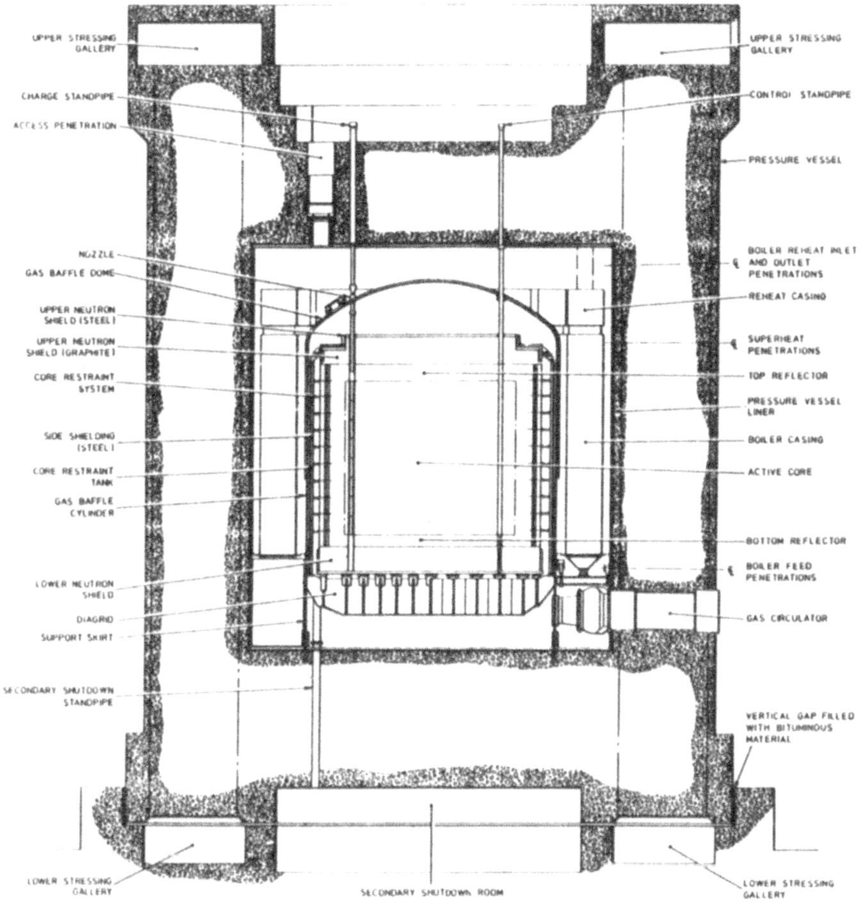

FIG. 2 - SECTIONAL ELEVATION ON AN AGR

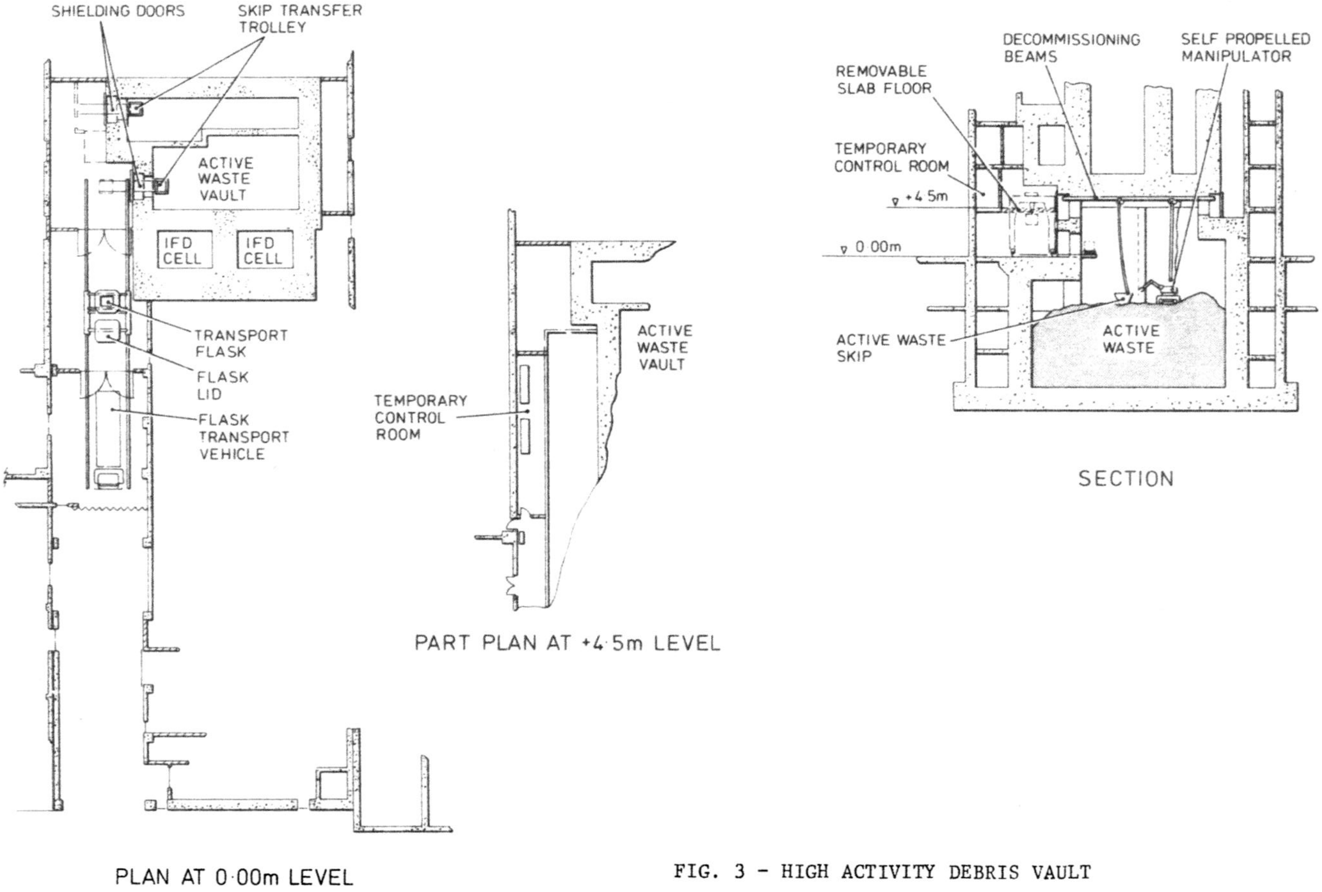

FIG. 3 - HIGH ACTIVITY DEBRIS VAULT

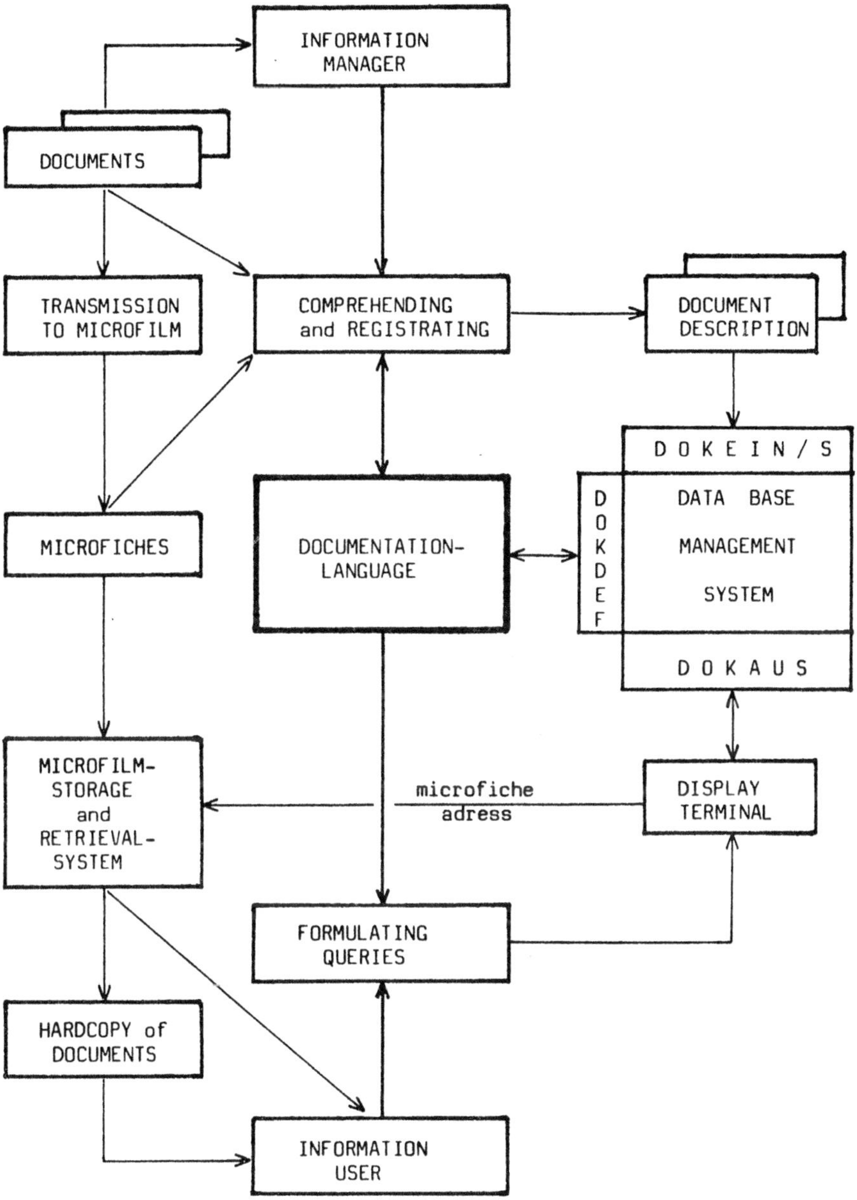

FIG. 4 - COMPONENTS AND PROCEDURES OF THE GENERAL SYSTEM

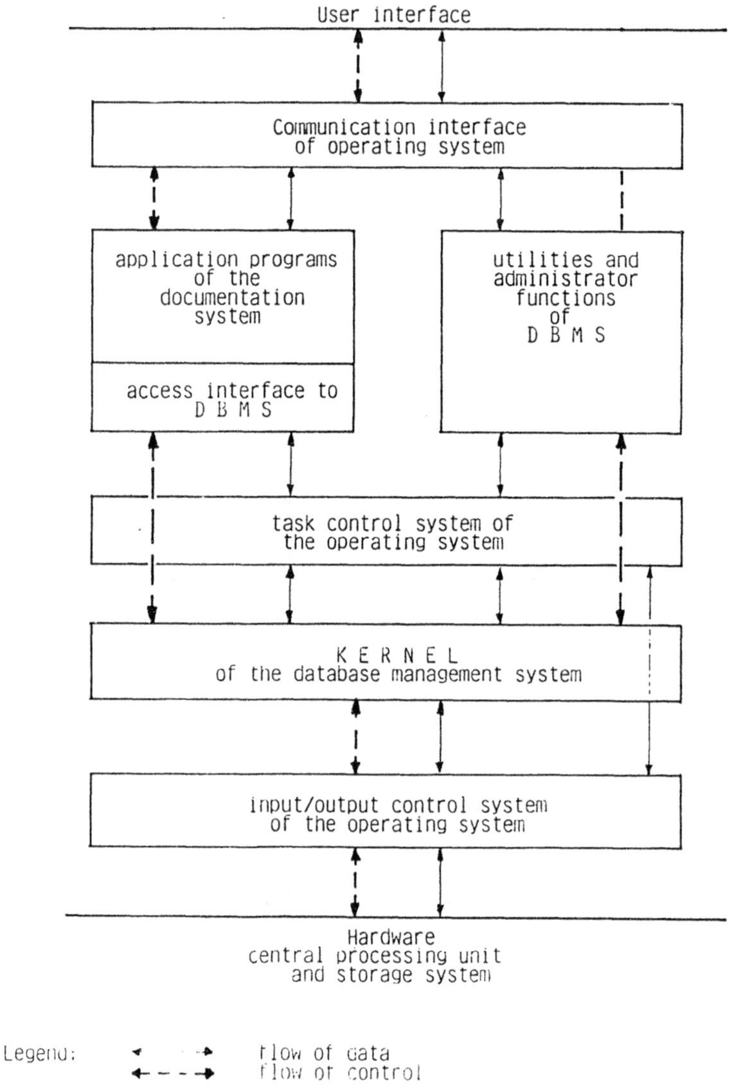

FIG. 5 - STRUCTURE OF LAYERS OF THE SOFTWARE SYSTEM

SESSION V : SUMMARY OF DISCUSSION

Paper : "Concepts aimed at minimizing the activation and contamination of concrete"

An FRG participant asks, firstly, if the reinforcement steel bars have been considered in the activation calculation and, secondly, what the cooling requirements for the borone-carbide layer are. Mr. Bittner confirms that the presence of 120 kg/m^3 of reinforcing steel has been included in calculations. The production of heat by the Borone-Lithium reaction under neutron irradiation has been calculated and the results are available in the full paper. The heat production is very low and can be neglected in comparison to the heat flow from the reactor vessel. This statement is valid for a standard PWR's neutron flux distribution at the chosen location of the layer.

An FRG participant asks if the ORIGEN-2 computer programme has been used with an adapted reactor spectrum or with the non-modified spectrum. Mr. Bittner indicates that the programme has been used with values for this part of the reactor as employed by the licensing authorities for calculations for reactors Grafenrheinfeld and Unterweser. The calculations reached from the inner diameter of the reactor vessel to the outer biological shield surface.

Paper : "Reduction of radiation exposure by replacement of Co-alloys and control of Co, Nb and other trace elemtns in steel.

An FRG participant comments that activation of the reactor vessel is not a concern, as this component will have to be treated as radioactive waste even after decay times of more than 100 years. Activation of steel is of importance as a source for contamination of the primary loop. He asks if the composition of radionuclides produced by activation is reflected by the composition of the contamination layers. Mr. Wieling confirms that the main objective is to diminish the amount of hard gamma emitters like ^{60}Co embodied in the contamination layers. Actually, Cobalt-base alloys form only about 1 to 2.5 m^2 of surface in the primary loop of an LWR, but the alloys contain 60% of Cobalt. Substitution of these alloys will not eliminate activation of corrosion products containing Cobalt, for the simple reason that the large surfaces in contact with the cooling water are made of steel containing small amounts of this element. Consequently, it is also necessary to aim at a reduction in Co-content of these steels.

The Chairman asks what the supplementary cost would be for ordering steel with very low Co and Nb contents, in order to fabricate components subject to heavy activation, like reactor vessel internals, from this base material. Mr. Wieling expects overcharge of 20% or more, depending on the degree of purity required. He thinks a better way would be to convince the steel supplier to select from his stock of base materials those parties having the lowest content in undesirable elements; in this manner, no extra cost is to be expected.

A U.S. participant asks if the hardness factor has been considered in reduction of Cobalt contents. In components like control rods Co-base alloys are employed in order to meet requirements for hardness. Mr. Wieling answers that substitution of Co-base alloys has to be done meeting all functional requirements for the component. In the Convoy line of reactors, this substitution is done progressively, and substitution of Co-base alloys in control rods, which are the most sensible parts, will be done as the last step after careful testing.

Paper : "Review of plant design features facilitating decommissioning"

A UKAEA representative relates his experience in dealing with documents on the Dragon reactor, which is shut-down and will certainly not be dismantled for a long time. Already now drawings are brittle. Microfilms, magnetic tapes, etc., also have limited life times. He asks what the expected life time of the documentation system is, which is mentioned in the presentation.

Mr. Müller (TH Darmstadt) who has been involved in the contractual work for the documentation system, explains that the technology in this field is progressing rapidly. Already now video disks and scanning procedures are being discussed. Within the next five years, the problem of refreshment by micro-film reproduction will probably be solved through these video disks which can be stored and protected as data bank carriers like magnetic tapes, etc., already now. The documentation system has been conceived as consisting of individual layers; in this manner, technological progress in hardware or software development can be integrated by replacing the corresponding layer. A documentation has to be conceived for life times from 80 to 100 years; discussion with reactor operators show that a unique documentation is needed which would be in use during design, construction, production and decommissioning periods.

A U.K. participant adds that now large capacities for storing information are available. For instance, the laser disk which is used to store television pictures, stores about 30 000 sets of engineering type drawing and documentation. The problem of retaining information of this sort is not one of the carrier, it is about the multiplicity of the records held; such large amounts of information can be copied readily and dispersed to many centers; this has to be done for this kind of high security data.

CONCLUDING PANEL

DECOMMISSIONING OF NUCLEAR INSTALLATIONS - AN OUTLOOK

Chairman

H. DAVIES
Deputy Director-General, CEC, Directorate-General for Science, Research, Development and the Joint Research Centre, Commission of the European Communities

Secretary

B. HUBER
CEC, Directorate-General for Science, Research and Development

Members

A. BERTINI
Capo Unità Speciale "Decommissioning", Direzione Produzione e Trasmissione, Ente Nazionale per l'Energia Elettrica

J. ESSMANN
Bereichsleiter der Technisch-Wissenschaftlichen Abteilung der Preussischen Elektrizitätswerke AG, Vorsitzender des Arbeitskreises Stillegung der Vereinigung Deutscher Elektrizitätswerke (VDEW) Versorgungs Unternehmen

E.H. HUBERT
Professeur Honoraire à l'Université de l'Etat à Liège, Conseiller auprès de l'Union des Exploitations Electriques de Belgique (U.E.E.B.)

H. LAWTON
Head of Laboratories, Windscale Nuclear Laboratories of the United Kingdom Atomic Energy Authority

P. TANGUY
Directeur de l'Institut de Protection et de Sûreté Nucléaire, Commissariat à l'Energie Atomique de la République Française.

CONTRIBUTIONS BY THE PANEL MEMBERS

Introduction by the Chairman

The intention of this Session is to allow a few moments for us to consider not the present state of work and the immediately recent past achievements of the programme, but to think of some of the important issues and trends which are of relevance to the future.

In a first phase of the Session, each panel member will make a short statement. The statements made by each panel member are not to be regarded as that member's total view of the range of issues which are important, but are a selection of statements which attempt to be, to some extent, complementary to each other and to cover the span of the field which is quite wide : technical and policy, economic and practical. The second phase of the Session will be devoted to a discussion which will be open to the floor.

A. BERTINI

I feel compelled to reiterate the need for a solution to be provided at national level and by the public authorities, to the problem of radioactive waste management. Such a solution is a necessary preliminary to any thoroughly planned decommissioning. Type and location of waste are the two basic factors to be taken into consideration when conditioning wastes, resulting not only from normal operations but also from decommissioning. On what basis should one decide, for example, whether to use one large or two small containers? The techniques are available : what it really boils down to is cost and evaluations. Only the State can provide the regulations in this field; to draw an analogy, one is perfectly free to install the plumbing in one's own home, but only the local authority can plan the water and sewage systems to which it is linked. Without a firm base, engineers will be forced to cover all contingencies and therefore waste both money and effort.

With regard to the "de minimis" value - i.e., of the limit below which a body or surface is no longer considered radioactive for the purposes of its unrestricted release - this limit must exist and must be established and approved at international level. The European Community could play an important part in this. As far as the limit itself is concerned, one should avoid creating further problems by continually picking holes in radiation protection assessments. As the former head of a power station, I am particularly conscious of the importance of radiation protection : radiation exists in a power station with implications both for oneself and for one's employees. One must, however, be logical. Take the magic number 10^{-5} µCi/cm². Although this figure is based on evaluations of direct radiation, decontaminating to a level of 10^{-5} rather than 10^{-4} Ci/cm² has been regarded as producing a far greater quantity of secondary waste, giving rise to an additional safety problem and a possibly unnecessary restricting of decontamination activities. Safety is a habit of mind, a fact of life in the engineering world - it should not be an obsession.

It has been suggested that differentiated limits should be established for the various radionuclides, which is fair enough, provided that one is thinking in terms of pure reason. However, it should not be forgotten that Kant also wrote the "Critique of Practical Reason". The plant engineer, who is a front-line operator, needs clear-cut but simple rules. When it comes to decontaminating large quantities of material (e.g., when the turbine hall of a BWR is dismantled), imagine the endless series of radiochemical tests that would have to be performed, with the requisite quality guarantees, in order to check the limits of each of the numerous radionuclides. Practical reason imposes a limit, possibly one that is more restrictive, but one that is easy to apply and quick to check.

For these and many other reasons, the experience acquired through dealing with the problem of decommissioning right from planning level is of incomparable value.

H. LAWTON

We, the British, evidently like to be different. We drive on the left hand side of the road, the rest of the world on the right. The world now concentrates on water-cooled reactors, we reserve our main effort for gas-cooled. However, I am impressed at the relevance to both reactor systems of much of the work funded by the CEC and discussed in the last three days.

There are several general points I would like to make.

First, I am sure that the development of machines to cut and handle radioactive material remotely, is still in its infancy. Undoubtedly, robotic techniques supported by the computer technology, are going to develop quickly so that our machine, described yesterday and now subject to detailed design, may appear to be rudimentary in concept in ten years' time. But we remain convinced of the importance of simplicity in the machine's design, since that should ensure reliability and ease of maintenance.

Our British project with the Windscale Advanced Gas-cooled Reactor (WAGR) is currently dominated by the issue of waste disposal. This is an unresolved issue, heavily coloured by political considerations. The uncertainty as to the disposal site for the waste is now beginning to affect some technical aspects of the project.

I think we all see now the increasing role played by the national and international authorities who lay down the regulations by which we must work. It is to be hoped that these rules clarify in way which are helpful to the decommissioning of nuclear facilities. It is a very real encouragement to the British nuclear industries that our government has welcomed the WAGR decommissioning project, since that gesture surely implies a confidence that the unresolved political issues, particularly over waste disposal, will be resolved.

Next, I would strongly urge that we gear on CEC development work in this field, whenever possible, to a live decommissioning project. This will place precise demands on the development people and, occasionally, their findings will compel the projects's plan to be changed. The project director must be prepared for such disappointments and be ready to adapt his plan to meet the new circumstances.

Finally, I remain convinced that we all must do a good job of decommissioning, finding the best solutions for a reasonable financial outlay. Time is not particularly pressing and delays can be tolerated, but the responsibility to execute our projects efficiently and, above all, safely, is a grave one. All nations will benefit from the success of one of us.

J. ESSMANN

In the course of this conference it has been shown that, even as things stand today, the available technologies are such that, from the stand-point of radiation protection in particular, a nuclear power plant can be decommissioned without risk at the end of its service life. Whatever strategy is ultimately selected for the purpose, both the technical and the financial requirements are within the bounds of feasibility.

The research projects on decommissioning initiated by the Commission, as well as the procedures and methods employed by specialized companies in implementing prolonged interventions such as the extensive back fitting works on German boiling-water reactors, have shown that the range of techniques already available for the various tasks is so wide that procedures can be selected according to local requirements and the operational history of the plant. As a rule, moreover, the expenditure on interventions involving plants in operation is likely to be considerably greater than under decommissioning conditions, owing to the greater difficulty of access and the radiation proection measures needed. In the case of decommissioning, for instance, accessibility can be improved by providing new access facilities.

What we have also learnt from this conference, I feel, is that there is no longer the same need to modify the plant design in order to facilitate decommissioning. Since our plants are designed to facilitate inspection and maintenance in the interests of operational safety and availability, we are already ensuring that decommissioning can be effected at reasonable cost. The emphasis should now be placed much more on optimizing the procedures and techniques and on adapting them to the existing design features of nuclear power stations with a view to decommissioning.

Let me clarify this with an example : instead of requiring the plant to be designed so that the diamond saw can be used for cutting through the biological shield, we should ask for the saw to be modified so that it can also be used in the decommissioning of nuclear power plants under the given spatial conditions. It is in this way that we should utilize the available engineering and research potential. It should be used firstly to develop an economic procedure that takes account of the cost/benefit ratio, and secondly to minimize the quantities of radioactive waste that occur in these procedures and subsequently have to be taken to a final repository.

In conclusion, I should like to underline this point once more : in the light of present-day knowledge, the need is no longer for power plant design standards that allow for eventual decommissioning : rather there is a need for guidelines and standards that the procedures and technologies must satisfy if they are to be used for decommissioning purposes.

E.H. HUBERT

For the top decision makers in companies operating nuclear power plants, the problem of decommissioning boils down to two basic questions :

1. is decommissioning necessary, not to say indispensable?
2. does the current state of scientific and technical knowledge permit the operation to be carried out in a satisfactory manner, i.e., while keeping individual and collective doses to acceptable levels and expenditure within economically acceptable limits?

It is on the answers to these two questions that the decision on the choice and time-table of decommissioning levels may depend. The answer to the first question is affirmative both because of the shortage of possible sites in the industrialized countries and for reasons of public acceptance of nuclear energy.

Like my colleagues, I think that the answer to the second question is likewise affirmative, at least as regards the scientific and technical knowledge permitting decommissioning with due regard to the health of the public and of the workers involved. But on the economic side, the assessment criteria should be more sharply defined in order to be able to reply to the following subsidiary questions :

1. is it possible, at present, to evaluate the cost of decommissioning?
2. how will the necessary expenditure be scheduled?
3. will there be a need for provisional financing and, if so, what will be its impact on the price per kWh?

These are complex questions : it is certainly possible, in the present state of technical knowledge, to evaluate the cost of a technically satisfactory operation at 1984 rates and subsequently to up-date the evaluation; but this ignores the consequences of technical advances made in research carried out between now and the year in which decommissioning will start, a date which itself is still somewhat vague.

This impression is even greater in the case of future reactors, since design improvements may well modify and perhaps reduce the cost of decommissioning. The operator therefore has no alternative but to make assumptions which are likely to make up for a lack of available credit if they are too optimistic as regards the useful life-time of the power station or, if too pessimistic, will put up the price of the kWh unnecessarily and by the same token adversely affect economic recovery, employment and even the balance of payments.

At present, for want of firmer data, reactor operators generally allow for a plant life-time of 30 years and a probable decommissioning cost amounting to 10-15% of the original investment.

There are several possible ways of obtaining more precise values :

a. pilot experiments on industrial reactors;
b. systematic evaluation of each of the operations necessary for decommissioning, with due regard to the effect of the various techniques employed;
c. intensifying the exchange of information at international level.

The first of these approaches was brought up and recommended in the course of our conference. The second approach was touched upon in the USA in the NUREG reports, at a meeting of the NEA in Paris in 1980 and, finally, by Mr. Regan in the last of our conference reports. The third approach leads me to regret that our conference all too often confines itself to describing the work undertaken in the context of the European Community's programme, without the authors' making any comparisons with similar studies conducted outside the Community or outside the Community programme.

Finally, with a view to the future of decommissioning, one should not under-estimate some important questions in the administrative and political field, such as licensing procedures, the change-over from nuclear third-party liability to conventional third-party liability and the part played by insurance in the case of a normal or premature decommissioning : international bodies like the IAEA, the NEA (OECD) and the International Nuclear Law Association (INLA) are applying themselves to these questions. Therefore, apart from the worries about dismantling, treatment and storage of residues and evolution of design that have been the subject of our meetings, the operator responsible in the eye of the public is beset by other worries, and it is only through increased cooperation between all the international organizations that worthwhile solutions will be found.

P. TANGUY

In my opinion, there are three major constraints to be taken into account when decommissioning :

- it must be safe,
- it must be economical,
- it must have credibility.

Safety. Decommissioning is a nuclear activity which has to be carried out by teams which are well aware of the exigences of working in the nuclear field and the relevant procedures, and these teams will have to have confirmed their skills in the construction, operation or maintenance of nuclear installations. The nuclear industry is a different industry : if one overlooks this, one encounters problems.

Economy. There is, potentially, a large decommissioning market; however, with time, this market could shrink almost indefinitely and it will only really be opened up when the dismantlers are able to put forward rapid, safe and economical plans and procedures. Techniques with a much higher productivity level than those currently available must, therefore, be developed and this will require a research and development programme.

Credibility. The response to the problems of safety and economy must be credible, which is to say that they must have the support of governments, owners and public. This credibility will only be acquired by rapid action on the part of the nuclear community towards the solution of the problem posed by the many installations which have been, or are about to be, shut down.

There are then, clear directions for the future :

- an R&D programme needs to be laid down;
- embarrassing "corpses" need to be got rid of; however, in so doing, we must take the opportunity to run full-scale tests on the new technical solutions which are developed. There must be some interface between R&D and pilot decommissioning operations. This could give the owners financial problems; they may well fear that what they can afford may not be in step with the wish to implement new techniques, although these fears possibly reflect an under-evaluation of the costs and the risks, associated with the use of "conventional" techniques. It would, therefore, be advisable for additional financing machinery to be provided for, by the Community, for example. This financing must be explicitly linked to the implementation of new processes which hold out the promise of significant progress.

I will end on two problems mentioned by previous speakers, problems which we still have to overcome :

- in R&D it seems to me that remote handling is an essential area; it does pose the problem of financing major investment, which I believe can only be resolved if the programme is part of a much greater whole, an entity which covers the fields of maintenance and repair or even, as in France, the entire field of nuclear and non-nuclear automatic industrial plant;
- for demonstration activities, a clear policy on developments in the field of waste which gives rational responses to the problems of standards for re-cycling, discharge or immersion, final storage, etc., represents a sine qua non. Considerable emphasis was placed on this point this morning, but it can never be stressed too much, since it is basically a problem of political will and public acceptance.

CONCLUDING PANEL:

DECOMMISSIONING OF NUCLEAR INSTALLATIONS: AN OUTLOOK
SUMMARY OF DISCUSSION

A U.K. participant comments the discussion on simplicity and simple systems, especially with respect to the particular area of remote operations. There seems to be a certain degree of ignorance of the present state, particularly of information technology, but also of remote operations. There are a lot more applications of remote operations which are well ahead of the nuclear industry, especially submarine teleoperation in its many forms. Teleoperation systems are not made with the intention of them being complex, but to ensure that the operational tasks themselves are simple. The equipment itself has to be robust and it has been demonstrated in many applications under the sea that this can be done.

A Swiss participant asks if decommissioning is a problem or not. In his opinion it is not a problem, but there is a lack of money to demonstrate this.

Mr. Hubert means that the problem of decommissioning is principally not a technical problem; the technical problem is largely solved; one may improve, one can do many things, but basically the technical problem is solved. What is not solved, is the political problem.

Mr. Bertini adds a comment on the question of simplicity raised before. For work in an area where the radiation level is high, a machine is needed which is safe, and there are various machines in our plants which are not very simple. However, in decommissioning, for every operation performed in an area of high radiation level, i.e. of the order of rem/h, there are many operations performed in areas where no real risk of high dose rates exists. These are areas with a certain contamination level, where many operations are to be carried out and where the main problem is to ensure that no material that is still contaminated gets out without being controlled. He is therefore referring to simplicity of procedures, simplicity of measurement and control methods, simplicity of regulations.

Mr. Crégut comments on the view that decommissioning urgently needs the support of politics. If we appeal to politicians, what should we ask them? Do we, as responsible and competent technicians, already possess a decommissioning doctrine? Are we capable, presently, of deciding whether an installation finally shut down must be dismantled immediately and completely, or whether one must proceed by intermediate waiting stages? Are we capable of deciding on the destination of all types of waste produced? When all these elements will be available, when we possess a doctrine, this will allow us to request a regulation, and it is only at this stage that we should turn to the politicians for support. Up to that time, what we can ask the politicians is to give us the funds.

A Belgian participant intervenes concerning the decisions to be taken by politicians and public acceptance of the solutions adopted in the field of decommissioning, an important aspect not mentioned so far is the role of the press. At least in certain countries it is very hard for a politician to take a decision against the current of the press: it is therefore very important to have good contacts with the press in order to gain public acceptance before turning to the politicians. There lies in his view the main problem which has delayed the clean-up of TMI. For instance, it has taken nine months to obtain the authorization for a load test of a crane bridge, because of opposition in the public.

A Swiss participant asks, after having recalled that radionuclides do not know borders, why it would be necessary to wait until national or international regulations are set up. He believes the European Community should take this problem in hand.

Mr. Tanguy comments on the question of international regulations that already now competent firms can commit themselves to carry out decommissioning operations within the range of acceptable risks for workers and without significative risk for the public and the environment. This is the present situaiton. Regulations, and in particular international regulations, which by definition must result from a consensus, which is generally reached by an addition of requirements, may at a certain stage of technical development act as a brake and tend to produce results which are contrary to the objectives.

What he is taking from this conference is that we must first furnish proof of our capabilities by carrying out demonstration projects. To do this, we need a certain political will, in order to obtain the funds and to raise certain barriers. These barriers appear to relate essentially to the destination of the wastes and concern, e.g., the "de minimis" limits and the question of sea disposal. These are points which need answers if we wish to accomplish demonstration operations. On the other side, it does not seem desirable to him, to embark at the present time in an international regulation of decommissioning operations.

LIST OF PARTICIPANTS

ACCORONA, C.
Researcher
E.N.E.A. - TERM- COASE
CRE Casaccia
C.P. 2400
I - 00100 ROMA A.D.

AHLFAENGER, W.
Dipl.-Ing.
VEW - AG
Postfach 941
D - 4600 DORTMUND 1

ALLEGRE, R.
Ingénieur
CEA/CEN-Cadarache
SEER- BT 208
B.P. 1
F - 13115 ST PAUL-LEZ-DURANCE

ALLIBERT, M.
Chercheur
INPG
ENSEEG, B.P. 75
F - 38402 ST MARTIN D'HERES

ANDREANI, A.
Chef du Département Nucléaire
Société NEYPRIC
75, rue Général Mangin
F - 38100 GRENOBLE

ANJUM, A.
Senior Engineer
Foster Wheeler Power
Products Ltd
Greater London House
Hampstead Road
UK - LONDON NW1 7QN

ANSELIN, F.
Ingénieur
CEA/IRDI-N
33, rue de la Fédération
F - PARIS

ANTOINE, P.
Ingénieur
C.E.A. IPSN/UDIN
C.E.N. Marcoule
B.. 171
F - 30205 BAGNOLS-SUR-CEZE

ARMSTRONG, P.J.
Design Engineer
Ontario Hydro
700 University Ave
CANADA - TORONTO, Ontario M5G IX6

ARNDT, K.-D.
Dipl.-Ing.
Salzgitter AG
Postfach 41 11 29
D - 3320 SALZGITTER 41

AUDE, G.
Ingénieur
C.E.N. Grenoble
C.E.A.
B.P. 85x
F - 38041 GRENOBLE Cedex

AUGUSTIN, X.
Ingénieur
TECHNICATOME
B.P. 18
F - 91190 GIF-SUR-YVETTE

AULER, I.
Ing. (grad.)
NIS - Nuklear-Ingenieur-
Service GmbH
Lyonerstr. 22
D - 6000 FRANKFURT 71

AVANZINI, P.G.
Manager
NIRA S.p.A.
Via dei Pescatori 35
I - 16129 GENOVA

AZUMA, K.
Senior Engineer
Engineering Sec., Nuclear Power Dept.
Okumura Corporation
1-3-10, Motoakasaka, Minato-ku
JAPAN - TOKYO 107

BACH, W.
Leiter Bereich Technologie
der Werkstoffe - Universität Hannover
Institut für Werkstoffkunde
Appelstr. 11a
D - 3000 HANNOVER 1

BAIER,
Dornier System GmbH
D - FRIEDRICHSHAFEN/IMMENSTAAD

BANTLE, S.
Dipl.-Ing.
Gesellschaft für Technisch
Orientiertes Rechnen mbH
Zerzabelhofstr. 76
D - 8500 NUERNBERG 30

BARBIER, R.
Chef Fabrications Nucléaires
CREUSOT LOIRE
Division Energie
B.P. 31 Cedex
F - 71208 LE CREUSOT

BARKER, A.
Chartered Engineer
United Kingdom Atomic
Energy Authority
Risley
UK - WARRINGTON, CHESHIRE, WA3 6AT

BARNIER, M.
C.E.A. - C.E.N. Saclay
IRDI/DERPE/SPS
F - 91191 GIF-SUR-YVETTE Cedex

BAROU, J.
Ingénieur
NOVATOME
La Boursidière, RN 186
F - 92357 LE PLESSIS ROBINSON Cedex

BAUMANN, J.
Techniker
Quadrex International
Alois-Senefelderstr. 1-3
D - 6800 MANNHEIM

BAURPER, C.
Ingénieur
Société NEYPRIC
75, rue Général Mangin
F - 38100 GRENOBLE

BEIER, B. G.
Dipl.-Ing.
Techn. Ueberwachungsverein Baden E.V.
Fachbereich Kerntechnik
und Strahlenschutz
Dudenstrasse 28
D - 6800 MANNHEIM 1

BERGAMASCHI, C.
Chef de départment nucléaire
Renault Automation
B.P. 30
F - 78390 BOIS d'ARCY

BERGEMANN, L.
Dipl.-Physiker
Kernkraftwerke Gundremmingen
Betriebsgesellschaft mbH
D - 8871 GUNDREMMINGEN

BERGMAN, C.
Radiation Protection Officer
National Institute of
Radiation Protection
P.O. Box 60204
S - 104 01 STOCKHOLM

BERGMAN, H.
Netherlands Energy
Research Foundation
P.O.B. 1
NL - 1755 ZG PETTEN

BERTINI, A.
Manager
Decommissioning Unit
ENEL - Ente Nazionale
per l'Energia Elettrica
3, Via G.B. Martini
I - 00198 ROMA

BESSON, B.
Ingénieur
Seri Renault Ingénierie
B.P. 19
F - 78391 BOIS D'ARCY Cedex

BIANCALE, P.
CEA/CEN Saclay
DEMT/SYST/EXPE
F - 91191 GIF-SUR-YVETTE

BIRSS, I.R.
Research Manager
U.K.A.E.A.
Windscale Nuclear Lab.
UK - SELLAFIELD, CUMBRIA CA20 IPF

BISCI, R.
Commission of the European
Communities, Directorate
General "Science, Research
and Development"
200, rue de la Loi
B - 1049 BRUSSELS

BITTNER, A.
Dipl.-Ing.
DYWIDAG
Dyckerhoff & Widmann AG
Erdinger Landstr. 1
D - 8000 MUENCHEN 81

BIVER, E.
Administrateur Délégué
GRADEL S.A.
B.P. 54
L - 8401 STEINFORT

BLACKWELL, W.Ch.
Principal Engineer (Nuclear)
GEC Energy Systems Ltd.
Cambridge Road
UK - WHETSTONE, LEICESTER LE

BOBLETTER, A.
Ingénieur
Bertin & Cie
Service Documentation
B.P. 3
F - 78373 PLAISIR Cedex

BOEDEKER, B.
Maschinenbau-Ingenieur
Gg. Noell GmbH
Postfach 6260
D - 8700 WUERZBURG

BOISSONNEAU, J.-F.
Ingénieur
Technicatome
B.P. 18
F - 91190 GIF-SUR-YVETTE

BONGERS, H.J.G.
KEMA
Utrechtse weg 310
NL - 6812 AR ARNHEM

BONNAIRE, B.
Cadre Commercial
S.N.P.E.
12, quai Henri IV
F - 75004 PARIS

BORRONI, P.
Researcher
CISE SpA
P.O. Box 12081
I - MILAN 20134

BOULITROP, D.
Ingénieur
C.E.A. - SPR/SIDS
C.E.N. Saclay
Bât. 120
F - 91191 GIF-SUR-YVETTE Cedex

BRAGG, K.
Radiation Protection and
Waste Management Division
Nuclear Energy Agency
38, bvd Suchet
F - 75016 PARIS

BRAMBILLA, G.
Engineer
AGIP S.p.A.
Via Sabbionara 611
I - 40059 MEDICINA
(Bologna)

BRAMY, W.
Ingénieur
NOVATOME
La Boursidière, RN 186
F - 92357 LE PLESSIS ROBINSON

BREGANI, F.
Research Engineer
ENEL
Centro di Ricerca
Termico e Nucleare
Via Rubattino 54
I - 20134 MILANO

BRICTEUX, J.
Chef de Département
I N I E X
200, rue du Chéra
B - 4000 LIEGE

BUONO, A.
Engineer
E.N.E.L.
Via G.B. Martini 3
I - 00198 ROMA

BUXBAUM, H.
Mechanical Engineer
Swedish State Power Board
S - 16287 VAELLINGBY

CARAMELLE, D.
Ingénieur
C.E.A. - C.E.N. de Saclay
DPT/SPIN - BAT 93
F - 91191 GIF-SUR-YVETTE Cedex

CAVELLI, F.
Dipartimento di Progettazione
Architettonica Eurbana
Via M. Pantaleoni, 23
I - 00191 ROMA

CETINCELIK, M.
Editor-in-Chief
Energy World - Press
P.O. Box 680 - Kizilay
Turquie - ANKARA

CHAPUIS, M.
Ingénieur
C.E.A. - Deptartement
Protection Technique
Sect. d'Etud. de Protection
pour les Déchets
B.P. 6
F - 92260 FONTENAY-AUX-ROSES

CHARAMATHIEU, A.
Ingénieur
S.T.M.I.
9, rue Fernand Léger
F - 91190 GIF-SUR-YVETTE

CHARLES, T.
Ingénieur
Ministère "Industrie et Recherche"
Serv. Central de Sûreté
des Inst. Nucl.
278, rue A. Pajeaud
F - 92160 ANTONY

CHAUDON, L.
Ingénieur
C.E.A. - Centre d'Etudes Nucléaires
de la Vallée du Rhône
B.P. 171
F - 30205 BAGNOS-SUR-CEZE Cedex

CHIESA, G.
Castagnetti SpA
Via Fabbrichetta 65
I - GRUGLIASCO (To)

CHIRON, A.
Ingénieur
Bertin & Cie
Service Documentation
B.P. 3
F - 78373 PLAISIR Cedex

CHNEIDER, G.
Collaborateur Université
Université de Liège
Laboratoire thermodynamique
Rue Ernest Solvay
B - 4000 LIEGE

CHUNG, B.H.
General Manager
Nuclear Power Generation Department
Korea Electric Power Cooperation
87 Samsungdong Kangnamgu
Korea - SEOUL

CIANI, V.
Commission of the European Communities
Directorate-General "Environment,
Consumer Protection and Nuclear Safety"
200, rue de la Loi
B - 1049 BRUXELLES

CLARK, M.
Scientist
National Radiological
Protection Board
Chilton
GB - DIDCOT Oxon OXII ORQ

CODEE, H.D.K.
Coordinator-Administrateur
Ministry of Housing, Physical
Planning and Environmental Management
Koopmansstraat 1
NL - 2280 HV RIJSWIJK

COHEN, P.
Directeur du Développement
et de la Prospective
Soc. de Travaux en
Milieu Ionisant (STMI)
9, rue Fernand Léger
F - 91190 GIF-SUR-YVETTE

CONSTANT, J.
Deputy Manager
Mercantile-Beliard N.V.
P.O. Box 269
B - 2000 ANTWERPEN

CONTI, M.
Engineer
E.N.E.A. - PAS/SMAIMP
CRE Casaccia
C.P. 2400
I - 00100 ROMA A.D.

CORDERO, G.
Ingénieur
Gagneraud Père et Fils
7/9, rue Auguste Maquet
F - 75016 PARIS

CORNET, G.
Ingénieur
Belgonucléaire
25, rue du Champ de Mars
B - 1050 BRUXELLES

COSSAR, A.E.
Engineer
Babcock Power Ltd
Construction Division
9 Bream Place
GB - HUNSTON Scotland

COSTE, G.
Ingénieur
C.E.A. - IPSN/UDIN
C.E.N. Marcoule
B.P. 171
F - 30205 BAGNOLS-SUR-CEZE

COUEZ, H.
Directeur de la Division Nucléaire
A.C.E.C.
B.P. 4
B - 6000 CHARLEROI

COURTOIS, C.
CEA/CEN Cadarache
SEDFMA/SETED - B.P. 1
F - 13115 SAINT PAUL LEZ DURANCE

CREGUT, A.
Ingénieur
C.E.A. - IPSN/UDIN
C.E.N. Marcoule
B.P. 171
F - 30205 BAGNOLS-SUR-CEZE

CROXFORD, J.S.
Director
Wimpey Gilbert
Beaconsfield Road
UK - HAYES, MIDDLESEX UB4 OLS

CURTIS, L.M.
Manager, Safety Audit
British Nuclear Fuels PLC
Risley
UK - WARRINGTON, CHESHIRE WA3 6A3

D'ANNA, C.
Manager of Nuclera Op. Dept.
Ente Nazionale per l'Energia
Elettrica
Viale Regina Margherita
I - 00198 ROMA

DA COSTA CABRAL GANAO, L.
Engineer
A.C.E.C.
Nuclear Division
P.O.Box 4
B - 6000 CHARLEROI

DAVIES, D.H.
Deputy Director General
Commission of the European
Communities, Directorate General
"Science, Research and Development"
200, rue de la Loi
B - 1049 BRUSSELS

DAVIES, I.
Chief Consultant
Taylor Woodrow Construction Ltd.
345 Ruislip Road
UK - SOUTHALL, MIDDLESEX UB1 2QX

DAVIES, J.
Manager R & D
Foster Wheeler Power
Products Limited
Greater London House, Hampstead Road
UK - LONDON NW1 7QN

DE BUCK, F.
ACEC
Division Nucléaire
B - 6000 CHARLEROI

DE TASSIGNY, Ch.
Ingénieur
C.E.A. - C.E.N. Grenoble
SPR - CENG
B.P. 85x
F - 38041 GRENOBLE Cedex

DE WAHA,
Ministère de la Santé
Division de la Radioprotection
L - LUXEMBOURG

DECKERS, J.
Project Engineer
nv. ENI s.a.
Kontichsesteenweg 25
B - 2630 AARTSELAAR

DEHON, Ch.
Ingénieur
Belgonucléaire S.A.
25, rue du Champ de Mars
B - 1050 BRUXELLES

DELANEY, E.G.
Manager for Decommissioning
United States Department
of Energy
6, Doolittle Court
USA - GAITHERSBURG, MD 20879

DELATTRE, B.
TECHNICATOME
B.P. 1
F - 91190 GIF-SUR-YVETTE

DELISLE, J.-P.
Ingénieur
C.E.A. - C.E.N. Cadarache
SEER - BT 208
B.P. 1
F - 13115 ST PAUL-LEZ-DURANCE

DETILLEUX, E.
Directeur général
Organisme national des déchets
radioactifs et matières fissilles
54, bvd du Régent, bte 5
B - 1000 BRUXELLES

DEUBEL, S.
Presse
DEUTSCHE WELLE
Aktueller Dienst
Postfach 100444
D - 5000 KOELN 1

DEVRESSE, M.
Ingénieur civil
A.C.E.C.
Division nucléaire 130
B.P. 4
B - 6000 CHARLEROI

DIEFENBACHER, W.
Dipl.-Ing.
Kernforschungszentrum Karlsruhe GmbH
Postfach 3640
D - 7500 KARLSRUHE 1

DUBOIS, F.
CEA/CEN Saclay
F - 91191 GIF- SUR- YVETTE

DUBOURG, M.
Assistant du directeur technique
FRAMATOME
Tour Fiat - Cedex 16
F - 92084 PARIS-LA-DEFENSE

EBELING, W.
Dipl.-Ing.
Salzgitter AG - Hauptabteilung
Forschung und Entwicklung
Postfach 41 11 29
D - 3320 SALZGITTER 41

EIRICH, A.J.
Manager, Office of
Surplus Facilities Management
UNC Nuclear Industries
P.O.B. 490 Met. Dept.
USA - RICHLAND, WASHINGTON 99352

ESSMANN, J.
Dipl.-Physiker
Preussische Elektrizitäts AG
Hauptverwaltung
Postfach 48 49
D - 3000 HANNOVER 1

FAVENNEC, J.
Ingénieur
FRAMATOME
Tour Fiat
F - 92084 PARIS LA DEFENSE

FELDMAN, C.
Nuclear Engineer
U.S. Nuclear Regulatory
Commission
USA - WASHINGTON, D.C. 20555

FERADAY, M.A.
Nuclear Engineer
International Atomic
Energy Agency
Wagramer Str. 5
A - 1400 WIEN

FICKEL, O.
Dipl.-Ing.
Brown Boveri Reaktor GmbH
Dudenstrasse 44
D - 6800 MANNHEIM

FINZI, S.
Director
Commission of the European
Communities, Directorate General
"Science, Research and Development"
200, rue de la Loi
B - 1049 BRUSSELS

FISH, G.F.
Project Group Manager
Strachan & Henshaw
P.O. Box 103
UK - BRISTOL BS99 7TJ

FITZPATRICK, J.
Consultant Engineer
Electrowatt Engineering
Services (UK) Ltd
Grandford House, Carfax
UK - HORSHAM, WEST SUSSEX RH12 1UP

FLEISCHER, C.
Senior Research Engineer
Taylor Woodrow Construction Ltd.
345 Ruislip Road
UK - SOUTHALL, MIDDLESEX UB1 2QX

FORSSTROEM, H.
Swedish Nuclear Fuel
Supply Co (SKBF/KBS)
P.O. Box 5864
S - 102 48 STOCKHOLM

FRANCE, P.G.W.
General Manager
Bird Group of Companies
Birmingham Road
GB - STRATFORD UPON AVON,
Warwickshire CV37 80A

FRANCIONI, W.M.
Masch. Ing.
Eidg. Institut für Reaktorforschung
CH - 5303 WUERENLINGEN

FREUND, H.U.
Physiker
Battelle-Institut
Am Römerhof 35
D - 6000 FRANKFURT 90

FRICKE, H.-D.
Wirtschaftsingenieur
NUKEM GmbH
Postfach 11 00 80
D - 6450 HANAU 11

FRIDMAN, G.
Journaliste
Energie Nucléaire Magazine
142, rue Montmartre
F - 75002 PARIS

FUCHS, R.
Dipl.-Ing.
Badenwerk AG
Abteilung PP
Postfach 16 80
D - 7500 KARLSRUHE 1

GARBAY, H.
C.E.A. - Dépt. Protection Technique
Sect. d'Etud. de Protection
pour les Déchets
B.P. 6
F - 92260 FONTENAY-AUX-ROSES

GATINEAU, J.-P.
Ingénieur
Electricité de France
Service Etudes & Projets
Thermiques et Nucléaires
Tour EdF-GdF
F - 92080 PARIS LA DEFENSE, Cedex 8

GAUCHON, J.P.
CEA/CEN Cadarache SETED
B.P. 1
F - 13115 ST. PAUL-LEZ-DURANCE

GAUFFRE, G.
FRAMATOME
Tour Fiat/Cedex 16
F - 92084 PARIS LA DEFENSE

GENOVA, M.
Researcher
CISE - Centro Informazioni
Studi Esperienze
C.P. 12081
I - 20134 MILANO

GODDARD, A.J.H.
Reader
Imperial College of
Science and Technology
Dept. of Mechanical Eng.
UK - LONDON SW7 2BX

GODFREY, D.G.
Chartered Engineer
Taylor Woodrow Construction Ltd
Taywood House, 345 Ruislip Road
UK - SOUTHALL, MIDDLESEX UB1 2QX

GOERTZ, R.
Physiker
Brenk Systemplanung
Heinrichsallee 38
D - 5100 AACHEN

GOMER, Ch.R.
Physicist
British Steel Corporation
Sheffield Laboratories
UK - MOORGATE, ROTTERHAM S60 3AR

GOUVRAS, G.
Commission of the European Communities,
Directorate General "Employment, Social
Affairs and Education"
Health and Safety Directorate
B.P. 1907
L - 2920 LUXEMBOURG

GRAS, M.
Ingénieur
Commissariat à l'Energie Atomique
31-33, rue de la Fédération
F - 75015 PARIS

GRASSI, G.
EURATOM CCR Ispra
Essor 80
I - 21020 ISPRA

GREGORY, A.R.
Head of Nuclear Decommissioning Project
Central Electricity Generating Board
Generation Develop.
and Construction Div.
Barnett Way
UK - GLOUCESTER GL4 7RS

GROLADE, R.
Directeur de Marketing
USSI
116, ave Aristide Briand
F - 92220 BAGNEUX

GUETAT, Ph.
C.E.A.
Dépt. Protection Technique
Sect. d'Etud. de Protection
pour les Déchets
B.P. 6
F - 92260 FONTENAY-AUX-ROSES

GUILLARD, C.
Directeur général
S.T.M.I.
9, rue Fernand Léger
F - 91190 GIF-SUR-YVETTE

HAHN, V.
Dipl.-Ing.
Kraftanlagen Aktiengesellschaft
Postfach 10 34 20
D - 6900 HEIDELBERG

HALSALL, J.R.
Computer Scientist
University of Warwick
Dept. of Computer Science
UK - COVENTRY CV4 7AL

HAMSTRA, J.
Director
AVORA B.V.
Irenelaan 16
NL - 1862 EB BERGEN (NH)

HANDA, K.
Section Manager
Engineering Sec.
Nuclear Power Engineering Dept.
The Furukawa Electric Co., Ltd.
5-1-9, Higashi-yawata, Hiratsuka-shi
JAPAN - KANAGAWA-KEN 254

HARA, S.
Manager
Transport Operations Dept.
Nuclear Transport Services Corp.
1-1-3, Shiba-daimon, Minato-ku
JAPAN - TOKYO 105

HARBECKE, W.
Dipl.-Ing.
Kernkraftwerk Lingen GmbH
Postfach 1640
D - 4450 LINGEN

HARDIMAN, S.A.
Chief Mechanical Engineer
Taylor Woodrow Management
and Engineering Limited
Alpha House
Westmount Centre (Delamere Rd)
UK - HAYES, MIDDLESEX UB4 OHD

HASSELHOFF, H.
Projektleiter
Siempelkamp Giesserei GmbH & Co
Siempelkampstr. 45
D - 4150 KREFELD

HEBEL, W.
Commission of the European
Communities, Directorate-General
"Science, Research and Development"
200, rue de la Loi
B - 1049 BRUXELLES

HEMPELMANN, W.
Ingenieur
Kernforschungszentrum Karlsruhe
Moselstr. 4
D - 7514 EGGENSTEIN 2

HENDRIKSE, C.
Aannemer
Hendrikse Services
Ambachtstraat 12
NL - 1251 PZ LAREN

HENNART, D.M.J.G.
Ingénieur civil
S.C.K./C.E.N.
Blg. WST
Boeretang 200
B - 2400 MOL-Donk

HERBILLON, G.
Ingénieur Principal
Ministère de l'Emploi et du Travail
Service Sécurité Technique
des Installations Nucléaires
53, rue Belliard
B - 1040 BRUXELLES

HERZOG, C.
Dipl.-Ing.
Arbeitsgemeinschaft
Versuchsreaktor, AVR
Hambacher Forst
D - 5170 JUELICH

HESS, R.
Nordostschweiz. Kraftwerke AG
Dorfstr. 24
CH - 5430 WETTINGEN

HESS, E.
Ing. HTL
Nordostschweiz. Kraftwerke AG
CH - 5400 BADEN

HIRAO, S.
Group Manager, Nuclear Energy
Construction & Service Dept.
Toshiba Corporation
8, Shinsugita-cho, Isogo-ku
JAPAN - YOKOHAMA-SHI, KANAGAWA-KEN 235

HISCOCK, S.
Director
Lead Development Association
34 Berkeley Square
UK - LONDON W1X 6AJ

HOLLMANN, H.
Assessor, juristischer Sachbearb.
Kernforschungszentrum Karlsruhe GmbH
Postfach 3640
D - 7500 KARLSRUHE 1

HORIE, H.
Manager
Building Engineering Dept.
Ohbayashi-Gumi Ltd.
3-37, Kyobashi, Higashi-ku
JAPAN - OSAKA-SHI, OSAKA 540

HORITA, S.
Senior Engineer, HTR Project Office
Nuclear Power Promotion Div.
Fuji Electric Co., Ltd
1-12-1, Yurakucho, Chiyoda-ku
JAPAN - TOKYO 100

HØJGAARD, K.E.
Chief Engineer
ELSAM
DK - 7000 FREDERICIA

HUBER, B.
Engineer
Commission of the European
Communities, Directorate General
"Science, Research and Development"
200, rue de la Loi
B - 1049 BRUSSELS

HUBERT, E.
Ingénieur
Université de Liège
58a, Drève Soetkin
B - 1070 BRUXELLES

HUEBENTHAL, K.H.
Dipl.-Chem.
Bundesministerium für Forschung
und Technologie
Heinemannstr. 2
D - 5300 BONN 2

HULOT, M.
Technicien Principal
Commissariat à l'Energie
Atomique
B.P. 6
F - 92260 FONTENAY-AUX-ROSES

HUNGER, H.
Brown Boveri Reaktor GmbH
Dudenstrasse 44
D - 6800 MANHEIM

HYVONEN, C.W.
Director of Marketing and Planning
Morrison-Knudsen Company Inc.
P.O.B. 7808
USA - BOISE, IDAHO

ISSLER, K.
Presse
DEUTSCHE WELLE
Aktueller Dienst
Postfach 100444
D - 5000 KOELN 1

ISSLER, R.
Presse
DEUTSCHE WELLE
Aktueller Dienst
Postfach 100444
D - 5000 KOELN 1

ITO, N.
Researcher
Japan Atomic Energy
Research Institute
Tokai-Mura, Naka-Gun
JAPAN - IBARAKI-KEN 319-11

ITO, K.
Manager, Director of Administrative
Office of Western Branches
Genshi-Ryoku Daiko Co., Ltd.
2-1-8, Dojima-hama, Kita-ku
JAPAN - OSAKA-SHI, OSAKA 530

ITO, S.
General Manager - Nuclear Power Pla
Service Dept. - Shin Nippon Air
Conditioning Engineering Co., Ltd.
1-1-34, Nakahara, Isogo-ku
JAPAN - YOKOHAMA-SHI, KANAGAWA-KEN

JACQUIER, M.
Ingénieur
A.G.S.A.A.
Pool Atomique
118, rue de Tocqueville
F - 75850 PARIS Cedex 17

JAGER, W.
Physiker
GKSS-Forschungszentrum
Geesthacht GmbH
Max-Planck-Strasse
D - 2054 GEESTHACHT-TESPERHUDE

JOSSO, F.
CEA/CEN Cadarache
SEDFMA/SETED - B.P. 1
F - 13115 SAINT PAUL LEZ DURANCE

KAFROUNY, A.
Directeur
U.S.S.I.
116, ave Aristide Briand
F - 92220 BAGNEUX

KALIS, D.
Journalist
DEUTSCHE WELLE
Aktueller Dienst
Postfach 100444
D - 5000 KOELN 1

KAMINSKY, S.
Manager of Nuclear Business
Development - Power Group
Morrison-Knudsen Company, Inc.
P.O.B. 7808
USA - BOISE, IDAHO

KAMITSUMA, M.
Japan Nus Co., Ltd
16th Floor Shinjuku Daiichi Seimei Bldg
7-1 Nishi-Shinjuku 2-chome
JAPAN - SHINJUKU-KU, TOKYO 160

KAMIYA, Y.
Engineer - Structural Technology Dept.
Nuclear Development Div.
Kumagai Gumi Co., Ltd.
17-1, Tsukudo-cho, Shinjuku-ku
JAPAN - TOKYO 162

KANAZAWA, K.
Manager, No.2 Plant Design Dept.
Nuclear Power Div. - Ishikawajima
Harima Heavy Industries Co., Ltd.
1, Shin-nakaharacho, Isogo-ku
JAPAN - YOKOHAMA-SHI, KANAGAWA-KEN 235

KASAGI, T.
General Manager - Nuclear Power
Designing Dept.- Shin Nippon Air
Conditioning Engineering Co., Ltd.
1-1-34, Nakahara- Isogo-ku
JAPAN - YOKOHAMA-SHI, KANAGWA-KEN 235

KATO, Y.
Engineer - Construction Quarters
Construction Department
Nihon Kensetsu Kogyo Co., Ltd.
13-11, Shimbashi 5-chome
JAPAN - MINATO-KU, TOKYO 105

KAYSER,
Ministère de la Santé
Division de la Radioprotection
1, av. des Archiducs
L - LUXEMBOURG

KIENHOEFER, M.
Ingenieur
Kernforschungszentrum Karlsruhe
Fichtenweg 2
D - 7514 EGGENSTEIN

KIENLE, F.
Dipl.-Physiker
Vereinigung Deutscher
Elektrizitätswerke - VDEW - e.V.
Stresemannallee 23
D - 6000 FRANKFURT 70

KINOSHITA, T.
Manager/Nuclear Dept
Sato Kogyo Co Ltd
4-8, Nihonbashi Honcho, Chuo-ku
Japan - TOKYO 103

KLARE, J.
Dipl.-Ing.
Salzgitter AG
Abt. Forschung und Entwicklung
Postfach 41 11 29
D - 3320 SALZGITTER 41

KLEIN, E.J.
Dipl.-Ing.
Technischer Ueberwachungsverein
Rheinland
Am Grauen Stein
D - 5000 KOELN-POLL

KLOECKNER, J.
Ingenieur
GNS Gesellschaft
für Nuklear Service mbH
Goeth str. 88
D - 43 ESSEN

KNACKSTEDT, H.G.
Dipl.-Ing.
Transnuklear GmbH
Postfach
D - 6450 HANAU

KNOWLES, A.N.
UKAFA
Central Technical Service
GB - WARRINGTON, Cheshire WA3 6AT

KOFAHL, A.
Dipl.-Ing.
Gesellschaft für
Reaktorsicherheit (GRS) mbH
Schwertnergasse 1
D - 5000 KOELN 1

KONING, C.
Netherlands Energy
Research Foundation
P.O.B. 1
NL - ZG PETTEN

KORN, K.
Dipl.-Ing.
Kernkraftwerk-Betriebs GmbH
D - 7514 EGGENSTEIN-LEOPOLDSHAFEN 2

KOUWENBERG, B.
Production Manager
Urenco Nederland Operations B.V.
P.O. Box 254
NL - 7600 AG ALMELO

KRAPP, J.
Ingenieur
Kernforschungszentrum Karlsruhe GmbH
Hauptabteilung Ingenieurtechnik
Postfach 3640
D - 7500 KARLSRUHE 1

KRUEGER, K.
Dipl.-Phys.
Arbeitsgemeinschaft
Versuchsreaktor, AVR
Hambacher Forst
D - 5170 JUELICH

KUKKOLA, T.
Chief Designer in Plant Engineering
Imatran Voima Oy
(Imatra Power Company)
P.O.B. 138
SF - 00101 HELSINKI 10

KUTSCH, W.
Dipl.-Ing.
Rheinisch-Westfälisches
Elektrizitatswerk AG, Hauptverwaltung
Kruppstrasse 5
D - 4300 ESSEN 1

KUTSUMIZU, A.
Manager
Nuclear Facilities Dept.
Ohbayashi-Gumi, Ltd.
2-3, Kanda-tsukusacho, Chiyoda-ku
JAPAN - TOKYO 101

KUYPERS, J.
Head of Dept. Techn. Development
N.V. KEMA
P.O. Box 9035
NL - 6800 ET ARNHEM

LAFONTAINE, I.
Directeur
TRANSNUBEL
Europalaan 20
B - 2480 DESSEL

LAMBLEY, J.
Physicist
British Steel Corporation
Sheffield Laboratories
Swinden House
UK - MOORGATE, ROTHERHAM S60 3AR

LARAIA, M.
Engineer
E.N.E.A. / DISP
Via Vitaliano Brancati 48
I - 00144 ROMA

LARCOMBE, M.H.E.
Computer Scientist
University of Warwick
Dept. of Computer Science
UK - COVENTRY CV4 7AL

LASCH, M.
Dipl.-Chemiker
Kernkraftwerke Gundremminen
Betriebsgesellschaft mbH
D - 8871 GUNDREMMINGEN

LAVALERIE, C.
Ingénieur
FRAMATOME
1, place de la Coupole (Tour Fiat)
F - 92084 PARIS LA DEFENSE

LAWTON, H.
Physicist
United Kingdom Atomic
Energy Authority
Sellafield
UK - SEASCALE, CUMBRIA CA20 1PF

LAWTON, W.
Managing Director
Thos. W. Ward (Nuclear
Decommissiong) Ltd. - Albion Works
Saville Street
Uk - SHEFFIELD S4 7UL

LE BER, J.
Ingénieur
C.E.A.
St. Guénolé - route de la Grande Grè
F - 29226 CARANTEC

LE GUELLEC, M.
TECHNICATOME
B.P. 1
F - 91190 GIF-SUR-YVETTE

LECERF, H.
Ets. Somalor-Ferrari
SOMAFER
Route de Vitry
F - 57270 UCKANGE

LECOFFRE, Y.
Ingénieur
Ets. de Grenoble de la
Sté ALSTHOM-ATLANTIQUE
B.P. 61x
F - 38041 GRENOBLE Cedex

LECONTE, M.
Ingénieur
A.C.E.C.
Division nucléaire 130
B.P. 4
B - 6000 CHARLEROI

LEFEVRE, J.
Directeur
C.E.A. - C.E.N.-FAR
DED.
B.P. 6
F - 92260 FONTENAY-AUX-ROSES

LELAIDIER, M.
Directeur
SOGEDEC
12bis, bvd Pèbre
F - 13008 MARSEILLE

LINSTER, R.
Commission of the European
Communities, Directorage General
"Personnel and Administration"
B.P. 1907
L - 2920 LUXEMBOURG

LIPSCHUTZ, E.
Project Manager
Comprimo België n.v.
Noorderlaan 139
B - 2030 ANTWERPEN

LOERCHER, G.
Dr.-Ing.
NIS - Nuklear-Ingenieur-
Service GmbH
Lyonerstr. 22
D - 6000 FRANKFURT/MAIN 71

LOESCHHORN, U.
Physiker
Kernforschungszentrum Karlsruhe GmbH
Postfach 3640
D - 7500 KARLSRUHE 1

LOURME, P.
Ingénieur
TECHNICATOME
B.P. 18
F - 91190 GIF-SUR-YVETTE

LUEDENBACH, B.
Dipl.-Ing.
Internationale Atomreaktorbau GmbH
Friedrich-Ebert-Str.
D - 5060 BERGISCH-GLADBACH 1

LYNN, J.
Nuclear Inspector
Health & Safety Executive
Thames House North, Millbank
UK - LONDON SN1P 4QJ

MABBS, E.G.
Chief Mechanical Engineer
Sir Robert McAlpine & Sons Ltd
P.O. Box 10
UK - KETTERING, NORTHANTS NN15 GJN

MACKENZIE, R.
Contracts Manager
Taylor Woodrow Construction Ltd
345 Ruislip Road
UK - SOUTHALL, MIDDLESEX UB1 2QX

MACK, J.
Research Chemist
Fairey Engineering Ltd
P.O. Box 41, Crossley Road
GB - HEATON CHAPEL, Stockport SK4 5BD

MALIGE, J.
Ingénieur
Seri Renault Ingénierie
B.P. 19
F - 78391 BOIS D'ARCY Cedex

MANASSERO, G.
FIAT Research Centre
Strada Torino 50
I - 10043 ORBASSANO-TORINO

MANDAHL, B.
Manager Protection Technology
OKG AKTIEBOLAG
P.O.B. 1746
S - 11187 STOCKHOLM

MANEVY, G.
Ingénieur
C.E.A.
Centre d'Etudes Nucleaires
CEN/FAR - IRDI/CRD, B.P. 6
F - 92260 FONTENAY AUX ROSES

MANTEGA, F.
Researcher
CISE - Centro Informazioni
Studi Esperienze
C.P. 12081
I - 20134 MILANO

MARCAILLOU, J.
Ingénieur
C.E.A.
C.E.N. Cadarache
B.P. 1
F - 13115 SAINT PAUL LEZ DURANCE

MARCUS, F.R.
Nordic Liaison Committee for
Atomic Energy
P.O. Box 49
DK - 4000 ROSKILDE

MARQUE, Y.
Ingénieur
C.E.A./ANDRA
31-33, rue de la Fédération
F - 75015 PARIS

MARSHALL, J.D.
Department of Environment
Room 19517
Romney House, Marsham St. 43
UK - LONDON SW1P 3PY

MARTIN, A.
Ingénieur, Chef de Service
Cockerill Mechanical Industries
Département Energie
374, rue des Vennes
B - 4020 LIEGE

MASUDA, H.
Chief Engineer
Tokai Works
Taihei Dengyo Kaisha, Ltd
2-4, Kandajinbocho, Chiyoda-ku
JAPAN - TOKYO 101

MATSUNAGA, K.
Engineer
Mitsubishi Heavy Industries, Ltd.
4-1, Shibakoen 2-chome, Minato-ku
JAPAN - TOKYO 105

MAY, S.
Directeur C.E.A. du Lab. Pierre Süe
C.E.A. - C.E.N. Saclay
Laboratoire Pierre Süe
F - 91191 GIF-SUR-YVETTE

MAYLIN, J.-A.
Ingénieur
C.E.A./ANDRA
31-33, rue de la Fédération
F - 75015 PARIS

MEERWALD, K.
Dipl.-Ing.
Hochtief AG
Abt. Kerntechnischer Ingenieurbau
Postfach 3189
D - 6000 FRANKFURT/MAIN 1

MEIER, F.
MAN - Maschinenfabrik Augsburg-Nürnb
Unternehmensbereich GHH Sterkrade
Bahnhofstr. 66
D - 4200 OBERHAUSEN 11

MENON, Sh.
Engineer
Studsvik Energiteknik AB
S - 61182 NYKOEPING

MEYER, A.-W.
Dipl.-Ing.
Kraftwerk Union
Berliner Str. 295-303
D - 6050 OFFENBACH

MIGLIORATI, B.
Engineer
FIAT TTG
Via Cuneo 20
I - 10100 TORINO

MIIKE, T.
Deputy Manager
CGR Research Office
The Japan Atomic Power Company
1-6-1, Otemachi, Chiyoda-ku
JAPAN - TOKYO 100

MILLIES, P.
Coordonnateur pour la robotique
Commissariat à l'Energie
Atomique
29-33, rue de la Fédération
F - 75015 PARIS

MILLOY, Ch. J.
Structural Engineer
Ove Arup and Partners
13 Fitzroy Street
UK - LONDON W1P 6BQ

MITTEMPERGHER, M.
President
NUCLECO S.p.A.
V.le Regina Margherita, 125
I - 00198 ROMA AD

MIYASUGI, T.
Assistant Principal Engineer
Nuclear Energy Project Dept.
Chiyoda Chemical Engineering
& Construction Co., Ltd.
2-12-1, Tsurumi-chuo, Tsurumi-ku
JAPAN - YOKOHAMA-SHI, KANAGAWA-KEN 230

MIZOGUCHI, T.
Sales Engineer, Business Develop. Sec.
Nuclear Energy Dept.
Nissho Iwai Corporation
2-4-5, Akasaka, Minato-ku
JAPAN - TOKYO 107

MOCHON MUNOZ, J.L.
Physicist
Hidroelectrica Espanola S.A.
Hermosilla 3
E - MADRID 1

MOODY, B.R.
Director & General Manager
Fairey Engineering Ltd
Nuclear & General Eng. Div.
P.O. Box 41, Heaton Chapel
UK - STOCKPORT, CHESHIRE SK4 5BD

MORANT, M.
Strahlenschutz
Fa. Morant & Buchen GmbH
Zellerhornstr. 73
D - 8213 ASCHAU/CHIEMGAU

MORIYA, Y.
Manager, Nuclear & Process
Engineering Dept., Nuclear Project Div
JGC Corporation
1-14-1, Bessho, Minami-ku
JAPAN - YOKOHAMA-SHI, KANAGAWA-KEN 232

MOTTE, F.
Ingénieur Civil
C.E.N. - S.C.K.
Centrale BR3
B - 2400 MOL

MOUREAU, J.-C.
Ingénieur en Chef-Directeur
S.P.R.I., Ministère de la Santé Public
Cité Administrative de l'Etat
Quartier Vésale 2-3, bureau 23
B - 1010 BRUXELLES

MUELLER-LYDA, I.
Diplom-Physiker
GSF - Institut für
Tieflagerung
Theodor-Heuss-Str. 4
D - 3300 BRAUNSCHWEIG

MUELLER, F.
Bauingenieur
IVB / Technische
Hochschule Darmstadt
Petersenstr. 13
D -- 6100 DARMSTADT

MUELLER, W.
Dipl.-Phys.
Gesellschaft für Reaktorsicherheit MbH
Schwertnergasse 1
D - 5000 KOELN 1

MacLACHLAN, A.
European Editor
Nucleonics Week
17, rue Georges Bizet
F - 75116 PARIS

MacMAHON, D.
University Teacher
Imperial College Reactor Centre
Silwood Park
UK - ASCOT, BERSHIRE SL5 7PY

NAKAGAWA, M.
Marubine
Hoffmannstr. 95
D - 8520 ERLANGEN

NAKAJIMA, N.
Manager, Chemistry Sec. I
Mitsubishi Atomic Power Ind., Inc.
1-297, Kitabukuro-machi, Omiya-shi
JAPAN - SAITAMA-KEN 330

NEAL, R.A.
Vice President of Marketing
and Contracts - Power Group
Morrison-Knudsen Company, Inc.
P.O.B. 7808
USA - BOISE, IDAHO

NEIDER, R.
Physiker
Bundesanstalt für Material-
prüfung (BAM)
Unter den Eichen 87
D - 1000 BERLIN 45

NICOLAY, D.
Commission of the European
Communities, Directorate General
"Information Market and Innovation"
B.P. 1907
L - 2920 LUXEMBOURG

NIEVELSTEIN, M.
Laboringenieur
Kernforschungszentrum Jülich GmbH
IRW/Heisse Zellen
Aachener-Str. 57
D - 5120 HERZOGENRATH

NODA, M.
Project Manager
Power Systems Engineering
Dept. - Power Systems Headquarters
Mitsubishi Heavy Industries, Ltd.
5-34-6, Shiba, Minato-ku
JAPAN - TOKYO 108

NOJIMA, H.
Deputy Manager
Nuclear Energy Department
Fujita Corporation
4-6-15, Sendagaya, Shibuya-ku
JAPAN - TOKYO 151

OGATA, N.
Principal Research Engineer
Nuclear Power Div.,
Research & Develop. Headquarters
Central Research Inst. of
Electric Power Industry
1-6-1, Otemachi, Chiyoda-ku
JAPAN - TOKYO 100

OLLIVIER, J.
Ingénieur
C.E.A. - C.E.N. Cadarache
SEER BT 208
B.P. 1
F - 13115 ST PAUL-LEZ-DURANCE

ORLOWSKI, S.
Head of division
Commission of the European
Communities, Directorate General
"Science, Research and Development"
200, rue de la Loi
B - 1049 BRUSSELS

OSE, K.
Department Manager
Nuclear Power Dept. - The Japan
Electrical Manufacturer's Association
2-4-15, Nagata-cho, Chiyoda-ku
JAPAN - TOKYO 100

PALENBERG, P.
C.E.N./S.C.K.
Operations BR2
Boeretang 200
B - 2400 MOL

PARKER, F.
Engineer
Vanderbilt University
P.O. Box 1596 Station B
USA - NASHVILLE, Tennessee 37235

PASCALI, R.
Research Engineer
ENEL - Centro Ricerca
Termica Nucleare
Via Rubattino 54
I - 20134 MILANO

PATON, A.
Civil Engineer
Taylor Woodrow Construction Ltd.
345 Ruislip Road
UK - SOUTHALL, MIDDLESEX UB1 2QX

PATTERSON, M.
National Nuclear Corporation Ltd
Warrington ad, Risley
GB - WARRINGTON, Cheshire WA3 6BZ

PAVLOVIC, Z.
Licencing Manager
KRSKO Nuclear
Power Plant
Nuklearna Elektrana
YU - 68270 KRSKO

PAWLYTSCH, St.
Dipl.-Ing. für Kerntechnik
Reaktorwartungsdienst &
Apparatebau
Eleonorenstr. 1
D - 5170 JUELICH

PELLECCHIA, V.
Mechanical Engineer
SNIA TECHINT
Tecnologie Energetiche Avanzate
Via A. Bargoni 34
I - 00153 ROMA

PETTERSSON, S.
Mechanical Engineer
Swedish State Power Board
S - 16287 VAELLINGBY

PFLUGRAD, K.
Commission of the European
Communities, Directorate General
"Science, Research and Development"
200, rue de la Loi
B - 1049 BRUSSELS

PICINI, P.
Scientist
E.N.E.A. Casaccia Ed. C.III.27
Via Anguillarese Km. 1,300
I - 00600 ROMA

PILOT, G.
Ingénieur
Commissariat à l'Energie Atomique
Saclay Département de Protection
Technique
SPIN/Bt 93
F - 91191 GIF SUR YVETTE Cedex

PLANQUART, J.
Professeur
Conseiller de la direction générale
Centre d'Etude de l'Energie Nucléaire
Boeretang 200
B - 2400 MOL

POCAI, M.
Ricercatore
CAMEN
REATTORE RTS-1
I - 56010 SAN PIERO A GRADO (Pisa)

POTTER, W.
Nuclear Development Division
Nuclear Energy Agency
38, bvd Suchet
F - 75016 PARIS

POURPRIX, M.
Ingénieur
C.E.A. - C.E.N. Saclay
Département de Protection Technique
SPIN/Bte 93
F - 91191 GIF-SUR-YVETTE Cedex

PREAT, M.
Ingénieur-Conseil
TRACTIONEL S.A.
31, rue de la Science
B - 1040 BRUXELLES

PRICE, M.S.T.
Scientist
United Kingdom Atomic
Energy Authority
Atomic Energy Establishment
UK - WINFRITH near DORCHESTER, Dorset

PRICE, L.
Journalist
ATOM
U.K.A.E.A.
Charles II St. 11
UK - LONDON S10 1Y

PUGH, J.
Design Engineer
National Nuclear Corporation Ltd
Warrington Road, Risley
UK - WARRINGTON, CHESHIRE WA3 6BZ

RAMCKE, K.
Ing. grad
Preussische Elektrizitäts-AG
Aktiengesellschaft
Postfach 4849
D - 3000 HANNOVER 91

RANTANEN, J.
Research Scientist
Technical Research Centre of Finland
Reactor Laboratory
Otakaari 3A
SF - 0215 ESPOO 15

RECHTLICH, R.
Dipl.-Ing.
Techn. Ueberwachungsverein
Baden E.V.
Fachbereich Kerntechnik
u. Strahlenschutz
Dudenstr. 28
D - 6800 MANNHEIM 1

REGAN, J.D.
Chartered Engineer
National Nuclear Corporation Ltd.
Booths Hall
Chelford Road
UK - KNUTSFORD, CHESHIRE WA16 8QZ

REITERS, J.
Secrétaire de la Ville de Huy
Administration communale
de Huy
4, ave Godin-Parnajon
B - 5200 HUY

REMOVILLE, J.
Scientific Consultant
C.E.E. Brussels, detached
at GRADEL S.A.
38, rte de Luxembourg
L - 8401 STEINFORT

RIECH, H.
Dipl.-Ing.
Hochtief AG
Abt. Kerntechnischer Ingenieurbau
Postfach 3189
D - 6000 FRANKFURT/MAIN 1

RIPPON, S.
Journalist
Nuklear News
Ruiguy Mansions Embankment, Putney
GB - LONDON SW15 1LE

ROBERT, J.,
Ingénieur
C.E.A. - SPR/SIDS
F - 91191 GIF-SUR-YVETTE

ROGER, J.
TECHNICATOME
B.P. 1
F - 91190 GIF-SUR-YVETTE

ROSENBAUM, F.
Betriebs-Ing.
Kernforschungsanlage Jülich
Postfach 19 13
D - 5170 JUELICH

ROSE, K.
Dipl.-Ing.
Gg. Noell GmbH
Postfach 6260
D - 8700 WUERZBURG

ROTTENBERG, G.
Civil engineer
C.E.N./S.C.K.
Boeretang 200
B - 2400 MOL

ROUSSEAU,
Belgian Building Research
Institute (CSTC-WTCB)
41, rue du Lombard
B - 1000 BRUSSELS

RUSSELL, G.
V-P General Manager
NPM Nuclear Project Managers
Canada Inc.
2000 Argentia Rd, Plaza 3
CANADA - MISSISSAUGA, ONTARIO L5N 1P7

SAITOH, T.
Section for Industrial
Program and Technology
Japan Atomic Industrial Forum, Inc.
1-13, 1-chome, Shimbashi, Minato-ku
JAPAN - TOKYO

SAKAKIBARA, Y.
Deputy General Manager
Tokyo Electric Power Co., Ind.
30 Charles II Street
UK - LONDON SW1Y 4AE

SANDALLS, J.
Scientist
U.K.A.E.A.
AERE Harwell
UK - DIDCOT, OXON OX11 ORA

SAPPOK, M.
Geschäftsführer
Siempelkamp Giesserei GmbH & Co
Siempelkampstr. 45
D - 4150 KREFELD

SAROUDIS, J.I.
Decommissioning Services
Atomic Energy of Canada Ltd.
1600 Dorchester, Blv. West
CANADA - MONTREAL, QUEBEC H3H 1P9

SCHALLER, K.
Commission of the European
Communities, Directorate General
"Science, Research and Development"
200, rue de la Loi
B - 1049 BRUSSELS

SCHEIDHAUER, J.
Directeur de la DQSR
Société de Travaux en
Milieu Ionisant (STMI)
9, rue Fernand Léger
F - 91190 GIF-SUR-YVETTE

SCHENKER, E.
Chem.-Ing.
Eidg. Institut für
Reaktorforschung
CH - 5303 WUERENLINGEN

SCHILLING, F.G.
Direktor
Schilling Abbautechnik GmbH
Pirmasenser Str. 20
D - 3000 HANNOVER 71

SCHILLINGS, K.L.
Interatom GmbH
Friedrich-Ebert- Str.
D - 5060 BERGISCH-GLADBACH 1

SCHLENKER, H.-V.
Dr.-Ing.
Rheinisch-Westfälisches
Elektrizitätswerk AG
Hauptverwaltung - Abt. KB-AK
Kruppstrasse 5
D - 4300 ESSEN 1

SCHMIDT, W.P.
DI
Kernforschungszentrum Karlsruhe GmbH
Hauptabt. Ingenieurtechnik
Postfach 3640
D - 7500 KARLSRUHE 1

SCHNEIDER, G.
Kernforschungszentrum Karlsruhe GmbH
Postfach 3640
D - 7500 KARLSRUHE 1

SCHUCHARDT, M.C.
Dipl.-Ing.
Kernforschungszentrum Karlsruhe GmbH
Postfach 3640
D - 7500 KARLSRUHE 1

SCHUETZLE, R.
Diplom-Ingenieur
MC-W Energie-Consult GmbH
Maybachstr. 8
D - 7000 STUTTGART 30

SCHUMACHER, J.
Technischer Angestellter
Physikalische-Technische-
Bundesanstalt
Bundesallee 100
D - 3300 BRAUNSCHWEIG

SELBY, R.
Project Manager
Vickers PLC
Design & Projects Division
Wessex House, Market Street
UK - EASTLEIGH, HAMPSHIRE

SHARPE, B.W.
Commission of the European Communities
Directorate General "Energy"
B.P. 1907
L - 2920 LUXEMBOURG

SIMS, J.
Business Development Manager
Balfour Beatty Gravatom
Ack Lane, Bramhall
GB - STOCKPORT, Cheshire SK7 1AW

SIWEK, H.
Dipl.-Ing.
Sozialministerium Baden-Württemberg
Postfach 1250
D - 7000 STUTTGART 1

SLEGERS, M.G.M.
Ingénieur, Chef de Secteur
BBC AG BROWN, BOVERI & Cie
Corporate Staff Marketing KLM 21
P.O. Box 58
CH - 5401 BADEN

SMITH, G.
Health Physicist
National Radiological
Protection Board
Chilton
GB - DIDCOT Oxon OXII ORQ

SPANO, A.M.
Ricercatore
CAMEN - REATTORE RTS-1
I - 56010 SAN PIERO A GRADO (Pisa)

SPENCER, R.H.J.
Chartered Engineer
U.K.A.E.A.
Safety Reliability Directorate
Wigshaw Lane
UK - CULCHETH, WARRINGTON WA3 4NE

STANG, R.
Dipl.-Ing.
Kraftwerk Union AG
Berliner Str. 295-299
Postfach 962
D - 6050 OFFENBACH/MAIN

STANG, W.
Dipl.-Ing.
Kernkraftwerke Gundremmingen
Betriebsgesellschaft mbH
D - 8871 GUNDREMMINGEN

STIENNON, G.
Directeur général adjoint
C.E.N./S.C.K.
Boeretang 200
B - 2400 MOL

STIEVENART, P.
Ing. Civ. Mines
Cabinet du Secrétaire
d'Etat à l'Energie
123. rue du Commerce
B - 1050 BRUXELLES

STORER, G.
Civil Engineer
Taylor Woodrow Construction Ltd
345 Ruislip Road
UK - SOUTHALL, MIDDLESEX UB1 2QX

STRADAL, K.-A.
Betriebsleiter
Kernforschungszentrum Jülich GmbH
IRW/Heisse Zellen
Driesch 31
D - 5163 LANGERWEHE

SUAREZ, J.
Director General Studies
Union Electrica-Fenosa
Capitan Haya, 53
E - MADRID 20

SUCHOWITZ, M.
Dipl.-Ing.
Kraftanlagen AG
Postfach 10 34 20
D - 6900 HEIDELBERG

TAKAHASHI, M.
Chief Engineer
Fukushima Works No. 1
Taihei Dengyo Kaisha, Ltd.
2-4, Kanda-jinbochi, Chiyoda-ku
JAPAN - TOKYO 101

TANAKA, K.
Assistant Manager,
New Energy & Nuclear
Eng. Dept. - Technology
& Develop. Headquarters
Hitachi Zosen Corporation
1-1-1, Hitotsubashi, Chiyoda-ku
JAPAN - TOKYO 100

TANGUY, P.
Ingénieur
C.E.A.
IPSN/C.E.N.-FAR
B.P. 6
F - 92260 FONTENAY-AUX-ROSES

TANIHIRO, Y.
Japan Nus Co., Ltd.
16th Floor Shinjuku Daiichi Seimei Blc
7-1 Nishi-Shinjuku 2-chome
JAPAN - SHINJUKU-KU, TOKYO 160

TANIKOSHI, T.
Department Manager, Nuclear Plant
Construction Engineering Dept.,
Hitachi Ltd
3-1-1, Saiwai-cho, Hitachi-shi
JAPAN - IBARAKI-KEN 317

TANIS, G.
CEA/CEN Grenoble
85 X
F - 38041 GRENOBLE CEDEX

TEGETHOFF, H.
MAN - Maschinenfabrik Augsburg-Nürnberg
Unternehmensbereich GHH Sterkrade
Bahnhofstr. 66
D - 4200 OBERHAUSEN 11

TESTEMALE, G.
Ingénieur
Commissariat à l'Energie
Atomique - SPR/SASR
B.P. 6
F - 92260 FONTENAY-AUX-ROSES

THOME, P.
Directeur
Field Automation
46, rue Sainte Anne
F - 75002 PARIS

TOGO, Y.
Professor
Faculty of Engineering
University of Tokyo
7-3-1, Hongo, Bunkyo-ku
JAPAN - TOKYO 112

TOKIOKA, S.
Engineer/Technical Div.
Kumaigumi Co Ltd
2-1, Honohara, Toyokawa-shi
Japan - AICHI-KEN 442

TOMASSINI, T.
Scientist
E.N.E.A. Casaccia Ed.C.III.27
Via Anguillarese Km. 1,300
I - 00060 ROMA

TONON, P.
Vice-président du CEN/SCK
Conseiller au Cabinet du
Secrétaire d'Etat à l'Energie
123, rue du Commerce
B - 1040 BRUXELLES

TORIKAI, K.
Professor
Tokyo Science University
2641, Higashi Kameyama,
Noda-shi, Chiba-ke
Japan - YAMAZAKI 278

TSUYUKI, T.
Deputy General Manager
Tokyo Electric Power Co., Inc.
30 Charles II Street
UK - LONDON SW1Y 4AE

TURLUER, G.
Ingénieur
CEN/CEA Fontenay-aux-Roses
Service S.C.E.C.F.
B.P. 6
F - 92260 FONTENAY-AUX-ROSES

TURVEY, F.
Engineer
Nuclear Energy Board
Lower Hatch St. 22
IRL - DUBLIN 2

UCHIDA, M.
Engineer - Engineering & Develop. Sec.
Nuclear Plant Service Integration Dept
Misubishi Heavy Industries Ltd
Kobe Shipyard & Engine Works
1-1-1, Wadasaki-cho, Hyogo-ku
JAPAN - KOBE-SHI, HYOGO-KEN 652

V.D.GEVEL, M.A.J.M.
Civil Engineer
NUCON B.V.
Weesperzijde 150
NL - 1097 DS AMSTERDAM

VAN DEN DAMME, R.
Directeur à la Direction Générale
INTERCOM S.A.
Place du Trône 1
B - 1000 BRUXELLES

VAN DER PLAS, T.
Chemist
KEMA
P.O. Box 9035
NL - 6800 ET ARNHEM

VAN DONGEN, A.C.
Process System Engineer
NUCON B.V.
Weesperzijde 150
NL - 1097 DS AMSTERDAM

VAN GESTEL, J.
S.A. SMET JET Industrial Cleaning
Services, Technical Department
Nijverheidsstraat 3
B - 2431 OEVEL

VAUNOIS, Ph.
Ingénieur Chimiste
CEA - CEN/Cadarache
B.P. 1
F - 13115 ST PAUL LEZ DURANCE

VERDEAU, J.-J.
C.E.N. Saclay
F - 91190 GIF-SUR-YVETTE

VERTUT, J.
Ingénieur
Commissariat à l'Energie
Atomique - DPT
B.P. 6
F - 92260 FONTENAY-AUX-ROSES

VIECENZ, H.-J.
Dipl.-Ing.
Gg. NOELL GmbH
Schweinfurter Strasse 28
D - 8700 WUERZBURG

VIGNERON, L.J.
Courtier d'Assurances
N.V. Thilly Van Eessel S.A.
58, ave des Arts, bte 6
B - 1040 BRUXELLES

VIROLLE, J.
Chargé de mission à
Electricité de France
2, rue Louis Murat
F - 75008 PARIS

VOGL, K.
Physiker
BGA/FRG
Ingolstädterlandstr. 1
D - 8042 NEUHERBERG

VOLLNER, H.-G.
Kernkraftwerke Gundremmingen
Betriebsgesellschaft mbH
D - 8871 GUNDREMMINGEN

VON DER DECKEN,
Vice-président de la
section de l'énergie,
des affaires nucléaires
et de la recherche
Comité Economique et Social des C.E.
2, rue Ravenstein
B - 1000 BRUXELLES

VON DER DECKEN, C.-B.
Universitätsprofessor
Kernforschungsanlage Jülich
Institut für Reaktorbauelemente
Postfach 1913
D - 5170 JUELICH

VUKOVIC, Z.
Dr. Sc.
Boris Kidric Institute
of Nuclear Sciences - Vinca
P.O.B. 522
YU - 11001 BELGRADE

WARREN, G.
Group Leader
Rolls Royce & Associates Ltd
P.O. Box 31
UK - DERBY DE2 8BJ

WEIR, C.
Chartered Engineer
British Nuclear Fuels PLC
Sellafield
UK - SEASCALE, CUMBRIA CA20 1PG

WENGER, J.-P.
Ing. HTL
NOK
Kernkraftwerk Beznau
CH - 5312 DOETTINGEN

WHITE, I.F.
Scientist
National Radiological
Protection Board
UK - CHILTON, DIDCOT OX11 ORQ

WIELAND, B.
Bundesamt für Energiewirtschaft
Kapellenstr.14
CH - 3003 BERN

WIELING, N.
Kraftwerk Union AG
Hammerbacherstrasse 12-14
D - 8520 ERLANGEN

WOOD, I.A.
Principal Engineer
GEC - Energy Systems Ltd
Cambridge Road
GB - WHETSTONE, Leicester

WOOLLAM, P.B.
Nuclear Physicist
Central Electricity
Generating Board
Berkeley Nuclear Laboratories
UK - BERKELEY, GLOS. GL13 9PB

WRIGHT, B.
Manager, Decommissioning
Babcock Woodhall Duckham Ltd.
The Boulevard
UK - CRAWLEY SUSSEX RH10 14X

YATSUHASHI, M.
Manager/Nuclear Plant Cable
Fujikura Ltd
2-11-20, Nishi Gotanda, Shinagawa-ku
Japan - TOKYO 141

YONEDA, M.
Ingénieur nucléaire
Kansai Electric Power Co.
3, ave Hoche
F - 75008 PARIS

ZETTWOOG, P.
Ingénieur
CEA
CENFAR IPSN - DPT B.P. 2
F - 92260 FONTENAY AUX ROSES

ZIMMERMANN, W.
Dipl.-Ing.
Kernforschungszentrum Karlsruhe GmbH
Hauptabteilung Ingenieurtechnik
Postfach 3640
D - 7500 KARLSRUHE 1

INDEX OF AUTHORS

GPSR Compliance

The European Union's (EU) General Product Safety Regulation (GPSR) is a set of rules that requires consumer products to be safe and our obligations to ensure this.

If you have any concerns about our products, you can contact us on ProductSafety@springernature.com

In case Publisher is established outside the EU, the EU authorized representative is:

Springer Nature Customer Service Center GmbH
Europaplatz 3
69115 Heidelberg, Germany

Batch number: 10823811

Printed by Printforce, the Netherlands